United States Nuclear Regulatory Commission

Protecting People and the Environment

I0482687

Safety Evaluation Report

Related to the License Renewal of Indian Point Nuclear Generating Unit Nos. 2 and 3

Docket Nos. 50-247 and 50-286

Entergy Nuclear Operations, Inc.

Office of Nuclear Reactor Regulation

AVAILABILITY OF REFERENCE MATERIALS
IN NRC PUBLICATIONS

U.S.NRC

United States Nuclear Regulatory Commission

Protecting People and the Environment

Safety Evaluation Report

Related to the License Renewal of Indian Point Nuclear Generating Unit Nos. 2 and 3

Docket Nos. 50-247 and 50-286

Entergy Nuclear Operations, Inc.

Manuscript Completed: October 2009
Date Published: November 2009

Office of Nuclear Reactor Regulation

ABSTRACT

This safety evaluation report (SER) documents the technical review of the Indian Point Nuclear Generating Unit Nos. 2 and 3 (IP2 and IP3), license renewal application (LRA) by the U.S. Nuclear Regulatory Commission (NRC) staff (the staff). By letter dated April 23, 2007, and as supplemented by letters dated May 3 and June 21, 2007, Entergy Nuclear Operations, Inc., (Entergy or the applicant) submitted the LRA in accordance with Title 10, Part 54, of the *Code of Federal Regulations*, "Requirements for Renewal of Operating Licenses for Nuclear Power Plants." Entergy requests renewal of the IP2 and IP3 operating licenses (Facility Operating License Numbers DPR-26 and DPR-64, respectively) for a period of 20 years beyond the current expirations at midnight on September 28, 2013, for IP2, and at midnight on December 12, 2015, for IP3.

Indian Point is located approximately 24 miles north of the New York City boundary line. The NRC issued the construction permits on October 14, 1966, for IP2, and on August 13, 1969, for IP3. The NRC issued the operating licenses on September 28, 1973, for IP2, and on December 12, 1975, for IP3. IP2 and IP3 employ a pressurized water reactor design with a dry ambient containment. Westinghouse Electric Corporation supplied the nuclear steam supply system and Westinghouse Development Corporation originally designed and constructed the balance of the plant with the assistance of its agent, United Engineers and Constructors. The licensed power output of each unit is 3216 megawatts thermal (MWt) with a gross electrical output of approximately 1080 megawatts electric (MWe).

On January 15, 2009, the staff issued an SER with Open Items Related to the License Renewal of Indian Point Nuclear Generating Unit Nos. 2 and 3, in which the staff identified 20 open items necessitating further review. This SER presents the status of the staff's review of information submitted through August 6, 2009, the cutoff date for consideration in this SER. The 20 open items that had been identified in the previous SER were resolved before the staff made a final determination on the LRA. SER Section 1.5 summarizes these items and their resolution. Section 6.0 provides the staff's final conclusion on its review of the IP2 and IP3 LRA.

TABLE OF CONTENTS

Appendices

List of Tables

ABBREVIATIONS

AC	alternating current
ACAR	aluminum conductor aluminum-reinforced
ACI	American Concrete Institute
ACRS	Advisory Committee on Reactor Safeguards
ACSR	aluminum core steel-reinforced
ADAMS	Agencywide Document Access and Management System
ADV	atmospheric dump valve
AEIC	Association of Edison Illuminating Companies
AERM	aging effect requiring management
AFW	auxiliary feedwater
AISC	American Institute of Steel Construction
AMP	aging management program
AMR	aging management review
AMSAC	ATWS Mitigating System Actuation Circuitry
ANSI	American National Standards Institute
APCSB	Auxiliary and Power Conversion Systems Branch
ART	adjusted reference temperature
ASME	American Society of Mechanical Engineers
ASTM	American Society for Testing and Materials
ATWS	anticipated transient without scram
B&PV	Boiler and Pressure Vessel
BADGER	boron-10 areal density gauge for evaluating racks
BIL	basic impulse level
BMI	bottom mounted instrumentation
BOP	balance of plant
BTP	branch technical position
BVS	building vent sampling
BWR	boiling water reactor
C	Celsius
CASS	cast austenitic stainless steel
CB	core barrel
CCW	component cooling water
CEA	control element assembly
CETNA^{TM}	core exit thermocouple nozzle assembly
CFR	*Code of Federal Regulations*
CII	containment inservice inspection
CL	chlorination system
CLB	current licensing basis
CO$_2$	carbon dioxide
CR	condition report

CRD	control rod drive
CRDM	control rod drive mechanism
Cr-Mo	chromium-molybdenum
CS	containment spray
CST	condensate storage tank
Cu	copper
CUF	cumulative usage factor
CVCS	chemical and volume control
C_vUSE	Charpy upper-shelf energy
CW	circulating water
CWM	IP3 city water system code
CYW	IP2 city water system code
DBA	design basis accident
DBD	design basis document
DBE	design basis event
DC	direct current
ECCS	emergency core cooling system
ECT	eddy current testing
EDG	emergency diesel generator
EFPY	effective full-power year
EMA	equivalent margin analysis
EN	shelter or protection
EPRI	Electric Power Research Institute
EQ	environmental qualification, environmentally qualified
EQAP	Entergy Quality Assurance Program
ER	Environmental Report (Applicant's Environmental Report Operating License Renewal Stage)
ESF	engineered safety features
F	Fahrenheit
FAC	flow accelerated corrosion
F_{en}	environmental fatigue life correction factor
FERC	Federal Energy Regulatory Commission
FLB	flood barrier
FLT	filtration
FMP	Fatigue Monitoring Program
FR	Federal Register
FRV	feedwater regulating valve
FSAR	final safety analysis report
ft-lb	foot-pound
FW	feedwater
FWST	fire water storage tank
GALL	Generic Aging Lessons Learned Report
GDC	general design criteria or general design criterion
GEIS	Generic Environmental Impact Statement
GL	generic letter

GSI	generic safety issue
GT	gas turbine
H_2	hydrogen
HELB	high-energy line break
HEPA	high efficiency particulate air
HPSI	high pressure safety injection
HVAC	heating, ventilation, and air conditioning
HX	heat exchanger
I&C	instrumentation and controls
IA	instrument air
IASCC	irradiation assisted stress corrosion cracking
IEEE	Institute of Electrical and Electronics Engineers
IGA	intergranular attack
IGSCC	inter-granular stress corrosion cracking
ILRT	integrated leak rate testing
IN	information notice
INPO	Institute of Nuclear Power Operations
IP1	Indian Point Nuclear Generating Unit 1
IP2	Indian Point Nuclear Generating Unit 2
IP3	Indian Point Nuclear Generating Unit 3
IP	Indian Point (site)
IPA	integrated plant assessment
IPEC	Indian Point Energy Center
ISG	interim staff guidance
ISI	inservice inspection
ISO	International Standards Organization
ksi	kip per square inch
KV or kV	kilo-volt
lb	pound
LBB	leak before break
LO	lube oil
LOCA	loss of coolant accident
LRA	license renewal application
μmhos/cm	micromhos per centimeter
MB	missile barrier
MC	ASME Class for metal containment components
MEB	metal-enclosed bus
MFW	main feedwater
MIC	microbiologically influenced corrosion
MOV	motor-operated valve
MPa	megapascal
MRP	Materials Reliability Program
MS	main steam
MSIV	main steam isolation valve

MWe	megawatts-electric
MWt	megawatts-thermal
n/cm^2	neutrons per square centimeter
NaOH	sodium hydroxide
NDE	nondestructive examination
NEI	Nuclear Energy Institute
NESC	National Electric Safety Code
NFPA	National Fire Protection Association
Ni	nickel
NPS	nominal pipe size
NRC	US Nuclear Regulatory Commission
NSAC	Nuclear Safety Analysis Center
NSAS	nonsafety system affecting safety system
NSSS	nuclear steam supply system
NYPA	New York Power Authority
O_2	oxygen
ODSCC	outside-diameter stress corrosion cracking
OI	open item
P&ID	piping and instrumentation diagram
PAB	primary auxiliary building
PB	pressure boundary
PBD	program basis document
pH	potential of hydrogen
PM	preventive maintenance
PORV	power-operated relief valve
ppb	parts per billion
ppm	parts per million
psi	pound per square inch
psig	pound-force per square inch gauge
PSPM	periodic surveillance and preventive maintenance
P-T	pressure-temperature
PTS	pressurized thermal shock
PVC	polyvinyl chloride
PW	primary water makeup
PWR	pressurized water reactor
PWSCC	primary water stress corrosion cracking
QA	quality assurance
RAI	request for additional information
RCCA	rod cluster control assembly
RCIC	reactor core isolation cooling
RCP	reactor coolant pump
RCPB	reactor coolant pressure boundary
RCS	reactor coolant system
RG	regulatory guide

RHR	residual heat removal
RI-ISI	risk-informed inservice inspection
RO	refueling outage
RPV	reactor pressure vessel
RT_{NDT}	reference temperature nil ductility transition
RT_{PTS}	reference temperature for pressurized thermal shock
RTD	resistance temperature detector
RVCH	reactor vessel closure head
RVI	reactor vessel internals
RVID	Reactor Vessel Integrity Database
RVLIS	reactor vessel level indication system
RW	river water
RWST	refueling water storage tank
S&PC	steam and power conversion
S_A	stress allowables
SAR	safety analysis report
SBO	station blackout
SC	structure and component
SCC	stress-corrosion cracking
SER	safety evaluation report
SFP	spent fuel pool
SFPC	spent fuel pit/pool cooling
SG	steam generator
SGBD	steam generator blowdown
SI	safety injection
SMP	structures monitoring program
SO_2	sulfur dioxide
SOC	statement of consideration
SOV	solenoid-operated valve
SPU	stretch power uprate
SR	surveillance requirement
SRP-LR	Standard Review Plan for Review of License Renewal Applications for Nuclear Power Plants
SS	stainless steel
SSC	system, structure, and component
SSE	safe-shutdown earthquake
SSFS	safety system function sheets
SW	service water
TLAA	time-limited aging analysis
TS	technical specification(s)
TSC	technical support center
UFSAR	Updated Final Safety Analysis Report
USE	upper-shelf energy
UT	ultrasonic testing
UV	ultraviolet

V	volt
VCT	volume control tank
WCAP	Westinghouse Commercial Atomic Power
WOG	Westinghouse Owners Group
XLPE	cross-linked polyethylene
yr	year
Zn	zinc
1/4 T	one-fourth of the way through the vessel wall measured from the internal surface of the vessel

SECTION 1

INTRODUCTION AND GENERAL DISCUSSION

1.1 Introduction

This document is a safety evaluation report (SER) on the license renewal application (LRA) for Indian Point Nuclear Generating Unit Nos. 2 and 3 (IP2 and IP3), as filed by Entergy Nuclear Operations, Inc. (Entergy or the applicant). By letter dated April 23, 2007, and as supplemented by letters dated May 3 and June 21, 2007, Entergy submitted its application to the U.S. Nuclear Regulatory Commission (NRC) for renewal of the Indian Point (IP) operating licenses for an additional 20 years. The NRC staff (the staff) prepared this report to summarize the results of its safety review of the LRA for compliance with Title 10, Part 54, "Requirements for Renewal of Operating Licenses for Nuclear Power Plants," of the *Code of Federal Regulations* (10 CFR Part 54). The NRC project manager for the license renewal review is Kim Green. Ms. Green may be contacted by telephone at 301-415-1627 or by electronic mail at Kimberly.Green@nrc.gov. Alternatively, written correspondence may be sent to the following address:

Division of License Renewal
US Nuclear Regulatory Commission
Washington, D.C. 20555-0001
Attention: Kim Green, Mail Stop O11-F1

In its April 23, 2007, submission letter, the applicant requested renewal of the operating licenses issued under Section 104b (Operating License Nos. DPR-26 and DPR-64) of the Atomic Energy Act of 1954, as amended, for IP2 and IP3 for a period of 20 years beyond the current expirations at midnight on September 28, 2013, for IP2, and at midnight on December 12, 2015, for IP3. Indian Point is located approximately 24 miles north of the New York City boundary line. The NRC issued the construction permits on October 14, 1966, for IP2, and on August 13, 1969, for IP3. The NRC issued the operating licenses on September 28, 1973, for IP2, and on December 12, 1975, for IP3. IP2 and IP3 employ a pressurized water reactor design with a dry ambient containment. Westinghouse Electric Corporation supplied the nuclear steam supply system and Westinghouse Development Corporation originally designed and constructed the balance of the plant with the assistance of its agent, United Engineers and Constructors. The licensed power output of each unit is 3216 megawatt thermal (MWt) with a gross electrical output of approximately 1080 megawatt electric (MWe). The updated final safety analysis reports (UFSARs) contain details of the plants and the site.

During its docketing sufficiency review, the staff identified two areas which required clarification from the applicant. The first issue was related to the name by which the applicant referred to the plant and the operating units. As noted in the LRA, the applicant refers to the operating units as Indian Point Energy Center Unit 2 and Unit 3. By letter dated May 3, 2007, the applicant clarified that the name "Indian Point Energy Center Units 2 and 3" is synonymous with the name "Indian Point Nuclear Generating Unit Nos. 2 and 3." The second issue was related to the proposed installation of the IP2 station blackout (SBO)/Appendix R diesel generator. By letter dated June 18, 2007, the staff notified Entergy that the staff believed that the current licensing basis (CLB) for IP2 was not fully represented in accordance with Section 54.4(a)(3) of Title 10 of the *Code of Federal Regulations* (10 CFR 54.4(a)(3)). The staff determined that the applicant had not included within the scope of license renewal those systems, structures, and components relied on in the safety analyses or plant evaluations to perform a function that demonstrates compliance with the requirements for station blackout (SBO) per 10 CFR 50.63, and safe shutdown per 10 CFR 50.48. In this regard, the LRA did not include information on the gas turbines, which at the time of submittal, were credited as an alternate power supply for the Appendix R and SBO events. Therefore, the staff requested that Entergy inform the staff of its plans to resolve this issue. By letter dated June 21, 2007, Entergy supplemented the LRA, and committed that the IP2 SBO/Appendix R diesel generator would be installed and operational by April 30, 2008. The applicant determined that the committed change to the facility met the requirements of 10 CFR 50.59(c)(1) and, therefore, a license amendment pursuant to 10 CFR 50.90 was not required. By letter dated July 25, 2007, the staff notified Entergy that it had completed its sufficiency review and that the application was acceptable for docketing.

The license renewal process consists of two concurrent reviews, a technical review of safety issues and an environmental review. The NRC regulations in 10 CFR Part 54, and 10 CFR Part 51, "Environmental Protection Regulations for Domestic Licensing and Related Regulatory Functions," respectively, set forth requirements for these reviews. The safety review for the IP license renewal is based on the applicant's LRA, amendments to the LRA, and on its responses to the staff's requests for additional information. On January 15, 2009, the staff issued an SER with Open Items Related to the License Renewal of Indian Point Nuclear Generating Unit Nos. 2 and 3, in which the staff identified 20 open items necessitating further review. Thereafter, the applicant supplemented the LRA and provided clarifications through its responses to the staff's RAIs and docketed correspondence. Unless otherwise noted, the staff reviewed and considered information submitted through August 6, 2009. The staff reviewed certain information received after that date as necessary and appropriate. The public may view the LRA and all pertinent information and materials, including the UFSARs, at the NRC Public Document Room, located on the first floor of One White Flint North, 11555 Rockville Pike, Rockville, MD 20852-2738 (301-415-4737 / 800-397-4209). Copies of the LRA are also available at White Plains Public Library, 100 Martine Avenue, White Plains, NY 10601, at Field Library, 4 Nelson Avenue, Peekskill, NY 10566, and at Hendrick Hudson Free Library, 185 Kings Ferry Rd., Montrose, NY 10548. In addition, the public may find the LRA, as well as materials related to the license renewal review, on the NRC Web site at http://www.nrc.gov.

This SER summarizes the results of the staff's safety review of the LRA and describes the technical details considered in evaluating the safety aspects of the units' proposed operation for an additional 20 years beyond the term of the current operating licenses. The staff reviewed the LRA in accordance with NRC regulations and the guidance in NUREG-1800, Revision 1, "Standard Review Plan for Review of License Renewal Applications for Nuclear Power Plants" (SRP-LR), dated September 2005.

SER Sections 2 through 4 address the staff's evaluation of license renewal issues considered during the review of the application. SER Section 5 is reserved for the report provided of the Advisory Committee on Reactor Safeguards (ACRS), which is expected to be issued subsequent to the publication of this SER. The conclusions of this SER are in Section 6.

SER Appendix A is a table showing the applicant's commitments for renewal of the operating licenses. SER Appendix B is a chronology of the principal correspondence between the staff and the applicant regarding the LRA safety review. SER Appendix C is a list of principal contributors to the SER and Appendix D is a bibliography of the references in support of the staff's review.

In accordance with 10 CFR Part 51, the staff issued the draft, plant-specific Supplement 38 to NUREG-1437, "Generic Environmental Impact Statement for License Renewal of Nuclear Plants Regarding Indian Point Nuclear Generating Unit Nos. 2 and 3 Draft Report for Comment," Volumes 1 and 2, on December 22, 2008. The supplement discusses the environmental considerations related to license renewal for IP2 and IP3. The draft Supplement is available on the NRC website (http://www.nrc.gov/reactors/operating/licensing/renewal.html).

1.2 License Renewal Background

Pursuant to the Atomic Energy Act of 1954, as amended, and NRC regulations, operating licenses for commercial power reactors are issued for 40 years and can be renewed for up to 20 additional years. The original 40-year license term was selected based on economic and antitrust considerations rather than on technical limitations; however, some individual plant and equipment designs may have been engineered for an expected 40-year service life.

In 1982, the staff anticipated interest in license renewal and held a workshop on nuclear power plant aging. This workshop led the NRC to establish a comprehensive program plan for nuclear plant aging research. From the results of that research, a technical review group concluded that many aging phenomena are readily manageable and pose no technical issues precluding life extension for nuclear power plants. In 1986, the staff published a request for comment on a policy statement that would address major policy, technical, and procedural issues related to license renewal for nuclear power plants.

In 1991, the staff published 10 CFR Part 54, the License Renewal Rule (Volume 56, page 64943, of the *Federal Register* (56 FR 64943), dated December 13, 1991). The staff participated in an industry-sponsored demonstration program to apply 10 CFR Part 54 to a pilot plant and to gain the experience necessary to develop implementation guidance. To establish a scope of review for license renewal, 10 CFR Part 54 defined age-related degradation unique to license renewal; however, during the demonstration program, the staff found that adverse aging effects on plant systems and components are managed during the period of initial license and that the scope of the review did not allow sufficient credit for management programs, particularly the implementation of 10 CFR 50.65, "Requirements for Monitoring the Effectiveness of Maintenance at Nuclear Power Plants," which regulates management of plant-aging phenomena. As a result of this finding, the staff amended 10 CFR Part 54 in 1995. As published May 8, 1995, in 60 FR 22461, amended 10 CFR Part 54 establishes a regulatory process that is simpler, more stable, and more predictable than the previous 10 CFR Part 54. In particular, as amended, 10 CFR Part 54 focuses on the management of adverse aging effects rather than on the identification of age-related degradation unique to license renewal. The staff made these rule changes to ensure that important systems, structures, and components (SSCs) will continue to perform their intended functions during the period of extended operation. In addition, the amended 10 CFR Part 54 clarifies and simplifies the integrated plant assessment process to be consistent with the revised focus on passive, long-lived structures and components (SCs).

Concurrent with these initiatives, the staff pursued a separate rulemaking effort (61 FR 28467, June 5, 1996) and amended 10 CFR Part 51 to focus the scope of the review of environmental impacts of license renewal in order to fulfill NRC responsibilities under the National Environmental Policy Act of 1969.

1.2.1 Safety Review

License renewal requirements for power reactors are based on two key principles:

(1)   The regulatory process is adequate to ensure that the licensing bases of all currently operating plants maintain an acceptable level of safety with the possible exceptions of the detrimental aging effects on the functions of certain SSCs, as well as a few other safety-related issues, during the period of extended operation.

(2)   The plant-specific licensing basis must be maintained during the renewal term in the same manner and to the same extent as during the original licensing term.

In implementing these two principles, 10 CFR 54.4, "Scope," defines the scope of license renewal as including those SSCs that (1) are safety-related, (2) whose failure could affect safety-related functions, or (3) are relied on to demonstrate compliance with the NRC's regulations for fire protection, environmental qualification (EQ), pressurized thermal shock (PTS), anticipated transient without scram (ATWS), and station blackout (SBO).

Pursuant to 10 CFR 54.21(a), a license renewal applicant must review all SSCs within the scope of 10 CFR Part 54 to identify SCs subject to an aging management review (AMR). Those SCs subject to an AMR perform an intended function without moving parts or without change in configuration or properties and are not subject to replacement based on a qualified life or specified time period. Pursuant to 10 CFR 54.21(a), a license renewal applicant must demonstrate that the aging effects will be managed such that the intended function(s) of those SCs will be maintained consistent with the current licensing basis (CLB) for the period of extended operation. However, active equipment is considered to be adequately monitored and maintained by existing programs. In other words, detrimental aging effects that may affect active equipment can be readily identified and corrected through routine surveillance, performance monitoring, and maintenance. Surveillance and maintenance programs for active equipment, as well as other maintenance aspects of plant design and licensing basis, are. required throughout the period of extended operation.

Pursuant to 10 CFR 54.21(d), the LRA is required to include a UFSAR supplement with a summary description of the applicant's programs and activities for managing aging effects and an evaluation of time-limited aging analyses (TLAAs) for the period of extended operation.

License renewal also requires TLAA identification. During the plant design phase, certain assumptions about the length of time the plant may operate are incorporated into design calculations for several plant SSCs. In accordance with 10 CFR 54.21(c)(1), the applicant must either show that these calculations remain valid for the period of extended operation, project the analyses to the end of the period of extended operation, or demonstrate that the aging effects on these SSCs will be adequately managed for the period of extended operation.

In 2005, the NRC revised Regulatory Guide (RG) 1.188, "Standard Format and Content for Applications to Renew Nuclear Power Plant Operating Licenses." This RG endorses Nuclear Energy Institute (NEI) 95-10, Revision 6, "Industry Guideline for Implementing the Requirements of 10 CFR Part 54 - The License Renewal Rule," issued in June 2005. NEI 95-10 details an acceptable method of implementing 10 CFR Part 54. The staff also used the SRP-LR to review the LRA.

In the LRA, the applicant utilized the process defined in NUREG-1801, Revision 1, "Generic Aging Lessons Learned (GALL) Report," dated September 2005. The GALL Report summarizes staff-approved aging management programs (AMPs) for many SCs subject to an AMR. If an applicant commits to implementing these staff-approved AMPs, the time, effort, and resources for LRA review can be greatly reduced, improving the efficiency and effectiveness of the license renewal review process. The GALL Report summarizes the aging management evaluations, programs, and activities credited for managing aging for most of the SCs used by nuclear power plants. The report is also a quick reference for both applicants and staff reviewers with respect to AMPs and activities that can manage aging adequately during the period of extended operation.

1.2.2 Environmental Review

Part 51 of 10 CFR contains regulations pertaining to environmental protection. In December 1996, the staff revised the environmental protection regulations to facilitate the environmental review for license renewal. The staff prepared the GEIS to document its evaluation of possible environmental impacts associated with nuclear power plant license renewals. For certain types of environmental impacts, the GEIS contains generic findings that apply to all nuclear power plants and are codified in Appendix B, "Environmental Effect of Renewing the Operating License of a Nuclear Power Plant," to Subpart A of 10 CFR Part 51, as Category 1 issues. Pursuant to 10 CFR 51.53(c)(3)(i), a license renewal applicant may incorporate these generic findings in its environmental report. In accordance with 10 CFR 51.53(c)(3)(ii), an environmental report also must include analyses of environmental impacts that must be evaluated on a plant-specific basis (i.e., Category 2 issues).

In accordance with the National Environmental Policy Act of 1969 and 10 CFR Part 51, the staff reviewed the plant-specific environmental impacts of license renewal, including any new and significant information not considered in the GEIS. As part of its scoping process, the staff held a public meeting on September 19, 2007 at the Colonial Terrace in Cortlandt Manor, New York, to identify plant-specific environmental issues. The draft, plant-specific Supplement 38 to the GEIS documents the results of the environmental review and makes a preliminary recommendation as to the license renewal action, based on environmental considerations. The staff held additional public meetings on February 12, 2009, in Cortlandt Manor, New York, to receive comments on the draft, plant-specific GEIS Supplement 38. The staff received numerous comments concerning the draft supplement. The staff plans to issue the final supplement in February 2010.

1.3 Principal Review Matters

Part 54 of 10 CFR describes the requirements for renewal of operating licenses for nuclear power plants. The staff's technical review of the LRA was performed in accordance with

10 CFR Part 54 requirements and NRC guidance. Section 54.29, "Standards for Issuance of a Renewed License," of 10 CFR sets forth the license renewal standards. This SER describes the results of the staff's safety review of the Indian Point LRA.

Pursuant to 10 CFR 54.19(a), the NRC requires a license renewal applicant to submit general information, which the applicant provided in LRA Section 1. The staff reviewed LRA Section 1 and finds that the applicant has submitted the required information.

Pursuant to 10 CFR 54.19(b), the NRC requires that the LRA include "conforming changes to the standard indemnity agreement, 10 CFR 140.92, Appendix B, to account for the expiration term of the proposed renewed license." On this issue, the applicant stated in the LRA:

> The agreement shall terminate at the time of expiration of the license specified in Item 3 of the Attachment to the agreement, which is the last to expire. Item 3 of the Attachment to the indemnity agreement, as revised by Amendment No. 25, lists IPEC operating license numbers DPR-26 and DPR-64. The applicants request that conforming changes be made to Article VII of the indemnity agreement, and Item 3 of the Attachment to that agreement, specifying the extension of agreement until the expiration date of the renewed IPEC facility operating license sought in this application. In addition, should the license number be changed upon issuance of the renewal license, the applicants request that conforming changes be made to Item 3 of the Attachment, and other sections of the indemnity agreement as appropriate.

The staff intends to maintain the original license numbers upon issuance of the renewed licenses, if approved. Therefore, conforming changes to the indemnity agreement need not be made and the 10 CFR 54.19(b) requirements have been met.

Pursuant to 10 CFR 54.21, "Contents of Application - Technical Information," the NRC requires that the LRA contain (a) an integrated plant assessment, (b) a description of any CLB changes during the NRC's review of the LRA, (c) an evaluation of TLAAs, and (d) an FSAR supplement. LRA Sections 3 and 4 and Appendix B address the license renewal requirements of 10 CFR 54.21(a) and (c). LRA Appendix A satisfies the license renewal requirements of 10 CFR 54.21(d).

Pursuant to 10 CFR 54.21(b), the NRC requires that, each year following submission of the LRA and at least three months before the scheduled completion of the NRC's review, the applicant submit an LRA amendment identifying any CLB changes to the facility that materially affect the contents of the LRA, including the FSAR supplement. By letters dated June 11, 2008, and June 30, 2009, the applicant submitted LRA annual updates which summarize the CLB changes that have occurred during the staff's review of the LRA.

Pursuant to 10 CFR 54.22, "Contents of Application - Technical Specifications," the NRC requires that the LRA include changes or additions to the technical specifications (TS) that are necessary to manage aging effects during the period of extended operation. In LRA Appendix D, the applicant stated that it had not identified any TS changes necessary for issuance of the renewed IP operating licenses. This statement adequately addresses the 10 CFR 54.22 requirement.

The staff evaluated the technical information required by 10 CFR 54.21 and 10 CFR 54.22 in accordance with NRC regulations and regulatory guidance. SER Sections 2, 3, and 4 document the staff's evaluation of the LRA technical information.

As required by 10 CFR 54.25, "Report of the Advisory Committee on Reactor Safeguards," the SER was referred to the ACRS, and the ACRS issued a report documenting its evaluation of the staff's LRA review and SER. SER Section 5 contains the ACRS report. SER Section 6 documents the findings required by 10 CFR 54.29.

1.4 Interim Staff Guidance

The staff, industry, and other interested stakeholders gain experience and develop lessons learned with each renewed license. The lessons learned address the staff's performance goals of maintaining safety, improving effectiveness and efficiency, reducing regulatory burden, and increasing public confidence. Interim staff guidance (ISG) is documented for use by the staff, industry, and other interested stakeholders until incorporated into such license renewal guidance documents as the SRP-LR and GALL Report.

Table 1.4-1 shows the current set of ISGs and proposed ISGs, as well as the SER sections in which they are addressed.

Table 1.4-1 Current and Proposed Interim Staff Guidance

ISG Issue (Approved ISG Number)	Purpose	SER Section
Nickel-alloy components in the reactor coolant pressure boundary (LR-ISG-19B)	To address the cracking of nickel-alloy components in the reactor pressure boundary.

This ISG is currently under development. NEI and EPRI-MRP will develop an augmented inspection program for GALL AMP XI.M11-B. This AMP will not be completed until the NRC approves an augmented inspection program for nickel-alloy base metal components and welds as proposed by EPRI-MRP. | SER Section 3.0.3.3.5 |
| Changes to Generic Aging Lesson Learned (GALL) Report Aging Management Program (AMP) XI.E6, "Electrical Cable Connections Not Subject to 10 CFR 50.49 Environmental Qualification Requirements" (LR-ISG-2007-02) | To address the frequency of inspection of electrical cable connections not subject to 10 CFR 50.49 prior to the period of extended operation.

The staff issued the proposed ISG for public comment. A final ISG has not yet been issued. | SER Section 3.0.3.3.6 |

ISG Issue (Approved ISG Number)	Purpose	SER Section
Staff Guidance Regarding the Station Blackout Rule (10 CFR 50.63) Associated with License Renewal Applications (LR-ISG-2008-01)	To clarify the scoping boundary of the offsite recovery paths that must be included within the scope of license renewal for station blackout. The staff issued the proposed ISG for public comments. On July 7, 2009, the staff withdrew LR-ISG-2008-01. See 74 FR 33478, dated July 13, 2009.	Not applicable

1.5 Summary of Open Items

On January 15, 2009, at the time the SER with Open Items was issued, the staff identified the following open items (OIs). An item was considered open if, in the staff's judgment, it has not been shown to meet all applicable regulatory requirements at the time of the issuance of the SER. The staff assigned a unique identifying number to each OI. By letters dated January 27, May 1, and June 12, 2009, the applicant provided additional information which enabled the staff to close out the open items.

OI 2.3A.3.11-1: (SER Section 2.3A.3.11 – IP2 Fire Protection – Water)

In LRA Section 2.3.3.11, the applicant listed the component types that require aging management review. However, some components were not included in the list that are either referenced in the applicant's current licensing basis documents or are shown on the license renewal drawings. Therefore, in RAI 2.3A.3.11-2, the staff asked the applicant to determine whether the components listed in the RAI should be included as component types subject to an AMR, and if not, to justify the exclusion. By letter dated November 16, 2007, the applicant stated that yard hose houses and chamber housings are not subject to an aging management review (AMR) because failure of these components will not result in a failure of the fire suppression function of the associated fire hydrant and the sprinkler system, respectively. The yard hose houses and chamber housings are passive, long-lived components that were identified as within the scope of license renewal. Therefore, the staff considers that these components are subject to an AMR in accordance with 10 CFR 54.21(a)(1). The staff indicated that the applicant should justify why the yard hose houses and chamber housings are not subject to an AMR.

By letter dated January 27, 2009, the applicant stated that yard hose houses and chamber housings provide no function that supports 10 CFR 50.48 requirements; therefore, they are not within the scope of license renewal. The closure of this item is documented in SER Section 2.3A.3.11.2.

OI 2.3.4.2-1: (SER Sections 2.3A.4.2 and 2.3B.4.2 – Main Feedwater System)

IP2 (SER Section 2.3A.4.2 – IP2 Main Feedwater System):

UFSAR Section 14.2.5.6, Containment Peak Pressure for a Postulated Steam Line Break, indicates that for IP2 the applicant takes credit for the main feedwater stop valves, BFD-5's, to close within 120 seconds, in the event of the failure of the main feedwater control valve.

In its revised response to RAI 2.3A.4.2-1 regarding feedwater isolation valves, dated March 24, 2008, the applicant stated that the feedwater valves credited for feedwater isolation are safety-related and, although not highlighted on the license renewal drawing, these valves and the remainder of the feedwater system components on the associated license renewal drawing are within scope of license renewal and are subject to an AMR based upon meeting the requirements of 10 CFR 54.4(a)(2) because of their potential for spatial interaction with safety-related equipment. Based upon the staff's understanding of the applicant's UFSAR, the main feedwater stop valves (BFD-5's), have an intended function that meets the criteria of 10 CFR 54.4(a)(1); however, these valves are neither included within the "system intended function boundary," nor are they highlighted on the license renewal drawings for having an intended function in accordance with 10 CFR 54.4(a)(1).

By letter dated December 30, 2008, the staff requested the applicant to justify the exclusion of the main feedwater stop valves (BFD-5's), from the scope of license renewal in accordance with 10 CFR 54.4(a)(1).

By letter dated January 27, 2009, the applicant explained that the BFD-5 isolation valves are nonsafety-related components, and consistent with the requirements in 10 CFR 54.4(a)(2), the valves are included in the scope for license renewal. The closure of this item is documented in SER Section 2.3A.4.2.2.

IP3 (SER Section 2.3B.4.2 – IP3 Main Feedwater System):

UFSAR Section 14.2.5, Rupture of a Steam Pipe, states in the event of a main steam line break incident, the motor-operated valves (MOVs) associated with each of the feedwater regulating valves (FRVs) also will close. The mechanical stroke time of 120 seconds to close these associated MOVs has been analyzed and is acceptable. License renewal drawing 9321-20193 shows a "HIGH STEAM FLOW SI LOGIC" signal goes to these MOVs (BFD-90's). UFSAR Section 14.2.5.1 states that redundant isolation of the main feedwater lines is necessary, because sustained high feedwater flow would cause additional cooldown. Therefore, in addition to the normal control action which will close the main feedwater valves, any safety injection signal will rapidly close all feedwater control valves (including the motor-operated block valves and low-flow bypass valves), trip the main feedwater pumps, and close the feedwater pump discharge valves.

The motor-operated block valves shown on license renewal drawings are BFD-5s and BFD-90s for the main FRVs, and the low flow bypass regulating valves, respectively. The feedwater isolation valves, BFD-5's and BFD-90's, are not included within the "system intended function boundary," nor are they highlighted on the license renewal drawings for having an intended function in accordance with 10 CFR 54.4(a)(1).

By letter dated December 30, 2008, the staff requested the applicant to justify the exclusion of the isolation valves, BFD-5's and BFD-90's, from the scope of license renewal in accordance with 10 CFR 54.4(a)(1).

By letter dated January 27, 2009, the applicant explained that the BFD-5 and BFD-90 isolation valves are nonsafety-related components, and consistent with the requirements in 10 CFR 54.4(a)(2), the valves are included in the scope for license renewal. The closure of this item is documented in SER Section 2.3B.4.2.2.

OI 2.3A.4.5-1: (SER Section 2.3A.4.5 – IP2 Auxiliary Feedwater Pump Room Fire Event)

In LRA Section 2.3.4.5 the applicant describes systems not described elsewhere in the application credited for mitigating the consequences of a Unit 2 fire event in the auxiliary feedwater (AFW) pump room. Each system listed has an intended function of support safe shutdown in the event of a fire in the auxiliary feed pump room (10 CFR 50.48) in accordance with 10 CFR 54.4(a)(3). However, the applicant did not highlight the components or flowpaths needed to support this event on the license renewal drawings. In addition, the applicant did not identify and list the structures and components that are subject to an AMR in accordance with 10 CFR 54.21(a)(1). Therefore, based upon the information provided in the LRA, the staff was not able to verify which components needed to perform the stated function are included within the scope of license renewal and are subject to an AMR.

By letter dated December 30, 2008, the staff requested the applicant to: a) identify the system support function for the AFW pump room fire event, b) clearly identify the portions of the systems' flowpaths that support these functions that are subject to an AMR, and c) identify the portions of these flowpaths that are not already in scope for 10 CFR 54.4(a)(1) or (a)(2).

By letter dated January 27, 2009, the applicant explained that it has included the components required to support the safety function in the event of a fire in the AFW pump room within the scope of license renewal in accordance with 10 CFR 54.4(a)(3), and identified the passive long-lived components requiring an AMR in accordance with 10 CFR 54.21(a)(1). The closure of this item is documented in SER Section 2.3A.4.5.2.

OI 2.5-1: (SER Section 2.5.1 – Electrical and Instrumentation and Control Systems)

By letter dated November 16, 2007, the applicant responded to RAI 2.5-1 and revised LRA Figures 2.5-2 and 2.5-3, the "Offsite Power Scoping Diagram(s)" for IP2 and IP3, to address staff concerns regarding the IP2 and IP3 primary and secondary offsite power paths. By letter dated March 24, 2008, the applicant revised its response to RAI 2.5-1. In a subsequent letter dated August 14, 2008, the applicant further clarified its response to RAI 2.5-1.

At the time of issuance of the SER with Open Items, the staff was completing its review of the applicant's information on the SBO scoping boundary. As a result of its review, the staff identified a need for additional information, and by letter dated May 20, 2009, the staff requested the applicant to explain why certain components associated with the delayed access circuit were not included within the scope of license renewal. By letter dated June 12, 2009, the applicant provided additional information. The closure of this item is documented in SER Section 2.5.1.2.

OI 3.0.3.2.7-1: (SER Section 3.0.3.2.7 – Fire Protection Program)

During an audit, the staff reviewed program basis documents (for IP3) associated with the fire protection AMP. One of the basis documents states that 15 percent of the fire seals located in fire barriers are demonstrated to be operable by visual inspection on a frequency of 24 months. However, for those penetration seals that are inaccessible, the frequency of inspection is given as "not required." By letter dated April 29, 2008, the staff requested that the applicant justify the lack of visual inspections of inaccessible penetration seals.

In its response, dated May 28, 2008, the applicant stated that penetration seals are inspected at least once every seven operating cycles. However, IP3 site surveillance procedure provides provisions for cases where a penetration seal may become inaccessible for periodic inspection as result of plant configuration changes (i.e., installation of new plant equipment, walls, barriers, or other obstacles). In such cases, the IP3 site procedure includes guidance for the cessation of periodic surveillance of such penetration seals, subject to preparation of a formal fire protection engineering evaluation justifying the discontinuance of periodic visual surveillance.

As stated in the IP3 basis document, the visual inspection of inaccessible penetration seals is "not required" if justified by a supporting fire protection engineering evaluation, developed in accordance with the guidance of GL 86-10. On a case-by-case basis, the inaccessibility of any such penetration seal must be justified, and the fire protection adequacy of the configuration must be demonstrated. The evaluation, as stated in the basis document, must include assessment of proximate combustible loading, mitigating features, and the consequences of potential failure of the affected seal.

The staff reviewed the applicant's response and found that it did not address the fact that GL 86-10 evaluations exist for all inaccessible fire barrier penetration seals; the response only indicated that it is a part of the fire protection program to perform such analyses. The staff requested the applicant to confirm that these analyses do exist and are periodically reviewed and updated to ensure their continued applicability.

By letter January 27, 2009, the applicant stated that there are no IP3 fire barrier penetration seals excluded from periodic inspection due to inaccessibility. Therefore, there are no corresponding engineering evaluations. The closure of this item is documented in SER Section 3.0.3.2.7.

OI 3.0.3.2.15-1: (SER Section 3.0.3.2.15 – Structures Monitoring Program)

In response to Audit Item 359 regarding IP2 reactor cavity leakage into the containment, Entergy described the degraded conditions, summarized corrective actions taken, and identified the current status of the degradation. The reactor cavity at IP2 has a history of leakage at the upper elevations of the stainless steel cavity liner when flooded during refueling outages. Attempts have been made over the last several outages to mitigate this condition, with limited success. An action plan is being developed for a permanent fix to this issue. However, Entergy made no commitment for augmented inspection during the extended period of operation. In a follow-up discussion, the staff expressed its concern with regard to the potential for degradation of the underlying concrete and reinforcement rebar due to the leakage of borated water through the cavity liner and potential impact of the leakage on other adjacent structures. The staff requested Entergy to provide the technical basis as to why augmented inspection during the period of extended operation is not necessary, if the recurring leak condition is not permanently fixed.

In an August 14, 2008, supplemental response to the staff's request, the applicant provided further information regarding the matter and committed to perform a one-time inspection and evaluation of a sample of potentially affected refueling cavity concrete, including embedded reinforcing steel, prior to the period of extended operation, in order to provide additional assurance that the concrete walls have not degraded (Commitment 36).

The staff has concluded that Entergy's commitment to perform a one-time inspection and evaluation of a sample of potentially affected refueling cavity concrete, including embedded reinforcing steel, prior to the period of extended operation, is appropriate in order to assess the current state of the concrete and rebar. However, because the applicant does not plan to perform periodic inspections of the refueling cavity and affected area, the staff determined that for this structure/environment/aging effect combination, the LRA is not consistent with the GALL Report AMP. Additionally, the applicant's program did not address concrete exposed to borated water.

By letter dated November 6, 2008, the applicant submitted a supplemental response to Audit Question 359, describing its plan for implementing a permanent fix over the next three (3) scheduled IP2 refueling outages (2010, 2012, and 2014). At the time of the issuance of the SER with Open Items, the staff was reviewing the applicant's response, pertinent to the effects of the refueling cavity leakage on the affected structures during the period of extended operation. As a result of the review, the staff identified the need for additional information, and by letters dated April 3, 2009, and May 20, 2009, the staff requested the applicant to provide additional information on the leakage path from the refueling cavity to the collection point lower in containment, and to explain how the structures monitoring program will adequately manage potential aging effects in this region during the period of extended operation. By letters dated May 1, 2009, and June 12, 2009, the applicant responded to the staff's request for additional information. The closure of this item is documented in the "Operating Experience" section of SER Section 3.0.3.2.15.

OI 3.0.3.2.15-2: (SER Section 3.0.3.2.15 – Structures Monitoring Program)

In response to Audit Item 360 regarding IP2 spent fuel pool (SFP) crack/leak paths, Entergy described the degraded conditions in greater detail, summarized corrective actions taken, and identified the current status of the degradation. The leakage from the spent pool liner was originally observed and repaired in 1992. Spent fuel pool leakage was again observed during excavation for the IP2 Fuel Storage Building in 2005. Entergy stated its belief that the conditions leading to this leakage have been corrected.

Entergy made no commitment for augmented inspection during the period of extended operation. Due to the lack of a leak-chase channel system at IP2 to monitor, detect and quantify potential leakage through the SFP liner, the staff was concerned that there has been insufficient time following the corrective actions to be certain that the leakage problems have been permanently corrected. In a follow-up discussion, the staff requested Entergy to provide the technical basis as to why augmented inspection during the extended period of operation is not necessary.

In an August 14, 2008, supplemental response to the staff's request, the applicant committed to test the groundwater outside the IP2 spent fuel pool for the presence of tritium from samples taken from adjacent monitoring wells, every 3 months (Commitment 25). The presence of tritium in the groundwater could be indicative of a continuing leak from the spent fuel pool.

Although Entergy has taken corrective action and has committed to quarterly monitoring for tritium in the groundwater, the staff was concerned that hidden degradation of concrete and rebar may have resulted from prior leakage, and may be continuing if there is still an active leakage mechanism. The staff requested the applicant to submit additional relevant information on the condition of concrete and rebar in areas where leakage was detected, and the design margins in these areas.

By letter dated November 6, 2008, the applicant submitted the requested information, which provides a detailed description of (1) the design margins for the spent fuel pool concrete walls and (2) the results of prior concrete core sample testing and rebar corrosion testing. At the time of the issuance of the SER with Open Items, the staff was reviewing the applicant's response. As a result of its review, the staff identified the need for additional information. By letter dated April 3, 2009, the staff requested the applicant to explain how the Structures Monitoring Program will adequately manage potential aging effects in the inaccessible concrete of the IP2 spent fuel pool due to borated water leakage during the period of extended operation. By letter dated May 1, 2009, the applicant responded to the staff's request for additional information. The closure of this item is documented in the "Operating Experience" section of SER Section 3.0.3.2.15.

OI 3.0.3.3.2-1: (SER Section 3.0.3.3.2-1 – Containment Inservice Inspection Program)

In response to Audit Item 361 regarding areas of spalling of the exterior concrete containment, Entergy provided information about the areas and reasons for the spalling. The applicant stated that the spalls occur at locations where Cadweld sleeves have insufficient concrete cover, attributed to an original installation deficiency. Rusting is not active and spalls are in an area where the rebar stresses are low. Entergy indicated that Raytheon has evaluated the structural margins for the IP containments, and at the locations of the exposed rebar, there is sufficient margin to accommodate additional loss of material due to corrosion. The condition is being monitored under the containment inservice inspection program (CII-IWL). Entergy stated that remedial action will be taken if the spalls further degrade and affect structural integrity.

In an August 14, 2008 supplemental response to the staff's request, the applicant committed to enhance the CII-IWL inspections during the period of extended operation through enhanced characterization of the degradation (i.e., quantifying the dimensions of noted indications through the use of optical aids), and that this quantification will allow for more effective trending of degradation following future inspections (Commitment 37). However, since the degraded areas will remain exposed to the environment during the period of extended operation, the staff needed additional clarification of how Entergy plans to implement aging management during the period of extended operation.

The staff requested additional relevant information for the IP2 and IP3 containments on the design margins at the locations of observed degradation, identifying the specific locations and dimensions of the damage.

By letter dated November 6, 2008, the applicant submitted the requested information, describing the design margins for the IP containment structures at the locations of existing concrete degradation. At the time of the issuance of the SER with Open Items, the staff was reviewing the applicant's response. As a result of its review, the staff identified the need for additional information. By letter dated April 3, 2009, the staff requested the applicant to explain how the existing degradation and design margin will be considered in performing periodic inspections to monitor degradation that would ensure that there is no loss of containment intended function during the period of extended operation. By letter dated May 1, 2009, the applicant responded to the staff's request for additional information. The closure of this item is documented in the "Operating Experience" section of SER Section 3.0.3.3.2.

OI 3.0.3.3.3-1: (3.0.3.3.3 Heat Exchanger Monitoring Program)

LRA Section B.1.17 states that the minimum acceptable tube wall thickness for each heat exchanger inspected is based upon a component-specific engineering evaluation. Wall thickness is acceptable if greater than the minimum wall thickness for the component.

The applicant stated that the existing program will be enhanced to include the minimum wall thickness for the new heat exchangers added to the scope of the program, and to specify that if visual examination is performed, the acceptance criterion is "no unacceptable signs of degradation." The acceptance criteria for the eddy current tests based on minimum wall thicknesses are acceptable. However, the acceptance criteria for visual examination were not clear and appeared to be subjective. By letter dated December 30, 2008, the staff requested that Entergy define the visual inspection acceptance criteria.

By letter dated January 27, 2009, the applicant stated that visual inspections are performed on heat exchangers that cannot be inspected by quantitative non-destructive examination due to design limitations. The applicant further stated that visual inspection of external portions of heat exchanger tubes focuses on detecting the extent of tube erosion, vibration wear, corrosion, pitting, fouling, and scaling. Any unacceptable signs of degradation will be evaluated through the corrective action process. The closure of this item is documented in the "Acceptance Criteria" section of SER Section 3.0.3.3.3.

OI 3.0.3.3.4-1: (SER Section 3.0.3.3.4 Inservice Inspection Program)

The staff noted that the applicant indicated it plans to enhance the Inservice Inspection Program to provide for periodic visual inspections of lubrite sliding supports used in the SG supports and reactor coolant pump (RCP) supports in order to confirm the absence of aging effects. By letter dated December 30, 2008, the staff requested the applicant to establish and justify its selection of the inspection methods, inspections frequencies, sample sizes, and acceptance criteria that are applicable to the lubrite components, and the corrective actions that would be implemented if these acceptance criteria are exceeded.

By letter dated January 27, 2009, the applicant stated that the Inservice Inspection Program will be enhanced prior to the period of extended operation to include explicit provisions for periodic inspections of the lubrite sliding supports. The closure of this item is documented in the "Detection of Aging Effects" section of SER Section 3.0.3.3.4.

OI 3.0.3.3.4-2: (SER Section 3.0.3.3.4 Inservice Inspection Program)

The staff noted that the "corrective actions" program element for AMP B.1.18, Inservice Inspection Program, credits only the corrective actions in the ASME Code, Section XI, Articles IWA-4000 and IWA-7000 as the corrective action criteria for the program. The ASME Code, Section XI editions of record for IP are the 2001 Edition of the ASME Code, Section XI inclusive of the 2003 Addenda for IP2, and the 1989 Edition of the ASME Code, Section XI, with no addenda for IP3. The staff noted that Entergy did not credit component-specific corrective action criteria in ASME Section XI, Article IWB-4000/7000 for Class 1 components, Article IWC-4000/7000 for Class 2 components, Article IWD-4000/7000 Class 3 components, or Article IWF-4000/7000 for ASME Code Class component supports as being within the scope of the "corrective action" program element for this AMP. By letter dated December 30, 2008, the staff asked the applicant to clarify whether the content of the "corrective actions" program element was intended to mean that Entergy will implement the corrective action provisions in the ASME Code, Section XI, Subsections IWA, IWB, IWC, IWD, and IWF that are applicable to the component Code Class in the applicable ASME Code, Section XI code of record.

By letter dated January 27, 2009, the applicant stated that it will implement the corrective action provisions in the ASME Code, Section XI, Subsections IWA, IWB, IWC, IWD, and IWF that are applicable to the component Code Class in the applicable ASME Code, Section XI edition of record. The closure of this item is documented in the "Corrective Actions" section of SER Section 3.0.3.3.4.

OI 3.0.3.3.7-1: (SER Section 3.0.3.3.7 – Periodic Surveillance and Preventive Maintenance)

In LRA Appendix B, Section B.1.29, the applicant described the existing Periodic Surveillance and Preventive Maintenance Program as an existing, plant-specific program. The staff reviewed the applicant's program using the review criteria and guidance in the SRP-LR, Appendix. During its review, the staff determined that additional information regarding certain program elements was needed. By letter dated December 30, 2008, the staff issued an RAI to obtain information in the following areas:

1.   The "scope of program" program element for the Periodic Surveillance and Preventive Maintenance Program did not specify which components were within the scope of the program.

2.   The applicant appeared to be crediting visual examinations, in part, to manage cracking but did not identify the visual techniques to be used.

3.   The "monitoring and trending" program element discussion only mentioned that the activities within the scope of the AMP provided for adequate monitoring and trending; there was no discussion on how the data from the inspections performed under the

"detection of aging effects" program element would be collected, quantified, or evaluated against applicable acceptance criteria, and used to make predictions related to degradation growth or to schedule re-inspections of the components.

4.   For the majority of the elastomeric or polymeric components within the scope of the AMP, the applicant credited both visual examinations and manual flexing of the components to manage changes in material properties of these elastomeric or polymeric components. However, material properties are intrinsic thermodynamic properties that cannot be monitored by direct visual or NDE inspection methods, and changes in material properties (such as loss of fracture toughness, hardening, or increases or reductions in strength) are more appropriately managed through appropriate material property analyses (including destructive analyses) or though performance of physical tests (such as flexing, etc.) that could provide some indication of whether the material properties for the components were changing.

5.   Certain statements regarding "operating experience" were ambiguous in that the applicant did not indicate clearly whether aging had been detected but that the amount of aging was determined to be acceptable when compared to the acceptance criteria for the aging effect, or whether the inspections did not identify the presence of aging effects in the components being inspected.

By letter dated January 27, 2009, the applicant responded to the staff's request for additional information. The closure of this item is documented in SER Section 3.0.3.3.7.

OI 3.1.2-1: (SER 3.1.2.1.3 Cracking Due to Cycling Loading, Stress Corrosion Cracking, and Primary Water Stress Corrosion Cracking)

During its review of the nickel alloy components and the Nickel Alloy Program, the staff determined that for some component types, the applicant: (1) did not indicate which base metal is used at IP (i.e., Alloy 600); (2) did not include any AMR entries for reactor vessel bottom head drains; and (3) did not credit the Inservice Inspection Program to manage cracking in steam generator primary nozzle closure rings. By letter dated December 30, 2008, the staff requested the applicant to:

Part A - Clarify whether the following components at IP2 or IP3 are fabricated from Alloy 600 base metal materials or welded with Alloy 182 or Alloy 82 filler metal materials: (1) control rod drive (CRD) housing-CRD nozzle welds, (2) upper reactor vessel closure head (RVCH) head vent nozzle-to-RVCH welds, and (3) CRD housing penetration core exit thermocouple nozzle assembly (CETNA™) components.

Part B - The staff notes that in the applicant's response to Audit Item 208, dated December 18, 2007, the applicant stated that the LRA Tables 3.1.2-1-IP2 through 3.1.2-4-IP2 and LRA Tables 3.1.2-1-IP3 through 3.1.2-4-IP3 include numerous AMR items for nickel-alloy components. The applicant stated that these AMR items are compared to GALL Report Items IV.A2-18 and IV.A2-19, which correspond to LRA table entries 3.1.1-31 and 3.1.1-65. The applicant stated that the AMR in LRA AMR 3.1.1-69 is only for management of cracking in the RV inlet and outlet nozzle safe-ends and the RV bottom head drain safe-ends. With respect to the AMRs on cracking of nickel alloy bottom mounted instrumentation (BMI) nozzle components, the staff notes that the response to Audit Item 208 stated that the RV bottom head safe-ends at IP2 and IP3 are those for the RV bottom head drains, but LRA Tables 3.1.2-1-IP2 and 3.1.2-1-IP3 do not include any AMR entries for RV bottom head drains. The staff requested the applicant to provide its basis on whether LRA Tables 3.1.2-1-IP2 and 3.1.2-1-IP3 need to be amended to include new AMRs for RV bottom head drains and their associated drain-to-bottom head welds, and if so to clarify whether the bottom head drains are fabricated from Alloy 600 base metal materials or are welded to the bottom RV heads using Alloy 82 or 182 nickel alloy filler metal materials.

Part C - AMRs of LRA Tables 3.1.2-4-IP2 and 3.1.2-4-IP3, which pertain to the management of cracking in the steam generator (SG) primary nozzle closure rings, credit only the Water Chemistry Control Program to manage cracking of the components. GALL Report Table IV.D1, Line Item D1-1 for these components recommends, in part, that the Inservice Inspection Program be credited for aging management of this effect in addition to Water Chemistry Control Program – Primary and Secondary. Given the information requested in Part A above, the staff requested the applicant to provide a basis for why the AMRs on cracking of the nickel alloy SG primary nozzle closure rings were aligned to GALL AMR Table VI.D1, Line Item D1-6, and why the Inservice Inspection Program is not also credited.

By letter dated January 27, 2009, the applicant responded to the staff's request for additional information. The closure of this item is documented in SER Sections 3.0.3.3.5, 3.1.2.1.3, 3.1.2.2.13, and 3.1.2.2.16.

OI 3.1.2.2.7-1: (SER Section 3.1.2.2.7 - Cracking Due to Stress Corrosion Cracking)

The Inservice Inspection Program is a plant-specific condition monitoring program for the management of cracking in ASME Code Class 1 components, including ASME Code Class 1 cast austenitic stainless steel (CASS) components. However, the staff noted that the inspections credited under this program might be either ultrasonic test (UT) examinations or enhanced VT-1 visual examinations. The staff also noted that the applicant's program includes a flaw evaluation methodology for CASS components that are susceptible to thermal aging embrittlement.

By letter dated December 30, 2008, the staff asked the applicant to: (a) clarify how current state of the art UT methods, as implemented through the Inservice Inspection Program or other programs, would be adequate to detect cracks in CASS materials, and (b) justify its basis for crediting the Thermal Aging Embrittlement of Cast Austenitic Stainless Steel (CASS) Program to manage and detect for cracking in the CASS pressurizer spray heads at IP2 and IP3.

By letter dated January 27, 2009, the applicant stated that because current volumetric examination methods are not adequate for reliable detection and evaluation of cracking in CASS components, ultrasonic testing examinations are not credited for use in aging management of reduction of fracture toughness in CASS components at Indian Point. Entergy also stated that listing the Thermal Aging Embrittlement of CASS Program on the line item for cracking may be unnecessary, but was included to demonstrate consistency with NUREG-1801 Item IV.C-3, which recommends a plant-specific program to address thermal aging embrittlement. The closure of this item is documented in SER Section 3.1.2.2.7.

OI 3.3-1: (SER Section 3.3A.2.3.1 – Service Water System - Summary of Aging Management Review – LRA Table 3.3.2-2-IP2)

The staff reviewed LRA Table 3.3.2-2-IP2, which summarizes the results of AMR evaluations for the service water system component groups. The LRA table referenced Note F for titanium heat exchanger shell externally exposed to condensation with no aging effect and no AMP. The staff noted that in LRA Table 3.3.2-9-IP2, the applicant used Note F for the same material/environment combination, but cited an aging effect of loss of material and stated that it will be managed by the Periodic Surveillance and Preventive Maintenance Program. This appears to be a discrepancy.

Similarly, the staff reviewed LRA Table 3.3.2-14-IP2, which summarizes the results of AMR evaluations for the emergency diesel generator system component groups. The LRA table referenced Note F for titanium heat exchanger tubes exposed to raw water (internal) having aging effects of fouling and loss of material which will be managed using the Service Water Integrity Program. The staff noted that in LRA Table 3.3.2-2-IP2, the applicant used Note F for the same material/environment combination but cites cracking as an additional aging effect. This appears to be a discrepancy.

The staff indicated that further information was required regarding the apparent discrepancies, before this item may be closed.

By letter dated January 27, 2009, the applicant stated that LRA Table 3.3.2-2-IP2 contains correct AMR results for the titanium heat exchanger shell externally exposed to condensation with no aging effect and no AMP, and that LRA Table 3.3.2-9-IP2 has been corrected. In addition, the applicant stated that the reason for the difference between cited aging effects in LRA Tables 3.3.2-14-IP2 and Table 3.3.2-2-IP2 is the difference between the grades of titanium used. In LRA Table 3.3.2-2-IP2, the grade of titanium installed in the service water system is unknown so it was conservatively assumed that the material was not grades 1, 2, 7, 11 or 12 and therefore, cracking was identified as an aging effect requiring management. The closure of this item is documented in SER Sections 3.3A.2.3.1 and 3.3A.2.3.11.

OI 3.4-1: (SER Section 3.4.2.1.9 – Auxiliary Feedwater Pump Room Fire Event)

In LRA Section 3.4.2, the applicant states that:

> The components in the systems required to supply feedwater to the steam generators during the short duration of the fire event are in service at the time the event occurs or their availability is checked daily. Therefore, integrity of the systems and components required to perform post-fire intended functions for at least one hour is continuously confirmed by normal plant operation. During the event these systems and components must continue to perform their intended functions to supply feedwater to the steam generators for a minimum of one hour. Significant degradation that could threaten the performance of the intended functions will be apparent in the period immediately preceding the event and corrective action will be required to sustain continued operation. For the minimal one hour period that these systems would be required to provide make up to the steam generators, further aging degradation that would not have been apparent

prior to the event is negligible. Therefore, no aging effects are identified, and no Summary of Aging Management Review table is provided.

Because these systems contain passive, long-lived components, the applicant must demonstrate that the effects of aging will be adequately managed so that the intended functions will be maintained consistent with the CLB for the period of extended operation. Based on the information contained in the LRA, Entergy did not appear to have demonstrated that the effects of aging for passive, long-lived components within the systems credited for providing flow to the steam generators during the fire event will be adequately managed.

By letter dated December 30, 2008, the staff issued an RAI to request that the applicant provide a list of passive, long-lived component types, material, environment, and aging effect combinations, and the programs that will be used to manage the aging effects for these SCs.

By letter dated January 27, 2009, the applicant responded to the staff's RAI and provided AMR results for the passive, long-lived components within the systems credited for providing flow to the steam generators during the fire event. For all component types, the applicant listed the aging effects and AMP as "none." The staff reviewed the response and determined that the systems contain passive, long-lived components made of materials that when exposed to the stated environments may experience aging effects, which must be managed during the period of extended operation in accordance with 10 CFR 54.21(a)(3).

By letter dated May 1, 2009, Entergy submitted a clarification response to RAI 3.4.2-1 as well as a new commitment to install a fixed automatic fire suppression system for IP2 in the AFW pump room prior to entering the period of extended operation. Entergy stated that this commitment will delete the requirement for IP2 to place reliance on certain portions of the secondary plant systems for alternate secondary heat sink measures to cope with potential AFW Pump Room fire scenarios.

The staff determined that because the planned installation is not yet part of the current licensing basis, it cannot make a finding consistent with the requirement in 10 CFR 54.29(a). Therefore, by letter dated May 20, 2009, the staff requested that the applicant provide information to demonstrate that the effects of aging will be adequately managed so that the intended function(s) will maintained consistent with the current licensing basis for the period of extended operation as required by 10 CFR 54.21(a)(3).

By letter dated June 12, 2009, the applicant responded to the staff's request and provided revised tables which include aging effects and AMPs to manage the aging effects for the component types that support the AFW pump room fire event that were not already included in scope and subject to aging management review for 10 CFR 54.4(a)(1) or (a)(2). The closure of this item is documented in SER Section 3.4.2.1.9

OI 3.5-1: (SER Section 3.5.2.2.1 – Containment Structures)

In LRA Sections 3.5.2.2.1.1 and 3.5.2.2.2.1, the applicant referenced an inconsistent combination of air entrainment and water-cement ratios. Per American Concrete Institute (ACI) 318-63, the water-cement ratio may be as high as 0.576 if there is no air entrainment. With air entrainment of four to six percent, the maximum water-cement should be 0.465. The staff asked the applicant to clarify if the correct value should be 0.465, and also to substantiate how it meets the code of record (i.e., ACI 318-63).

By letter dated November 6, 2008, the applicant stated that ACI 318-63 provides two methods for determination of concrete properties which will result in the required concrete strength. The applicant further stated that the concrete mixture at IP was established based on tests of concrete mixtures and actual tests for containment concrete showed compressive strengths above the required 20.7 MPa (3000 psi). In the SER with Open Items, the staff stated that it was reviewing the applicant's response, and that its evaluation of this matter would be included in the final SER.

The staff also noted that the applicant states in the LRA that the concrete also meets the requirements of a later ACI guide, ACI 201.2R-77. The staff asked the applicant to clarify the use of the later ACI 201.2R-77, since the editions of the ASTM standards may have changed between 1963 and 1977. In its letter dated November 6, 2008, the applicant stated that IP structures designed in accordance with ACI 318-63 align with many of the recommendations in ACI 201.2R-77. At the time of the issuance of the SER with Open Items, the staff was reviewing the applicant's response. As a result of the review, the staff identified the need for additional information. By letter dated April 3, 2009, the staff asked the applicant to describe the methodology used to establish the required concrete compressive strength of 3000 psi for the containment and other safety-related concrete structures, in accordance with ACI 318- 63, Method 2. By letter dated May 1, 2009, the applicant responded to the staff's request for additional information. The closure of this item is documented in SER Section 3.5.2.2.1.

OI 3.5-2: (SER Section 3.5.2.2.1, Subsection entitled "Reduction of Strength and Modulus of Concrete Structures Due to Elevated Temperature")

In LRA Section 3.5.2.2.1.3, the applicant stated that ACI 349 specifies long-term temperature limits of 66 °C (150 °F) for general areas and 93 °C (200 °F) for local areas. The effects of aging due to elevated temperature exposure are not significant below these temperatures.

The applicant also stated that the IP2 containment areas during normal operation are below 54 °C (130 °F) bulk average temperature. Penetrations through the containment cylinder wall for pipes carrying hot fluid are cooled by air-to-air heat exchangers and the pipes are insulated to maintain the temperature in the adjoining concrete below 121 °C (250 °F). The GALL Report provides for local area concrete temperatures higher than 93°C (200 °F) if tests or calculations evaluate the reduction in strength. The applicant also states that an evaluation of IP2 hot piping penetration concrete has found temperatures up to 121 °C (250 °F) acceptable.

The applicant further stated that the IP3 containment areas normally operate below a bulk average temperature of 54 °C (130 °F). Penetrations through the containment cylinder wall for pipes carrying hot fluid are cooled by air-to-air heat exchangers and the pipes are insulated to maintain the temperature in the adjoining concrete below 93 °C (200 °F).

The applicant concluded that these are not aging effects requiring management for IP.

In SRP-LR Section 3.5.3.2.1.3, it is stated that the GALL Report recommends further evaluation of programs to manage reduction of strength and modulus of concrete structures due to elevated temperature for PWR and BWR concrete and steel containments. The GALL Report notes that the implementation of ASME Section XI, Subsection IWL examinations and 10 CFR 50.55a would not be able to detect the reduction of concrete strength and modulus due to elevated temperature and also notes that no mandated aging management exists for managing this aging effect. The GALL Report recommends that a plant-specific evaluation be performed if any portion of the concrete containment components exceeds specified temperature limits, i.e., general temperature greater than 66 °C (150 °F) and local area temperature greater than 93 °C (200 °F).

The staff's review of operating experience did not identify any occurrences of concrete degradation at the IP2 hot penetrations. However, because concrete degradation at elevated temperatures is a slow process, there is a need to confirm that an additional 20 years of operation will not lead to significant degradation. The staff asked the applicant what the effects on the concrete will be during the period of extended operation for areas where the local temperature exceeds 93 °C (200 °F). By letter dated November 6, 2008, the applicant stated that an engineering evaluation of the effect of 121 °C (250 °F) temperatures on the hot piping penetration concrete was performed. The evaluation determined that a reduction in strength of 15 percent could be expected from the elevated temperatures. The applicant further stated that this reduction in strength was acceptable since the original concrete compressive strength tests showed an actual strength more than 15 percent greater than the design strength of 20.7 MPa (3,000 psi).

At the time of the issuance of the SER with Open Items, the staff was reviewing the applicant's response. As a result of its review, the staff identified the need for additional information. By letter dated April 3, 2009, the staff requested the applicant to clearly explain the role of the air-to-air heat exchangers in cooling the concrete around the hot piping penetrations. In addition the staff asked the applicant to describe the methodology used to arrive at the conclusion that the actual concrete strength is more than 15 percent greater than 20.7 MPa (3,000 psi), i.e., greater than 23.8 MPa (3,450 psi), to provide a summary of the results, and to explain how consideration was given to the reduction in modulus of elasticity in the high temperature concrete evaluation. By letter dated May 1, 2009, the applicant responded to the staff's request for additional information. The closure of this item is documented in SER Section 3.5.2.2.1.

OI 3.5-3: (SER Section 3.5.2.2.2 – Safety-Related and Other Structures and Component Supports)

Item 3.5.1-40 of LRA Table 3.5.1 addresses building concrete at locations of expansion and grouted anchors for the aging effect of reduction in concrete anchor capacity due to local concrete degradation/service-induced cracking or other concrete aging mechanisms. The GALL Report recommends the Structures Monitoring Program (SMP) for monitoring this concrete component for the stated aging effect. In the SER with Open Items, the staff found that the applicant had appropriately credited the SMP for Groups B2 through B5 component supports and surrounding concrete consistent with the GALL Report. However, for the Group B1 (ASME Class 1, 2, 3 & MC) supports, the applicant's reference to "IP concrete anchors and surrounding concrete" implies that the applicant is crediting the ISI-IWF AMP for both the supports and surrounding concrete. The staff found that, while ISI-IWF is appropriate for the Group B1 component supports themselves, ISI-IWF is not specifically applicable for concrete surrounding the anchors for these supports, because the code support boundary definition which extends to the surface of the building but does not include the building structure. Therefore, the staff indicated that the applicant should indicate which AMP it will use to manage the effects of aging for the concrete surrounding the B1 supports.

By letter dated January 27, 2009, the applicant stated that the applicable aging management program for concrete surrounding concrete anchors is the Structures Monitoring Program. The applicant also clarified the statement in LRA Section 3.5.2.2.2.6(1). The closure of this item is documented in SER Section 3.5.2.2.2.

OI 4.3-1: (SER Section 4.3.1 – Class 1 Fatigue)

In its review, the staff noted that the applicant used data from 1973 to 1995 to project the number of plant heatups and cooldowns from 1995 to March 31, 2006 (current cycles), rather than use actual data. As stated above, the applicant will track the number of transients under the Fatigue Monitoring Program. However, without the actual number of heatups and cooldowns from 1995 to March 31, 2006, the applicant may not be able to accurately predict when the number of analyzed cycles might be exceeded. The staff notes that changes in operating practices such as refueling (12-month refueling cycle vs. 24-month refueling cycle) would decrease the number of heatups and cooldowns experienced post 1995, which should yield a more conservative projection. Nonetheless, the applicant should have the actual data for the plant startups and shutdowns during this period of time. Therefore, the staff believes that the use of actual plant operating experience in lieu of a projection for the current number of cycles is appropriate.

By letter dated January 27, 2009, the applicant provided the actual number of cycles for IP3 plant heatups and cooldowns through March 31, 2006. This information was also provided in response to Audit Item 14. The closure of this item is documented in the "Staff Evaluation" section of SER Section 4.3.1.

1.6 Summary of Proposed License Conditions

Following the staff's review of the LRA, including subsequent information and clarifications from the applicant, the staff identified three proposed license conditions.

The first license condition requires the applicant to include the UFSAR supplement required by 10 CFR 54.21(d) in the first UFSAR update required by 10 CFR 50.71(e) following the issuance of the renewed licenses.

The second license condition requires future activities described in the UFSAR supplement to be completed prior to the period of extended operation.

The third license condition requires that all capsules in the reactor vessel that are removed and tested meet the requirements of American Society for Testing and Materials (ASTM) E 185-82 to the extent practicable for the configuration of the specimens in the capsule. Any changes to the capsule withdrawal schedule, including spare capsules, must be approved by the staff prior to implementation. All capsules placed in storage must be maintained for future insertion. Any changes to storage requirements must be approved by the staff, as required by 10 CFR Part 50, Appendix H.

SECTION 2

STRUCTURES AND COMPONENTS SUBJECT TO AGING MANAGEMENT REVIEW

2.1 Scoping and Screening Methodology

2.1.1 Introduction

Title 10 of the *Code of Federal Regulations* (10 CFR) 54.21, "Contents of Application—Technical Information," requires for each license renewal application (LRA) an integrated plant assessment (IPA) listing those structures and components (SCs) subject to an aging management review (AMR) for all of the systems, structures, and components (SSCs) within the scope of license renewal.

LRA Section 2.1, "Scoping and Screening Methodology," describes the methodology for identifying those SSCs at the Indian Point Nuclear Generating Unit Nos. 2 and 3 (IP2 and IP3) that are within the scope of license renewal and those SCs that are subject to an AMR. The staff reviewed the scoping and screening methodology of Entergy Nuclear Operations, Inc. (Entergy or the applicant), to determine whether it meets the scoping requirements of 10 CFR 54.4(a) and the screening requirements of 10 CFR 54.21.

In developing the scoping and screening methodology for the LRA, the applicant considered the requirements of 10 CFR Part 54, "Requirements for Renewal of Operating Licenses for Nuclear Power Plants" (the Rule), Statements of Consideration for the Rule, and the guidance of Nuclear Energy Institute (NEI) 95-10, Revision 6, "Industry Guideline for Implementing the Requirements of 10 CFR Part 54—The License Renewal Rule," issued June 2005. The applicant also considered the correspondence between the U.S. Nuclear Regulatory Commission (NRC) staff, other applicants, and NEI.

2.1.2 Summary of Technical Information in the Application

LRA Sections 2 and 3 detail the technical information required by 10 CFR 54.4, "Scope," and 10 CFR 54.21(a). This safety evaluation report (SER) with open items contains sections entitled "Summary of Information from the Application," which provide information taken directly from the LRA.

In LRA Section 2.1, the applicant described the process to identify the SSCs that meet the license renewal scoping criteria of 10 CFR 54.4(a) and the process used to identify the SCs that are subject to an AMR, as required by 10 CFR 54.21(a)(1). Additionally, LRA Section 2.2, "Plant Level Scoping Results," Section 2.3, "Scoping and Screening Results: Mechanical Systems," Section 2.4, "Scoping and Screening Results: Structures," and Section 2.5, "Scoping and Screening Results: Electrical and Instrumentation and Control Systems," provide the results of the process used to identify the SCs that are subject to an AMR. LRA Section 3.0, "Aging Management Review Results," presents information regarding the IP2 and IP3 AMR process in Section 3.1, "Reactor Vessel, Internals and Reactor Coolant System," Section 3.2, "Engineered

Safety Features Systems," Section 3.3, "Auxiliary Systems," Section 3.4, "Steam and Power Conversion Systems," Section 3.5, "Structures and Component Supports," and Section 3.6, "Electrical and Instrumentation and Controls." Section 4.0 of the LRA, "Time-Limited Aging Analyses," contains the applicant's identification and evaluation of time-limited aging analyses (TLAAs).

2.1.3 Scoping and Screening Program Review

The staff evaluated the LRA scoping and screening methodology in accordance with the guidance contained in NUREG-1800, Revision 1, "Standard Review Plan for Review of License Renewal Applications for Nuclear Power Plants" (hereafter referred to as the SRP-LR), Section 2.1, "Scoping and Screening Methodology." The following regulations form the basis for the acceptance criteria for the scoping and screening methodology review:

- 10 CFR 54.4(a), as it relates to the identification of plant SSCs within the scope of the Rule

- 10 CFR 54.4(b), as it relates to the identification of the intended functions of SSCs within the scope of the Rule

- 10 CFR 54.21(a)(1) and 10 CFR 54.21(a)(2), as they relate to the methods used by the applicant to identify plant SCs subject to an AMR

As part of the review of the applicant's scoping and screening methodology, the staff reviewed the activities described in the following sections of the LRA using the guidance contained in the SRP-LR:

- Section 2.1 to ensure that the applicant described a process for identifying SSCs that are within the scope of license renewal in accordance with the requirements of 10 CFR 54.4(a)

- Section 2.2 to ensure that the applicant described a process for determining SCs that are subject to an AMR in accordance with the requirements of 10 CFR 54.21(a)(1) and 10 CFR 54.21(a)(2)

In addition, the staff conducted a scoping and screening methodology audit at IP2 and IP3, located outside Buchanan, NY, during the week of October 8–12, 2007. The audit focused on ensuring that the applicant had developed and implemented adequate guidance to conduct the scoping and screening of SSCs in accordance with the methodologies described in the LRA and the requirements of the Rule. The staff reviewed implementation of the project-level guidelines and topical reports describing the applicant's scoping and screening methodology. The staff conducted detailed discussions with the applicant on the implementation and control of the license renewal program and reviewed the administrative control documentation used by the applicant during the scoping and screening process, the quality practices used by the applicant to develop the LRA, and the training and qualification of the LRA development team. The staff evaluated the quality attributes of the applicant's Aging Management Program activities described in Appendix A, "Updated Final Safety Analysis Report Supplement," and Appendix B, "Aging Management Programs and Activities," to the LRA. On a sampling basis, the staff performed a system review of the service water (SW) system and the turbine building, including a review of the scoping and screening results reports and the supporting design

2-2

documentation used to develop the reports, to ensure that the applicant had appropriately implemented the methodology outlined in the administrative controls and to verify that the results are consistent with the current licensing basis (CLB) documentation.

2.1.3.1 *Implementing Procedures and Documentation Sources for Scoping and Screening*

The staff reviewed the applicant's scoping and screening implementing procedures as documented in the Scoping and Screening Methodology Audit Trip Report (Agencywide Documents Access and Management System [ADAMS] Accession No. ML083540648) to verify that the process for identifying SCs subject to an AMR was consistent with the SRP-LR. Additionally, the staff reviewed the scope of CLB documentation sources and the process used by the applicant to ensure that the applicant's commitments, as documented in the CLB and relative to the requirements of 10 CFR 54.4 and 10 CFR 54.21, were appropriately considered and that the applicant adequately implemented its procedural guidance during the scoping and screening process.

2.1.3.1.1 Summary of Technical Information in the Application

In LRA Section 2.1, the applicant addressed the following information sources for the license renewal scoping and screening process:

- updated final safety analysis reports (UFSARs)
- technical specifications and bases
- technical requirements manual
- design-basis documents (DBDs)
- licensing commitment database
- Maintenance Rule bases documents
- fire hazards analysis
- Appendix R safe-shutdown analysis
- station blackout (SBO) analysis
- SERs
- docketed correspondence
- plant drawings

The applicant stated that it used this information to identify the functions performed by plant systems and structures. It then compared these functions to the scoping criteria in 10 CFR 54(a)(1)–(3) to determine whether the associated plant system or structure performed a license renewal intended function. It also used these sources to develop the list of SCs subject to an AMR.

2.1.3.1.2 Staff Evaluation

Scoping and Screening Implementation Procedures. The staff reviewed the applicant's scoping and screening methodology implementation procedures, including license renewal guidelines, documents, reports, and AMR reports, as documented in the audit report, to ensure that the guidance is consistent with the requirements of the Rule, the SRP-LR, and NEI 95-10. The staff finds that the overall process used to implement the 10 CFR Part 54 requirements described in the implementing documents and AMRs is consistent with the Rule, the SRP-LR, and industry guidance. The applicant's implementing documents contain guidance for determining plant SSCs within the scope of the Rule and for determining which SCs within the scope of license renewal are subject to an AMR (see ADAMS Accession No. ML080730399). During the review of the implementing documents, the staff focused on the consistency of the detailed procedural guidance with information in the LRA, including the implementation of staff positions documented in the SRP-LR, and the information in responses, dated February 13, 2008, to the staff's requests for additional information (RAIs).

After reviewing the LRA and supporting documentation, the staff determined that the scoping and screening methodology instructions are consistent with the methodology description provided in LRA Section 2.1. The applicant described its methodology in sufficient detail to provide concise guidance on the scoping and screening implementation process to be followed during the LRA activities.

Sources of Current Licensing Basis Information. The staff reviewed the scope and depth of the applicant's CLB review to verify that the methodology is sufficiently comprehensive to identify SSCs within the scope of license renewal, as well as SCs requiring an AMR. As defined in 10 CFR 54.3(a), the CLB is the set of NRC requirements applicable to a specific plant and a licensee's written commitments for ensuring compliance with, and operation within, applicable requirements of the NRC and the plant-specific design bases that are docketed and in effect. The CLB includes applicable NRC regulations, orders, license conditions, exemptions, technical specifications, and design-basis information (documented in the most recent final safety analysis report). The CLB also includes licensee commitments remaining in effect that were made in docketed licensing correspondence, such as licensee responses to NRC bulletins, generic letters, and enforcement actions, and licensee commitments documented in NRC safety evaluations or licensee event reports.

During the audit, the staff reviewed pertinent information sources used by the applicant including the UFSARs, license renewal boundary diagrams, and Maintenance Rule information. In addition, the applicant's license renewal process identified additional potential sources of plant information pertinent to the scoping and screening process, including the equipment database, system safety function sheets, safety classification documents, design-basis references, piping and instrumentation diagrams (P&IDs), electrical drawings, docketed correspondence, technical specifications and bases, the fire hazards analysis, and safety evaluations. The staff confirmed that the applicant's detailed license renewal program guidelines specify the use of the CLB source information in developing scoping evaluations.

The IP2 and IP3 equipment database and the system safety function sheets are the applicant's primary repository for component safety classification information. During the audit, the staff reviewed the applicant's administrative controls for the IP2 and IP3 equipment database, the system safety function sheets, and safety classification data. Plant administrative procedures describe these controls and govern their implementation. Based on a review of the administrative controls and a sample of the safety classification information contained in the IP2 and IP3 equipment database and system safety function sheets, the staff concludes that the applicant established adequate measures to control the integrity and reliability of IP2 and IP3 safety classification data and, therefore, the IP2 and IP3 equipment database and system safety function sheets provide a sufficiently controlled source of system and component data to support scoping and screening evaluations.

During the staff's review of the applicant's CLB evaluation process, the applicant explained the incorporation of updates to the CLB and the process used to ensure that those updates are adequately incorporated into the license renewal process. The staff determined that LRA Section 2.1 describes the CLB and related documents used during the scoping and screening process consistently with the guidance contained in the SRP-LR.

In addition, the staff reviewed the implementing procedures and results reports used to support identification of SSCs relied on to demonstrate compliance with the safety-related criteria, nonsafety-related criteria, and the regulated events criteria detailed in 10 CFR 54.4(a). The applicant's license renewal program guidelines provide a comprehensive listing of documents used to support scoping and screening evaluations. The staff finds these design documentation sources useful for ensuring that the initial scope of SSCs identified by the applicant is consistent with the plant's CLB.

2.1.3.1.3 Conclusion

Based on its review of LRA Section 2.1, the detailed scoping and screening implementation procedures, and the results from the scoping and screening audit, the staff concludes that the applicant's scoping and screening methodology considers CLB information consistently with the Rule, the SRP-LR and NEI 95-10 guidance and, therefore, is acceptable.

2.1.3.2 *Quality Controls Applied to the Development of the License Renewal Application*

2.1.3.2.1 Staff Evaluation

The staff reviewed the quality controls used by the applicant to ensure that scoping and screening methodologies used in the LRA were adequately implemented. The applicant applied the following quality assurance processes during the LRA development:

- The applicant developed written plans and procedures to direct implementation of the scoping and screening methodology, control LRA development, and describe training requirements and documentation.

- The applicant developed written requirements for developing, revising, and approving the guidelines and procedures.

- The applicant considered pertinent issues in previous LRAs and corresponding RAIs to determine their relevance to the IP2 and IP3 application.

- Industry peers and the site review committee examined the LRA before its submittal to the staff.

2.1.3.2.2 Conclusion

On the basis of its review of pertinent LRA development guidance, discussion with the applicant's license renewal staff, and a review of the applicant's documentation of the activities performed to assess the quality of the LRA, the staff concludes that the applicant's quality assurance activities meet current regulatory requirements and provide assurance that LRA development activities were performed in accordance with the applicant's license renewal program requirements.

2.1.3.3 Training

2.1.3.3.1 Staff Evaluation

The staff reviewed the applicant's training process for consistent and appropriate guidelines and methodology for the scoping and screening activities. As outlined in the implementing documents, the applicant requires training and documentation for all personnel participating in the LRA development. Personnel are required to complete the training before preparing and approving implementing procedures. Training materials include the applicant's project guidelines; pertinent industry documents; 10 CFR Part 54 and its Statements of Consideration; NEI 95-10, Revision 6; Regulatory Guide 1.188, Revision 1, "Standard Format and Content for Applications to Renew Nuclear Power Plant Operating Licenses," issued September 2005; SRP-LR; NUREG-1801, Revision 1, "Generic Aging Lessons Learned (GALL) Report" (hereafter referred to as the GALL Report); and attendance at a license renewal orientation session.

The applicant's procedures specify two levels of training—(1) training for the corporate project team personnel and (2) training for site personnel. Generally, the project team personnel review all training documents in order to identify those documents directly related to their specific scoping and screening responsibilities. The intent of the training for site personnel is to ensure that personnel understand the license renewal process and the materials specifically related to each individual's license renewal responsibilities. Completion of the training allows site personnel to evaluate and approve the license renewal documents for technical accuracy. Qualification and training records and a checklist serve as documentation for each individual's completed license renewal training. The staff reviewed completed qualification and training records and the completed checklists for several of the applicant's license renewal personnel. Additionally, after discussions with the applicant's license renewal personnel during the audit, the staff verified that the applicant's personnel are knowledgeable about the license renewal process requirements and specific technical issues within their areas of responsibility.

2.1.3.3.2 Conclusion

On the basis of discussions with the applicant's license renewal project personnel responsible for the scoping and screening process and its review of selected documentation in support of the process, the staff concludes that the applicant's personnel are adequately trained to implement the scoping and screening methodology as described in the applicant's implementing documents and the LRA.

2.1.3.4 Conclusion of Scoping and Screening Program Review

On the basis of a review of information provided in LRA Section 2.1, a review of the applicant's detailed scoping and screening implementing procedures, discussions with the applicant's license renewal personnel, and the results from the scoping and screening audit, the staff concludes that the applicant's scoping and screening program is consistent with the SRP-LR and the requirements of 10 CFR Part 54 and, therefore, is acceptable.

2.1.4 Scoping Methodology for Plant Systems, Structures, and Components

In LRA Section 2.1, the applicant described the methodology used to scope SSCs under the 10 CFR 54.4(a) scoping criteria. The applicant described the scoping process for the plant in terms of systems and structures. Specifically, the scoping process consisted of developing a list of plant systems and structures, identifying their intended functions, and determining which functions meet one or more of the three criteria detailed in 10 CFR 54.4(a). The applicant developed the list of systems using the equipment database; the list of plant structures was developed from a review of plant layout drawings, Maintenance Rule documentation, DBDs, and the UFSARs. Mechanical system functions were identified from the IP2 and IP3 safety system function sheets (SSFSs). The applicant obtained additional information on mechanical system functions from the UFSARs, the Maintenance Rule documents, piping flow diagrams, and DBDs. Structural functions were identified using the UFSARs, the Maintenance Rule basis documents for structures, the fire hazards analyses, DBDs, and structural drawings. According to the LRA, all electrical and instrumentation and control (I&C) systems, and electrical and I&C components in mechanical systems, are within the scope of license renewal.

2.1.4.1 Application of the Scoping Criteria in 10 CFR 54.4(a)(1)

2.1.4.1.1 Summary of Technical Information in the Application

LRA Section 2.1.1.1, "Application of Safety-Related Scoping Criteria," describes the scoping methodology as it relates to the safety-related criterion in accordance with 10 CFR 54.4(a)(1). With respect to the safety-related criterion, the applicant stated that safety-related system and structure functions are initially identified through a review of the SSFSs and then confirmed by a review of the UFSARs, Maintenance Rule documents, piping flow diagrams, and DBDs, as applicable. Systems and structures whose intended functions meet one or more of the criteria in 10 CFR 54.4(a)(1) were included within the scope of license renewal. The applicant confirmed that it considered all plant conditions, including conditions of normal operation, design-basis accidents (DBAs), external events, and natural phenomena for which the plant must be designed, for license renewal scoping under the 10 CFR 54.4(a)(1) criteria.

2.1.4.1.2 Staff Evaluation

Pursuant to 10 CFR 54.4(a)(1), the applicant must consider all safety-related SSCs relied on to remain functional during and following a design-basis event (DBE) to ensure the performance of certain functions. These functions are (1) the integrity of the reactor coolant pressure boundary (RCPB), (2) the ability to shut down the reactor and maintain it in a safe-shutdown condition, or (3) the capability to prevent or mitigate the consequences of accidents that could result in potential offsite exposures comparable to those described in 10 CFR 50.34(a)(1), 10 CFR 50.67(b)(2), or 10 CFR 100.11, "Determination of Exclusion Area, Low Population Zone, and Population Center Distance."

With regard to identification of DBEs, SRP-LR, Section 2.1.3, "Review Procedures," states the following:

> The set of DBEs as defined in the Rule is not limited to Chapter 15 (or equivalent) of the UFSAR. Examples of DBEs that may not be described in this chapter include external events, such as floods, storms, earthquakes, tornadoes,

or hurricanes, and internal events, such as a high energy line break. Information regarding DBEs as defined in 10 CFR 50.49(b)(1) may be found in any chapter of the facility UFSAR, the Commission's regulations, NRC orders, exemptions, or license conditions within the CLB. These sources should also be reviewed to identify SSCs relied upon to remain functional during and following DBEs (as defined in 10 CFR 50.49(b)(1)) to ensure the functions described in 10 CFR 54.4(a)(1).

During the audit, the applicant stated that it evaluated the types of events listed in NEI 95-10 (i.e., anticipated operational occurrences, DBAs, external events, and natural phenomena) that were applicable to IP2 and IP3. The applicant identified the documents (the UFSARs and the fire hazards analysis) that described the events. The applicant also reviewed licensing correspondence and DBDs. The staff determined that the applicant's evaluation of DBEs is consistent with the SRP-LR.

The applicant performed scoping of SSCs for the 10 CFR 54.4(a)(1) criterion in accordance with the license renewal implementing documents, which provide guidance for the preparation, review, verification, and approval of the scoping evaluations to ensure the adequacy of the results of the scoping process. The staff reviewed the implementing documents governing the applicant's evaluation of safety-related SSCs and sampled the applicant's reports of the scoping results to ensure that the applicant applied the methodology in accordance with those written instructions. In addition, the staff discussed the methodology and results with the applicant's personnel who were responsible for these evaluations.

The staff reviewed the applicant's evaluation of the Rule and CLB definitions pertaining to 10 CFR 54.4(a)(1). The IP2 and IP3 CLB definition of "safety-related" meets the definition in 10 CFR 54.4(a)(1). LRA Section 2.1.1.1 documents the applicant's definition of safety-related and exceptions to the definition in 10 CFR 54.4(a)(1). Based on its review, the staff confirmed that the applicant correctly identified the applicable dose criteria for IP2 and IP3 as set forth in 10 CFR 54.4(a)(1)(iii). The dose criteria are set forth in 10 CFR 50.67(b)(2) and 10 CFR 100.11 for IP2, as reflected in the LRA. Although the IP3 CLB definition of "safety-related" did not explicitly include reference to 10 CFR 50.67(b)(2), the requirements of 10 CFR 50.67(b)(2), which concern the use of an alternate source term in the dose analysis, are also applicable to IP3, which has been approved for the use of an alternate source term. The staff confirmed that the applicant reviewed the IP3 systems and components credited in the plant's dose analyses to ensure that the applicable systems and components were included in the scope of the license renewal. The applicant did not identify any additional SSC functional requirements, beyond those established to meet the requirements of 10 CFR Part 100, "Reactor Site Criteria," credited for the application of the alternate source term, and no additional SSCs for IP3 were required for inclusion in the scope of license renewal under 10 CFR 50.67(b)(2).

The staff reviewed a sample of the license renewal scoping results for the SW system and the turbine building to provide additional assurance that the applicant adequately implemented its scoping methodology in accordance with 10 CFR 54.4(a)(1). The staff verified that the applicant developed the scoping results for each of the sampled systems consistently with the methodology, identified the SSCs credited for performing intended functions, and adequately described the basis for the results, as well as the intended functions.

In order to verify that the applicant identified and used pertinent engineering and licensing information to identify the SSCs required by 10 CFR 54.4(a)(1) to be within the scope of license renewal, the staff determined that it would require additional information to complete its review of the applicant's scoping methodology.

In RAI 2.1-1(c), dated January 14, 2008, the staff stated that during the audit, it reviewed the applicant's technical evaluation and onsite documentation for nonsafety-related SSCs affecting safety-related SSCs, which indicate that certain similar SSCs were included within the scope of license renewal under 10 CFR 54.4(a)(1) for one unit, but under 10 CFR 54.4(a)(2) for the other unit. The staff requested that the applicant provide the rationale and basis for including similar SSCs within the scope of license renewal under 10 CFR 54.4(a)(1) for one unit, but under 10 CFR 54.4(a)(2) for the other unit, and describe how it performed the corresponding review of the adjacent or attached nonsafety-related SSCs (for inclusion within the scope of license renewal) for similar systems in the two units. In its February 13, 2008, response to RAI 2.1-1(c), the applicant stated the following:

> Because IP2 and IP3 were operated independently for an extended period of time, there are differences between IP2 and IP3 in terms of the number of systems, as well as system boundaries and intended functions for similarly named systems. The site component database along with system flow diagrams were used to define system boundaries and identify system intended functions. Consequently, certain similarly named SSCs were included within the scope of license renewal in accordance with the requirements of 10 CFR 54.4(a)(1) only for one unit and 10 CFR 54.4(a)(2) only for the other unit because the system boundaries were different.
>
> The IP2 city water system (CYW) is in-scope for 10 CFR 54.4(a)(1) and 10 CFR 54.4(a)(2) while the IP3 city water system (CWM) is in-scope only for 10 CFR 54.4(a)(2). IP2 piping assigned to the city water system provides containment isolation, a 10 CFR 54.4(a)(1) intended function, for supply to fire water hose reels inside the containment building. The IP3 city water system does not provide a similar intended function or any other (a)(1) functions and therefore is not in-scope for 10 CFR 54.4(a)(1). Since the city water systems are fluid-filled, all components not included for 54.4(a)(1) or (a)(3) in structures containing components with safety functions were reviewed for potential spatial impact. Appropriate LRA drawings were also reviewed to verify that no components required for structural support of components with safety functions were excluded. This review was performed for both systems regardless of system functions to ensure all in-scope components were identified.
>
> The IP2 instrument air closed cooling water system is in-scope only for 10 CFR 54.4(a)(2) while the IP3 instrument air closed-cooling system is in-scope for 10 CFR 54.4(a)(1) and 10 CFR 54.4(a)(2). IP3 instrument air closed-cooling heat exchangers SWN CLC 31 HTX, SWN CLC 32 HTX perform an intended function of providing service water system pressure boundary and are in-scope for 10 CFR 54.4(a)(1). The corresponding IP2 instrument air closed-cooling heat exchangers 21 CWHX, 22CWHX are assigned to the SW system and not instrument air closed-cooling, so the IP2 instrument air closed-cooling system has no components with a 10 CFR 54.4(a)(1) intended function. Since the

instrument air closed-cooling systems are fluid-filled, all components in structures containing components with safety functions were reviewed for potential spatial impact. Appropriate LRA drawings were also reviewed to verify that no components required for structural support of components with safety functions were excluded. This review was performed for both systems regardless of system functions to ensure all in-scope components were identified.

The IP2 river water service system (RW) is in-scope for 10 CFR 54.4(a)(1) to support the service water system pressure boundary. Both IP2 and IP3 RW systems are in-scope for 10 CFR 54.4(a)(2). The IP3 RW system has no components within its boundary that support the service water system pressure boundary or any other (a)(1) functions. Since the RW systems are fluid-filled, all system components in structures containing components with safety functions were included for potential spatial impact. Appropriate LRA drawings were also reviewed to verify that no components required for structural support of components with safety functions were excluded. This review was performed for both systems regardless of system functions to ensure all in-scope components were included.

The staff reviewed the applicant's response to RAI 2.1-1(c) and determined that the applicant's description of the process used to ensure that SSCs have been appropriately included within the scope of license renewal is in accordance with 10 CFR 54.4(a)(1) or (a)(2), as applicable, based on the intended function of the SSC for the unit that the system serves. The staff's concern described in RAI 2.1-1(c) is resolved.

2.1.4.1.3 Conclusion

On the basis of its review of systems (on a sampling basis), discussions with the applicant, review of the applicant's scoping process, and the applicant's response to RAI 2.1-1(c), the staff concludes that the applicant's methodology for identifying systems and structures is consistent with the SRP-LR and the requirements of 10 CFR 54.4(a)(1) and, therefore, is acceptable.

2.1.4.2 Application of the Scoping Criteria in 10 CFR 54.4(a)(2)

2.1.4.2.1 Summary of Technical Information in the Application

In LRA Section 2.1.1.2, "Application of Criterion for Nonsafety-Related SSCs Whose Failure Could Prevent the Accomplishment of Safety Functions," the applicant described the scoping methodology as it relates to the nonsafety-related criteria in 10 CFR 54.4(a)(2). The applicant based its 10 CFR 54.4(a)(2) scoping methodology on guidance provided in Appendix F of NEI 95-10, Revision 6. By considering functional failures and physical failures, the applicant evaluated the impacts of nonsafety-related SSCs that meet the 10 CFR 54.4(a)(2) criteria.

Functional Failure of Nonsafety-Related SSCs. LRA Section 2.1.1.2.1, "Functional Failures of Nonsafety-Related SSC," states that SSCs required to perform a function in support of safety-related components are generally classified as safety related and are included within the scope of license renewal in accordance with 10 CFR 54.4(a)(1). For the few exceptions where nonsafety-related components are required to remain functional to support a safety function, the applicant identified this intended system function and included the components within the scope of license renewal in accordance with the requirements of 10 CFR 54.4(a)(2).

Nonsafety-Related SSCs with the Potential for Spatial Interaction with Safety-Related SSCs. LRA Section 2.1.1.2.2, "Physical Failures of Nonsafety-Related SSCs," states that nonsafety-related systems and nonsafety-related portions of safety-related systems are identified as in scope under 10 CFR 54.4(a)(2) if there is a potential for spatial interactions with safety-related equipment. Spatial failures are defined as failures of nonsafety-related SSCs that are connected to or located in the vicinity of safety-related SSCs, creating the potential for interaction between the SSCs from physical impact, pipe whip, jet impingement, a harsh environment resulting from a piping rupture, or damage from leakage or spray that could impede or prevent the accomplishment of the safety-related functions of a safety-related SSC. In addition, the applicant included overhead handling systems and mitigative features, such as missile barriers, flood barriers, and spray shields, within the scope of license renewal in accordance with 10 CFR 54.4(a)(2).

The applicant used the preventive option described in NEI 95-10, Appendix F, to determine the scope of license renewal with respect to the protection of safety-related SSCs from spatial interactions that the CLB does not address. This scoping process, referred to as the "spaces" approach, involves an evaluation based on the location of nonsafety-related equipment and its proximity to safety-related SSCs, including the identification of fluid-filled system components located in the same space as safety-related equipment. The applicant defined a "space" as a room or cubicle that is separated from other spaces by substantial objects (such as walls, floors, and ceilings).

Nonsafety-Related SSCs Directly Connected to Safety-Related SSCs. LRA Section 2.1.1.2.2 states that the scope of license renewal includes the nonsafety-related piping and supports up to and including the first seismic anchor beyond the safety/nonsafety interface such that the safety-related portion of the piping will be able to perform its intended function. For piping in this structural boundary, pressure integrity is not required; however, piping within the safety class pressure boundary depends on the structural boundary piping and supports so that the system can fulfill its safety function. For IP2 and IP3, "structural boundary" is defined as the portion of a piping system that, although outside the safety class pressure boundary, is relied on to provide structural support for the pressure boundary.

2.1.4.2.2 Staff Evaluation

As detailed in 10 CFR 54.4(a)(2), the applicant must consider all nonsafety-related SSCs whose failure could prevent satisfactory accomplishment of safety-related SSCs relied on to remain functional during and following a DBE to ensure (1) the integrity of the RCPB; (2) the ability to shut down the reactor and maintain it in a safe shutdown condition; or (3) the capability to prevent or mitigate the consequences of accidents that could result in potential offsite exposures, comparable to those referred to in 10 CFR 50.34(a)(1), 10 CFR 50.67(b)(2), or 10 CFR 100.11.

Regulatory Guide 1.188, Revision 1, endorses the use of NEI 95-10, Revision 6. NEI 95-10 discusses the staff's position on the 10 CFR 54.4(a)(2) scoping criteria, including nonsafety-related SSCs typically identified in the CLB; consideration of missiles, cranes, flooding, and high-energy line breaks (HELBs); nonsafety-related SSCs connected to safety-related SSCs; nonsafety-related SSCs in proximity to safety-related SSCs; and the mitigative and preventive options related to nonsafety-related and safety-related SSC interactions.

In addition, the staff's position (as discussed in NEI 95-10, Revision 6) is that applicants should not consider hypothetical failures, but rather should base their evaluation on the plant's CLB, engineering judgment and analyses, and relevant operating experience. NEI 95-10 further describes operating experience as all documented plant-specific and industry-wide experience that can be used to determine the plausibility of a failure. Documentation would include NRC generic communications and event reports, plant-specific condition reports, industry reports such as safety operational event reports, and engineering evaluations. The staff reviewed LRA Section 2.1.1.2, in which the applicant described the scoping methodology for nonsafety-related SSCs under 10 CFR 54.4(a)(2). In addition, the staff reviewed the applicant's results report, which documents the guidance and corresponding results of the applicant's scoping review under 10 CFR 54.4(a)(2). The applicant stated that it performed this review in accordance with the guidance in NEI 95-10, Revision 6, Appendix F.

Nonsafety-Related SSCs Required To Perform a Function That Supports a Safety-Related SSC. The staff determined that nonsafety-related SSCs required to remain functional to support a safety-related function were included within the scope of license renewal as safety-related as if these SSCs were in scope under 10 CFR 54.4(a)(1). The applicant's scoping report discusses the evaluation criteria described in 10 CFR 54.4(a)(2). The staff finds that the applicant implemented an acceptable method for scoping of the nonsafety-related systems that perform a function that supports a safety-related intended function, as required by 10 CFR 54.4(a)(2).

Nonsafety-Related SSCs Directly Connected to Safety-Related SSCs. Based on a review of the information in the LRA and the applicant's implementing documents, the staff determined that, to identify the nonsafety-related SSCs connected to safety-related SSCs and which require structural soundness to maintain the integrity of the safety-related SSCs, the applicant used a combination of the information contained in the IP2 and IP3 structural analysis to identify the structural boundary. The applicant also applied the bounding approach as described in NEI 95-10, Appendix F. The applicant reviewed the safety-related to nonsafety-related interfaces for each mechanical system to identify the nonsafety-related components located between the interface and the structural boundary. The applicant included all nonsafety-related SSCs within the structural boundary that are within the scope of license renewal, in accordance with 10 CFR 54.4(a)(2).

If a seismic support could not be located using the structural boundary, the applicant identified the portion of the nonsafety-related piping up to, and including, a base-mounted component, flexible connection, or the end of the piping run, in accordance with the guidance of Appendix F to NEI 95-10. This guidance describes the use of bounding criteria as a method of determining the portion of nonsafety-related SSCs that an applicant should include within the scope of license renewal.

The staff noted during the scoping and screening methodology audit that the applicant included fluid-filled, nonsafety-related pipes located in a safety-related space within the scope of license renewal based on the spaces approach; the applicant separately addressed nonsafety-related piping attached to safety-related SSCs. However, the applicant did not provide sufficient information in either the LRA or the implementing procedures to demonstrate that, when the fluid-filled pipe was also attached to a safety-related SSC, an additional portion of the pipe, beyond the safety-related space, up to and including an appropriate seismic anchor, equivalent anchor, or bounding condition, was also included within the scope of license renewal. The staff determined that it needed additional information to complete the review of the applicant's scoping methodology.

In RAI 2.1-1(a), dated January 14, 2008, the staff requested that the applicant describe the process used to ensure that fluid-filled, nonsafety-related pipe, attached to safety-related SSCs and exiting the safety-related space, was included within the scope of license renewal, up to and including an appropriate seismic anchor, equivalent anchor, or bounding condition.

In its February 13, 2008, response to RAI 2.1-1(a), the applicant stated the following:

> The process for determining the components to be included for 10 CFR 54.4(a)(2) included a review of all passive mechanical components at IP2 and IP3 that were not already included in an AMR report under 10 CFR 54.4(a)(1) or (a)(3). The review began with a determination of which components need to be in-scope due to their potential for spatial interaction with components with a safety function. If piping and components for fluid-filled systems exit areas containing components with safety functions, further review was performed. This occurred in only limited locations. For those few locations, IPEC [Indian Point Energy Center] reviewed the component database and associated drawings and confirmed that those components required for structural support are within the safety-related space.

The staff reviewed the applicant's response to RAI 2.1-1(a) and determined that the applicant described an adequate process, which includes additional review of certain fluid-filled, nonsafety-related pipe to ensure that fluid-filled, nonsafety-related pipe attached to safety-related SSCs that exits the safety-related space is included within the scope of license renewal, up to and including an appropriate seismic anchor, equivalent anchor, or bounding condition. The staff's concern described in RAI 2.1-1(a) is resolved.

Nonsafety-Related SSCs with the Potential for Spatial Interaction with Safety-Related SSCs. The applicant considered physical impact (pipe whip, jet impingement), harsh environments, flooding, spray, and leakage when evaluating the potential for spatial interactions between nonsafety-related systems and safety-related SSCs. The applicant used a spaces approach, as described above, to identify the portions of nonsafety-related systems with the potential for spatial interaction with safety-related SSCs.

Physical Impact or Flooding. The applicant considered nonsafety-related supports for nonseismic piping systems and electrical conduit and cable trays with potential for spatial interaction with safety-related SSCs for inclusion within the scope of license renewal, in accordance with 10 CFR 54.4(a)(2). These supports and components are addressed in a commodity fashion, (i.e., grouping structural components that typically do not have unique identifiers based on common characteristics such as materials of construction, within civil/structural AMR reports). The applicant's review of earthquake experience revealed no occurrence of welded steel pipe segments falling during a strong motion earthquake. The applicant, using the guidance in NEI 95-10, concluded that, as long as the effects of aging on supports for piping systems are managed, collapse of piping systems is not credible (except

2-13

from flow-accelerated corrosion as considered in the HELB analysis for high-energy systems) and the piping sections are not within scope under 10 CFR 54.4(a)(2). The applicant evaluated the missiles that could be generated from internal or external events such as failure of rotating equipment or overhead-handling systems. The applicant included nonsafety-related design features that protect safety-related SSCs from such missiles within the scope of license renewal. In addition, the applicant included walls, curbs, dikes, doors, and similar structures that provide flood barriers to safety-related SSCs within the scope of license renewal in accordance with 10 CFR 54.4(a)(2).

Pipe Whip, Jet Impingement, and Harsh Environment. The applicant evaluated nonsafety-related portions of high-energy lines in accordance with 10 CFR 54.4(a)(2). The applicant based its evaluation on a review of documents including the UFSARs, DBDs, and relevant site documentation. The applicant evaluated its high-energy systems to ensure identification of components that are part of nonsafety-related, high-energy lines that can affect safety-related equipment. If the applicant's HELB analysis assumed that a nonsafety-related piping system did not fail, or assumed failure only at specific locations, then the applicant included that piping system (piping, equipment, and supports) within the scope of license renewal, in accordance with 10 CFR 54.4(a)(2), and designated it as subject to an AMR to ensure that those assumptions remain valid through the period of extended operation. Also, as discussed in the IP2 and IP3 scoping report (in accordance with 10 CFR 54.4(a)(2)), the applicant reviewed the reference documents (primarily DBDs) that contain HELB analyses for inside and outside containment and identified high-energy lines. Many of the identified systems are safety-related or are required for a regulated event and are included within the scope of license renewal in accordance with 10 CFR 54.4(a)(1) or (a)(3). The applicant included remaining nonsafety-related, high-energy lines, which were determined to have potential interaction with safety-related SSCs, within the scope of license renewal in accordance with 10 CFR 54.4(a)(2).

Spray and Leakage. The applicant evaluated moderate and low-energy systems that have the potential for spatial interactions from spray or leakage. Nonsafety-related systems and nonsafety-related portions of safety-related systems with the potential for spray or leakage that could prevent safety-related SSCs from performing their required safety function were considered within the scope of license renewal. The applicant used a spaces approach to identify the nonsafety-related SSCs located within the same space as safety-related SSCs, as described above. After identifying the applicable mechanical systems, the applicant reviewed the system functions to determine whether the system contained fluid, air, or gas. On the basis of plant and industry operating experience, the applicant excluded the nonsafety-related SSCs containing air or gas from the scope of license renewal. The applicant then determined whether the system had any components located within a space containing safety-related SSCs. The applicant included those nonsafety-related SSCs determined to contain fluid and located within a space containing safety-related SSCs within the scope of license renewal.

RAI 2.1-1(b), dated January 14, 2008, states that during the staff audit, the audit team reviewed the applicant's technical evaluation and onsite documentation for nonsafety-related SSCs affecting safety-related SSCs. This technical evaluation found that certain nonsafety-related SSCs affecting safety-related SSCs were not included within the scope of license renewal, based on the proximity of the nonsafety-related SSCs to the safety-related SSCs. The staff requested that the applicant provide the rationale and basis for not including nonsafety-related

SSCs in the vicinity of safety-related SSCs within the scope of license renewal, based on their proximity to safety-related SSCs.

In its February 13, 2008, response to RAI 2.1-1(b), the applicant stated the following:

> Within a structure that contains components with safety functions, the proximity of components to components with a safety function is not used as a criterion for exclusion of a system or component from (a)(2) scope due to spatial interaction. The wording in the original version of the AMR report reviewed during the license renewal scoping and screening audit did not clarify why fluid-filled components in locations with safety-related equipment were excluded. Some systems have fluid-filled nonsafety-related components located in structures that contain components with safety functions but cannot spatially affect components with safety functions due to physical barriers such as room separation within the structure. During the license renewal scoping and screening audit, a portion of the IP2 chlorination (CL) system was determined to be in proximity to service water system components which perform a safety function. The CL system had been excluded from 10 CFR 54.4(a)(2) scope. The CL system is added to the scope of license renewal for 10 CFR 54.4(a)(2) with components to be managed by the Periodic Surveillance and Preventive Maintenance, External Surfaces Monitoring, and Bolting Integrity Programs.

The staff reviewed the applicant's response to RAI 2.1-1(b) and determined that the applicant described an adequate process, including consideration of room boundaries to prevent interaction, to ensure that fluid-filled, nonsafety-related pipes were not excluded from the scope of license renewal based on the proximity of the nonsafety-related SSCs to safety-related SSCs. The applicant also concluded that an additional system, the chlorination system, is included within the scope of license renewal. SER Section 2.3A.3.19 documents the staff's review of the IP2 chlorination system that was added to the scope. SER Section 3.3A.2 documents the staff's evaluation of AMR results for the IP2 chlorination system components. The staff's concern described in RAI 2.1-1(b) is resolved.

Protective Features. The applicant evaluated protective features, such as whip restraints, spray shields, supports, and missile and flood barriers installed to protect safety-related SSCs against spatial interaction with nonsafety-related SSCs from fluid leakage, spray, or flooding. These protective features are credited in the plant design and included within the scope of license renewal.

2.1.4.2.3 Conclusion

On the basis of its review of the applicant's scoping process and systems (on a sampling basis), discussions with the applicant, and review of the information provided in the response to RAIs 2.1-1(a) and (b), the staff concludes that the applicant's methodology for identifying and including nonsafety-related SSCs that could affect the performance of safety-related SSCs within the scope of license renewal is consistent with the scoping criteria in 10 CFR 54.4(a)(2) and, therefore, is acceptable.

2.1.4.3 Application of the Scoping Criteria in 10 CFR 54.4(a)(3)

2.1.4.3.1 Summary of Technical Information in the Application

LRA Section 2.1.1.3, "Application of Criterion for Regulated Events," describes the methodology for identifying those systems and structures within the scope of license renewal in accordance with the Commission's criteria for five regulated events. These criteria appear in (1) 10 CFR 50.48, "Fire Protection," (2) 10 CFR 50.49, "Environmental Qualification of Electric Equipment Important to Safety for Nuclear Power Plants," (3) 10 CFR 50.61, "Fracture Toughness Requirements for Protection against Pressurized Thermal Shock Events," (4) 10 CFR 50.62, "Requirements for Reduction of Risk from Anticipated Transients without Scram (ATWS) Events for Light-Water-Cooled Nuclear Power Plants," and (5) 10 CFR 50.63, "Loss of All Alternating Current Power."

Fire Protection. LRA Section 2.1.1.3.1, "Commission's Regulations for Fire Protection (10 CFR 50.48)," describes the scoping of systems and structures relied on in safety analyses or plant evaluations to perform a function that demonstrates compliance with the fire protection criterion. The LRA stated that in-scope systems and structures for fire protection include equipment based on the functional requirements defined in 10 CFR 50.48. The applicant identified this equipment based on a review of the CLB for systems and structures relied on for compliance with 10 CFR 50.48. The applicant indicated in the LRA that those SSCs credited with fire prevention, detection, and mitigation in areas containing equipment important to the plant's safe operation and equipment credited to achieve safe shutdown in the event of a fire are within the scope of license renewal.

Environmental Qualification. LRA Section 2.1.1.3.2, "Commission's Regulations for Environmental Qualification (10 CFR 50.49)," describes the scoping of systems and structures relied on in safety analyses or plant evaluations to perform a function in compliance with the environmental qualification (EQ) criterion. The LRA states that the EQ program satisfies the requirements of 10 CFR 50.49 and that, because a bounding approach was used for scoping electrical equipment, the electrical and I&C systems and electrical equipment contained in mechanical systems are included within the scope of license renewal by default.

Pressurized Thermal Shock. LRA Section 2.1.1.3.3, "Commission's Regulations for Pressurized Thermal Shock (10 CFR 50.61)," describes the scoping of systems and structures relied on in safety analyses or plant evaluations to perform a function that demonstrates compliance with the pressurized thermal shock (PTS) criterion. The LRA states that for both IP2 and IP3, the only system relied on to comply with the PTS regulation is the reactor coolant system (RCS), specifically the reactor vessel.

Anticipated Transient without Scram. LRA Section 2.1.1.3.4, "Commission's Regulations for Anticipated Transients without Scram (10 CFR 50.62)," describes the scoping of systems and structures relied on in safety analyses or plant evaluations to perform a function that demonstrates compliance with the ATWS criterion. The LRA states that the applicant determined the mechanical system intended functions supporting anticipated transient without scram (ATWS) regulation based on CLB information for IP2 and IP3. The LRA also states that, because the applicant used a bounding approach for scoping electrical and I&C equipment, the electrical and I&C systems contained in mechanical systems are included within the scope of license renewal by default.

Station Blackout. LRA Section 2.1.1.3.5, "Commission's Regulations for Station Blackout (10 CFR 50.63)," describes the scoping of systems and structures relied on in safety analyses or plant evaluations to perform a function that demonstrates compliance with the SBO criterion. The LRA states that the applicant determined the system intended functions supporting 10 CFR 50.63 requirements based on information contained in the CLB. The LRA further states that, because the applicant used a bounding approach for scoping electrical and I&C equipment, the onsite electrical and I&C systems and electrical equipment contained in mechanical systems are included within the scope of license renewal by default.

2.1.4.3.2 Staff Evaluation

The staff reviewed the applicant's approach to identifying mechanical systems and structures relied on to perform functions meeting the requirements of the fire protection, EQ, PTS, ATWS, and SBO regulations. As part of its review, the staff (1) discussed the methodology with the applicant, (2) reviewed the documentation developed to support the approach, and (3) evaluated a sample of the mechanical systems and structures indicated as within the scope of license renewal in accordance with 10 CFR 54.4(a)(3).

The applicant's implementing procedures describe the process for identifying systems and structures within the scope of license renewal. The procedures state that all mechanical systems and structures that perform functions addressed in 10 CFR 54.4(a)(3) are to be included within the scope of license renewal and the results documented in scoping results reports. The results reports reference the information sources used for determining the systems and structures credited for compliance with the events listed in the specified regulations.

Fire Protection. The applicant's scoping results reports indicate that the applicant considered CLB documents to identify in-scope systems and structures. These documents include the (1) fire protection plan, which includes the fire protection program plan as required by 10 CFR 50.48; (2) IP2 and IP3 fire hazards analyses; and (3) safe-shutdown analyses for the requirements in Appendix R, "Fire Protection Program for Nuclear Power Facilities Operating Prior to January 1, 1979," to 10 CFR Part 50, "Domestic Licensing of Production and Utilization Facilities." The staff reviewed the scoping results reports in conjunction with the LRA and the IP2 and IP3 CLB information to validate the methodology for including the appropriate systems and structures within the scope of license renewal.

The staff found that the scoping results reports indicate which of the mechanical systems and structures are included within the scope of license renewal because they perform intended functions that meet 10 CFR 50.48 requirements. As an example, for a mechanical system, the applicant's IP2 fire hazards analysis report credits the reactor coolant pump (RCP) oil collection system, which is included under the IP2 fire protection—CO_2, Halon, and RCP oil collection systems. From this report, the applicant identified a license renewal intended function for the system as providing each RCP with an oil collection system designed to contain and direct the oil to remote storage containers if leakage occurs. The scoping results also identify structures within the scope of license renewal. For example, the foundation structures of the IP2 and IP3 fire water storage tanks are within the scope of license renewal because they maintain the structural integrity of the fire water storage tanks that support equipment credited in safe-shutdown capability analyses. The staff determined that the applicant's scoping methodology is adequate for including systems and structures credited in performing fire protection functions.

Environmental Qualification. The applicant employed a bounding approach for scoping plant electrical and I&C systems. All of these systems are included within the scope of license renewal, and electrical and I&C components in mechanical systems are included in the electrical systems. This method also includes within the scope of license renewal any equipment relied on to perform functions that demonstrate compliance with the EQ regulation.

The staff reviewed the LRA, implementing procedures, scoping results reports, and the IP2 and IP3 master EQ component equipment lists to verify that the applicant identified SSCs within the scope of license renewal that meet EQ requirements. The staff determined that the applicant's scoping methodology is adequate for identifying EQ SSCs within the scope of license renewal.

Pressurized Thermal Shock. The applicant addressed PTS requirements for the reactor vessels in a TLAA in LRA Section 4.2.5. This methodology is appropriate for identifying SSCs with functions credited for complying with the PTS regulation. For this requirement, the applicant identified the IP2 and IP3 reactor vessels as the only components within the scope of license renewal. SER Section 4.2.5 documents the staff's review of the applicant's PTS TLAA.

Anticipated Transient without Scram. The applicant's scoping results reports indicate that mechanical systems are included within the scope of license renewal because they perform intended functions that meet 10 CFR 50.62 requirements. The applicant determined the intended functions based on IP2 and IP3 CLB information and identified most in-scope components as electrical equipment in mechanical systems. For scoping electrical equipment, the applicant's bounding methodology included within the scope of license renewal all electrical and I&C systems in mechanical systems, by default. The applicant also conservatively included mechanical systems with ATWS intended functions based on CLB information from the SSFSs. The staff determined that this scoping methodology is adequate for identifying systems with functions credited for complying with the ATWS regulation.

Station Blackout. The scoping results reports identify the mechanical systems and structures credited with performing intended functions to comply with the SBO requirement. In its scoping effort, the applicant considered CLB information, including the UFSARs, SSFS, and the SBO report for electrical systems. The applicant used additional information (e.g., drawings and engineering judgment) to identify other systems that support SBO functions.

The applicant included within the scope of license renewal electrical equipment, mechanical systems, and structures with intended functions meeting SBO requirements. For scoping electrical equipment, the applicant's bounding methodology included within the scope of license renewal all electrical and I&C systems in mechanical systems by default. The mechanical systems and structures within the scope of license renewal are those relied on in the CLB for the 8-hour SBO coping duration phase and for the SBO recovery phase. The staff determined that this scoping methodology is adequate for identifying systems and structures with functions credited for complying with the SBO regulation. SER Section 2.5 documents the staff's review of the results of the implementation of the SBO scoping methodology.

2-18

2.1.4.3.3 Conclusion

The staff concludes that the applicant's methodology for identifying systems and structures meets the scoping criteria detailed in 10 CFR 54.4(a)(3) and, therefore, is acceptable. The staff based this conclusion on sample reviews, discussions with the applicant, and review of the applicant's scoping process.

2.1.4.4 Plant-Level Scoping of Systems and Structures

2.1.4.4.1 Summary of Technical Information in the Application

System- and Structure-Level Scoping. The applicant documented its methodology for performing the scoping of systems and structures in accordance with 10 CFR 54.4(a) in the LRA, guidance documents, and scoping and screening reports. The applicant's approach to system and structure scoping provided in the site guidance documents and implementing procedures is consistent with the methodology described in Section 2.1 of the LRA. Specifically, the implementing procedures require personnel performing license renewal scoping to use CLB documents, describe the system or structure, and include a list of functions that the system or structure is required to accomplish. Sources of information regarding the CLB for systems include the UFSARs, DBDs, P&IDs, Maintenance Rule information, drawings, and docketed correspondence. The applicant then compared identified system or structure function lists to the scoping criteria to determine whether the functions meet the scoping criteria of 10 CFR 54.4(a). The applicant documented the results of the plant-level scoping process in accordance with the implementing procedures. The results were provided in the systems and structures documents and reports that contain a description of the structure or system, a listing of functions performed by the system or structure, identification of intended functions, the 10 CFR 54.4(a) scoping criteria met by the system or structure, references, and the basis for the classification of the intended functions of the system or structure.

Insulation. LRA Section 2.1.1, "Scoping Methodology," states that insulation was treated as a bulk commodity for the purposes of scoping. LRA Section 2.4.4, "Bulk Commodities," discusses insulation and states that certain insulation has the specific intended functions of (1) controlling the heat load during DBAs in areas with safety-related equipment or (2) maintaining integrity such that falling insulation (such as reflective metallic-type reactor vessel insulation) does not damage safety-related equipment and was included within the scope of license renewal in accordance with 10 CFR 54.4(a)(1) or (a)(2) as applicable.

Consumables. In LRA Section 2.1.2.4, "Consumables," the applicant used the information in SRP-LR Table 2.1-3 to categorize and evaluate consumables. For the purpose of license renewal, consumables were divided into four categories (a) packing, gaskets, component seals, and O-rings, (b) structural sealants, (c) oil, grease, and component filters, and (d) system filters, fire extinguishers, fire hoses, and air packs.

Group (a) consumables (packing, gaskets, component mechanical seals, and O-rings) are typically used to provide a leakproof seal when components are mechanically joined together. These items are commonly found in components such as valves, pumps, heat exchangers, ventilation units or ducts, and piping segments. According to American National Standards Institute (ANSI) B31.1 and American Society of Mechanical Engineers (ASME) Boiler and Pressure Vessel (B&PV) Code Section III, the subcomponents of these pressure-retaining components are not pressure-retaining parts. Therefore, these subcomponents are not relied on to perform a pressure boundary intended function and are not subject to an AMR.

Group (b) consumables (elastomers and other materials used as structural sealants) are subject to an AMR if they are not periodically replaced and they perform an intended function, typically supporting a pressure boundary, flood barrier, or rated fire barrier. Seals and sealants, including pressure boundary sealants, compressible joints and seals, seismic joint filler, and waterproofing membranes, are included in the AMR of bulk commodities. Sealants with a pressure boundary function are included in the AMR of the containment buildings.

Group (c) consumables (oil, grease, and component filters) are treated as consumables because either (1) they are periodically replaced or (2) they are monitored and replaced based on condition. They are not subject to an AMR.

Group (d) consumables (system filters, fire hoses, fire extinguishers, self-contained breathing apparatus, and self-contained breathing apparatus cylinders) are considered consumables because they are routinely tested and inspected and they are replaced when necessary. Periodic inspection procedures specify the replacement criteria of these components that are routinely checked by tests or inspections. Therefore, while these consumables are in the scope of license renewal, they are not subject to an AMR.

2.1.4.4.2 Staff Evaluation

During the audit, the staff reviewed the applicant's methodology for performing the scoping of plant systems and structures to ensure that it is consistent with 10 CFR 54.4(a). The methodology used by the applicant to determine the systems and structures within the scope of license renewal is documented in implementing procedures and scoping results reports for mechanical systems. The scoping process defines the plant in terms of systems and structures. Specifically, the implementing procedures identify the systems and structures that are subject to review in accordance with 10 CFR 54.4(a) and describe the processes for capturing the results of the review. The procedures are used to determine whether the system or structure performs intended functions consistent with the requirements of 10 CFR 54.4(a). The implementing procedures indicate that the applicant completed this process for all systems and structures to ensure that the entire plant was addressed. During the audit, the staff reviewed a sampling of the documents and reports and concluded that the applicant's scoping results contain an appropriate level of detail to document the scoping process.

2.1.4.4.3 Conclusion

Based on its review of the LRA, site guidance documents, and scoping and screening implementing procedures, and based on a sampling of system scoping results reviewed during the audit, the staff concludes that the applicant's methodology for identifying systems and structures within the scope of license renewal, and their intended functions, is consistent with the requirements of 10 CFR 54.4 and, therefore, is acceptable.

2.1.4.5 Mechanical Scoping

2.1.4.5.1 Summary of Technical Information in the Application

LRA Section 2.1.1 describes the methodology for identifying license renewal evaluation boundaries. For mechanical systems, the mechanical components include those portions of the system that are necessary to ensure that the intended functions will be performed. The LRA states that components needed to support each of the system-level intended functions identified in the scoping process are included within the evaluation boundary.

The LRA states that, for mechanical system scoping, system boundaries were defined in part by the collection of components in the database assigned to the system code. The database represents all systems and contains the vast majority of system components. The database was useful in preparing the list of plant systems but could not be used alone to determine all system boundaries.

In addition, the LRA states that flow diagrams were used with the component database to help define system boundaries. System functions were determined based on the functions performed by the components within those boundaries. The LRA notes that, because of the differences in IP2 and IP3 system boundaries, the intended functions for the systems are often different, even for similarly named systems. The applicant evaluated structural commodities associated with mechanical systems, such as pipe hangers and insulation, with the structural bulk commodities, (i.e., grouping structural components that typically do not have unique identifiers that are common to in-scope systems and structures (e.g., anchors, embedments, equipment supports, insulation)), while it evaluated electrical and I&C components separately. The evaluation boundaries for mechanical systems were documented on license renewal drawings created by marking mechanical P&IDs to indicate the components within the scope of license renewal. The applicant evaluated mechanical systems against the criteria of 10 CFR 54.4(a)(1), (a)(2), and (a)(3).

2.1.4.5.2 Staff Evaluation

The staff evaluated LRA Section 2.1 and the guidance in the implementing procedures and reports in its review of the mechanical scoping process. The implementing procedures and reports provide instructions for identifying the evaluation boundaries. An understanding of system operations in support of intended functions is required to determine the mechanical system evaluation boundary.

This process is based on the review of Maintenance Rule basis documents, DBDs, SSFSs, the fire hazards analysis, the safe-shutdown analysis, internal flooding analyses, technical specifications, applicable sections of the UFSARs, and plant drawings. The evaluation boundaries for mechanical systems are documented on license renewal boundary drawings that were created by marking mechanical P&IDs to indicate the components within the scope of license renewal and subject to an AMR. Components within the evaluation boundary were reviewed to determine if they perform an intended function. Intended functions were established based on whether a particular function of a component is necessary to support the system functions that meet the scoping criteria.

The staff reviewed the implementing procedures and the CLB documents associated with mechanical system scoping and found that the guidance and CLB source information are acceptable to identify mechanical components and support structures in mechanical systems that are within the scope of license renewal. The staff conducted detailed discussions with the applicant's license renewal project management personnel and reviewed documentation pertinent to the scoping process. The staff assessed whether the applicant had appropriately applied the scoping methodology outlined in the LRA and implementing procedures and whether the scoping results are consistent with CLB requirements.

The staff determined that the applicant's procedure is consistent with the description in LRA Section 2.1 and the guidance in SRP-LR Section 2.1 and was adequately implemented.

On a sampling basis, the staff reviewed the applicant's methodology for identifying SW system mechanical component types meeting the scoping criteria of 10 CFR 54.4. The staff also reviewed the implementing procedures for the scoping methodology and discussed the methodology and results with the applicant. The staff verified that the applicant had identified and used pertinent engineering and licensing information to determine the SW system mechanical component types that fall within the scope of license renewal. As part of the review process, the staff evaluated each system intended function identified for the SW system, the basis for inclusion of the intended function, and the process used to identify each of the system component types. The staff verified that the applicant had identified and highlighted system P&IDs to develop the license renewal boundaries in accordance with the procedural guidance.

Based on its review of the LRA, scoping implementing procedures, the sample system, and review and discussions with the applicant, the staff verified that the applicant is knowledgeable about the process and conventions for establishing boundaries, as defined in the license renewal implementing procedures, and that the applicant independently verified the results in accordance with the implementing procedures. Specifically, other license renewal personnel knowledgeable about the system independently reviewed the marked-up drawings to ensure accurate identification of the system intended functions. The applicant performed additional cross-discipline verification and independent reviews of the associated drawings before approving the scoping effort.

2.1.4.5.3 Conclusion

Based on its review of the LRA, scoping implementing procedures, the sample system review, and discussions with the applicant, the staff concludes that the applicant's methodology for identifying mechanical systems and components within the scope of license renewal is consistent with the requirements of 10 CFR 54.4 and, therefore, is acceptable.

2.1.4.6 Structural Scoping

2.1.4.6.1 Summary of Technical Information in the Application

In LRA Section 2.1.1, the applicant described the methodology for identifying structures that are within the scope of license renewal. The applicant developed a list of plant structures from a review of plant layout drawings, Maintenance Rule documentation, DBDs, and the UFSARs. The structures list includes all structures that potentially support plant operations or could adversely impact structures that support plant operations. In addition to buildings and facilities, the list of structures includes other structures that support plant operation.

The applicant identified intended functions for structures and mechanical systems based on reviews of applicable plant licensing and design documentation. The applicant reviewed documents that included Maintenance Rule documents, DBDs, site SSFSs, the fire hazards analysis, the safe-shutdown analysis, internal flooding analyses, technical specifications, the UFSARs, and station drawing. The LRA states that the applicant evaluated each structure against the criteria of 10 CFR 54.4(a)(1), (a)(2), and (a)(3).

2.1.4.6.2 Staff Evaluation

The staff reviewed the applicant's approach for identifying structures relied on to perform the license renewal intended functions in accordance with 10 CFR 54.4(a). As part of this review, the staff discussed the methodology with the applicant, reviewed the documentation developed to support the applicant's review, and evaluated the scoping results for several structures that were identified as within the scope of license renewal.

The applicant identified and developed a list of plant structures and the structures' intended functions, through a review of the UFSARs, Maintenance Rule documents, the fire hazards analysis, DBDs, and structural drawings. The applicant determined that the primary structural safety functions applicable to the requirements of 10 CFR 54.4(a)(1) were to provide (1) containment of radioactive products to mitigate post-accident offsite doses and (2) to support or protect safety-related equipment. The applicant also included structures housing safety-related SSCs within the scope of license renewal in accordance with 10 CFR 54.4(a)(2).

The staff reviewed selected portions of the UFSARs, Maintenance Rule documents, the fire hazards analysis, DBDs, structural drawings, implementing procedures, and selected AMR reports to verify the adequacy of the methodology.

In addition, the staff reviewed the scoping results, including information contained in the source documentation for the turbine building, to verify that application of this methodology would provide the results as documented in the LRA. The staff verified that the applicant had identified and used pertinent engineering and licensing information to determine the turbine building structural component types that fall within the scope of license renewal. As part of the review process, the staff evaluated the intended functions identified for the turbine building, the basis for inclusion of the intended functions, and the process used to identify each of the component types.

2.1.4.6.3 Conclusion

Based on the staff's review of information in the LRA, the applicant's detailed scoping procedures, and a review of a sample of structural scoping results, the staff concluded that the applicant's methodology for identification of the structures within the scope of license renewal is consistent with the requirements of 10 CFR 54.4 and, therefore, is acceptable.

2.1.4.7 Electrical Scoping

2.1.4.7.1 Summary of Technical Information in the Application

LRA Section 2.1.1 states that, for the purposes of system-level scoping, all plant electrical and I&C systems are included in the scope of license renewal. The evaluation of electrical systems includes electrical and I&C components in mechanical systems.

LRA Section 2.5, "Scoping and Screening Results: Electrical and Instrumentation and Control Systems," states that the applicant included electrical and I&C components within the scope of license renewal unless they were specifically screened out. When used with the plant spaces approach, this method eliminates the need for unique identification of every component and its specific location and ensures that components are not improperly excluded from an AMR.

The applicant began the electrical and I&C scoping process by grouping the total population of components into commodity groups. The commodity groups include similar electrical and I&C components with common characteristics. The applicant identified component-level intended functions of the commodity groups. During the IPA, commodity groups and specific plant systems were eliminated from further review as the intended functions of commodity groups were examined.

2.1.4.7.2 Staff Evaluation

The staff reviewed the applicant's approach for identifying electrical and I&C SSCs relied on to perform the license renewal intended functions detailed in 10 CFR 54.4(a). As part of this review, the staff reviewed the implementing procedures and documentation developed to support the applicant's review and evaluated, on a sampling basis, the scoping results for several electrical systems identified as within the scope of license renewal.

The staff evaluated LRA Sections 2.1.1 and 2.5, scoping implementing procedures, scoping reports and aging management reports, as documented in the audit report, governing the electrical scoping methodology. The staff determined that the scoping phase for electrical components began with placing all electrical components from plant systems within the scope of license renewal. In addition, non-plant electrical systems including certain switchyard components required to support SBO and to restore offsite power were included within the scope of license renewal. The staff determined that the data sources used for scoping included the UFSARs, DBDs, cable database, component database, the station single-line drawings, cable procurement specifications, electrical drawings, the EQ master list, the IP2 and IP3 fuse list, and connection diagrams to identify the electrical and I&C components.

During the scoping methodology audit, the staff reviewed the applicant's process for identifying fuse holders within the scope of license renewal. The staff determined that the applicant had reviewed the plant fuse list and connection diagrams to identify fuses outside of complex assemblies and had determined that no fuses were within the scope of license renewal. During the scoping methodology audit, the staff reviewed the applicant's process for identifying tie wraps within the scope of license renewal. The staff determined that the applicant had reviewed the CLB for credit taken for tie wrap installation and reviewed operating experience to determine if the nonsafety-related tie wraps could affect a safety-related function, but did not identify any tie wraps within the scope of license renewal. The staff reviewed selected portions of the data sources and selected several examples of components for which the applicant demonstrated the process used to determine that electrical components were within the scope of license renewal.

2.1.4.7.3 Conclusion

On the basis of the review of information contained in the LRA, the applicant's scoping implementing procedures, and a review of a sample of electrical scoping results, the staff concludes that the applicant's methodology for identification of electrical components within the scope of license renewal is consistent with the requirements of 10 CFR 54.4 and, therefore, is acceptable.

SER Section 2.5 documents the results of the staff's review of the implementation of the SBO scoping methodology.

2.1.4.8 Conclusion for Scoping Methodology

On the basis of its review of the LRA and the scoping implementation procedures, the staff concludes that the applicant's scoping methodology is consistent with the guidance contained in the SRP-LR and identifies those SSCs (1) that are safety related, (2) whose failure could affect safety-related functions, and (3) that are necessary to demonstrate compliance with NRC regulations for fire protection, EQ, PTS, ATWS, and SBO. The staff concludes that the applicant's methodology is consistent with the requirements of 10 CFR 54.4(a) and, therefore, is acceptable.

2.1.5 Screening Methodology

2.1.5.1 General Screening Methodology

2.1.5.1.1 Summary of Technical Information in the Application

In LRA Section 2.1.2, "Screening Methodology," the applicant described its process for determining which components and structural elements require an AMR. Screening is the process by which the applicant identifies SCs within the scope of license renewal that perform an intended function, as described in 10 CFR 54.4, without moving parts or without a change in configuration or properties, and that are not subject to replacement based on a qualified life or specified time period.

LRA Section 2.1.6 states that the screening process for IP2 and IP3 followed the recommendations of NEI 95-10. For the group of systems and structures that were within the scope of license renewal, the applicant determined that passive long-lived components or structural elements that perform license renewal intended functions require an AMR. Components or structural elements that are either active or subject to replacement based on a qualified life do not require an AMR. Although the requirements for the IPA are the same for each system and structure, in practice, the screening process differed for mechanical systems, electrical systems, and structures.

2.1.5.1.2 Staff Evaluation

As required by 10 CFR 54.21, each LRA must contain an IPA that identifies SCs that are within the scope of license renewal and subject to an AMR. The IPA must identify components that perform an intended function without moving parts or a change in configuration or properties (passive components), as well as components that are not subject to periodic replacement based on a qualified life or specified time period (long-lived components). The IPA includes a description and justification of the methodology used to determine the passive and long-lived SCs and a demonstration that the effects of aging on those SCs will be adequately managed so that the intended function(s) will be maintained under all design conditions imposed by the plant-specific CLB for the period of extended operation.

The staff reviewed the methodology used by the applicant to determine whether mechanical and structural component types and electrical commodity groups within the scope of license renewal should be subject to an AMR. The applicant implemented a process for determining which SCs are subject to an AMR in accordance with the requirements of 10 CFR 54.21(a)(1). In LRA Section 2.1.2, the applicant discussed these screening activities as they relate to the component types and commodity groups within the scope of license renewal.

The applicant applied the screening process to evaluate the component types and commodity groups included within the scope of license renewal and to determine which ones were passive and long lived and, therefore, subject to an AMR. The staff reviewed LRA Sections 2.3, 2.4, and 2.5, which provide the results of the process used by the applicant to identify component types and commodity groups subject to an AMR. During the scoping and screening methodology audit, the staff discussed the processes used for each discipline, reviewed the implementing procedures describing the screening methodology, and reviewed documentation of the screening results. On a sampling basis, the staff also reviewed the screening results reports for the SW system and the turbine building. The following sections of this SER discuss specific methodologies for mechanical, electrical, and structural components.

2.1.5.2 Mechanical Component Screening

2.1.5.2.1 Summary of Technical Information in the Application

LRA Section 2.1.2.1, "Screening of Mechanical Systems," discusses the screening methodology for identifying passive and long-lived mechanical components and their support structures that are subject to an AMR.

License renewal drawings were prepared to indicate portions of systems that support system intended functions within the scope of license renewal, with the exception of those systems that are within scope under 10 CFR 54.4(a)(2) for physical interactions.

2.1.5.2.2 Staff Evaluation

The staff evaluated the mechanical screening methodology discussed and documented in LRA Section 2.1.2.1, implementing procedures, AMR reports, and the license renewal drawings. The mechanical system screening process began with the results from the scoping process. The applicant reviewed each system evaluation boundary as illustrated on P&IDs to identify passive and long-lived components. Within the system evaluation boundaries, all passive, long-lived components that perform or support a license renewal intended function are subject to an AMR. The results of the review are documented in the AMR reports. The AMR reports contain information such as the sources reviewed and the intended functions of the system.

During the scoping and screening methodology audit, the staff reviewed the results of the boundary evaluations and discussed the process with the applicant. The staff verified that mechanical system evaluation boundaries were established for each system within the scope of license renewal and that the boundaries were determined by mapping the system intended function boundaries onto P&IDs. The applicant reviewed the components within the system intended function boundary to determine whether the component supported the system's intended function. Components that supported the system's intended function were reviewed to determine whether the component was passive and long lived and, therefore, subject to an AMR.

The staff reviewed selected portions of the UFSARs, Maintenance Rule documents, the fire hazards analysis, DBDs, structural drawings, implementing procedures, and selected AMR reports. The staff conducted detailed discussions with the applicant's license renewal team and reviewed documentation pertinent to the screening process. The staff assessed whether the mechanical screening methodology outlined in the LRA and procedures was appropriately implemented and whether the scoping results are consistent with CLB requirements. The staff also reviewed the mechanical screening results for the SW system to verify proper implementation of the screening process. These audit activities revealed no discrepancies between the methodology documented and the implementation results.

2.1.5.2.3 Conclusion

Based on its review of the LRA, the implementing procedures, and a sample of the SW system screening results, the staff concludes that the applicant's mechanical component screening methodology is consistent with the SRP-LR guidance. The staff concludes that the applicant's methodology for identification of passive, long-lived mechanical components within the scope of license renewal and subject to an AMR is consistent with the requirements of 10 CFR 54.21(a)(1) and, therefore, is acceptable.

2.1.5.3 Structural Component Screening

2.1.5.3.1 Summary of Technical Information in the Application

LRA Section 2.1.2.2, "Screening of Structures," states that, for each structure within the scope of license renewal, the screening process identified those structural components that are subject to an AMR. LRA Section 2.4 presents the results for structures. The screening process for structural components involved a review of DBDs, design drawings, general arrangement drawings, penetration drawings, the UFSARs, plant modifications, system descriptions, and plant walkdowns to identify specific structural components and commodities that make up the structure. The LRA states that structures are inherently passive and, with few exceptions, are long lived. Therefore, the screening of structural components and commodities was based primarily on whether they perform an intended function. The applicant grouped structural components as commodities based on materials of construction (steel, bolted connections, concrete, and other materials). The applicant evaluated structural components and commodity groups to identify intended functions as they relate to license renewal.

2.1.5.3.2 Staff Evaluation

The staff reviewed the applicant's methodology for identifying structural components that are subject to an AMR as required by 10 CFR 54.21(a)(1). As part of this review, the staff discussed the methodology with the applicant, reviewed the documentation developed to support the activity, and evaluated the screening results for several structures that were identified within the scope of license renewal.

The staff reviewed the applicant's methodology used for structural screening described in the LRA sections noted above and in the applicant's implementing procedures and AMR reports. The applicant performed the screening review in accordance with the implementing procedures and identified pertinent structure design information, components, materials, environments, and aging effects. The staff verified that the applicant determined that structures are inherently passive and long lived, such that the screening of structural components and commodities was based primarily on whether they perform an intended function. During the scoping and screening methodology audit, the staff discussed the screening methodology and, on a sampling basis, reviewed the screening reports for a selected group of structures.

The staff reviewed selected portions of the UFSARs, Maintenance Rule documents, the fire hazards analysis, DBDs, structural drawings, implementing procedures, and selected AMR reports. The staff conducted detailed discussions with the applicant's license renewal team and reviewed documentation pertinent to the screening process. The staff assessed whether the screening methodology outlined in the LRA and implementing procedures was appropriately implemented and whether the scoping results are consistent with CLB requirements.

The staff also reviewed structural screening results for SCs contained in the turbine building to verify proper implementation of the screening process. Based on these audit activities, the staff identified no discrepancies between the methodology documented and the implementation results.

2.1.5.3.3 Conclusion

On the basis of the staff's review of information contained in the LRA, the applicant's detailed implementing procedures, and a review of a sample of structural screening results, the staff concludes that the applicant's methodology for identifying structural components within the scope of license renewal and subject to an AMR is consistent with the requirements of 10 CFR 54.21(a)(1) and, therefore, is acceptable.

2.1.5.4 Electrical Component Screening

2.1.5.4.1 Summary of Technical Information in the Application

In LRA Section 2.1.2.3, "Electrical and Instrumentation and Control Systems," the applicant discussed the screening of electrical and I&C system components. For each electrical system within the scope of license renewal, the screening process identified those electrical components and commodity groups that are subject to an AMR. Electrical components in mechanical systems were included in the scope of license renewal and were addressed under the electrical screening process.

The LRA states that the process of electrical screening differs from the mechanical and structural processes because the electrical components were addressed completely within their respective commodity groups. The applicant assigned each electrical component within the scope of license renewal to an electrical component commodity group for the screening evaluation. An electrical commodity group is a collection of electrical components grouped by type of equipment or function.

In the LRA, the applicant indicated that for the electrical equipment within the scope of license renewal, the passive, long-lived components that perform or support an intended function are subject to an AMR. Appendix B to NEI 95-10 identifies the electrical commodity groups considered to be passive and potentially requiring an AMR. For IP2 and IP3, electrical commodity groups were identified and cross-referenced to the appropriate NEI 95-10 commodity, which identifies the passive commodity groups. Electrical commodity groups determined to be active were not subject to an AMR. Electrical commodity groups that are not subject to replacement based on a qualified life or specified time period were considered long lived. The applicant further stated that components subject to replacement and addressed in replacement programs, such as the EQ Program, are not subject to an AMR.

2.1.5.4.2 Staff Evaluation

The staff reviewed the applicant's methodology for electrical screening, described in LRA Section 2.1.2.3, and the applicant's implementing procedures and AMR reports. The applicant used the screening process described in these documents to identify the electrical commodity groups subject to AMR.

The applicant used the component database, the station's single-line drawings, and cable procurement specifications as data sources to identify the electrical and I&C commodity groups subject to an AMR. The applicant also reviewed additional IP2 and IP3 documents such as electrical drawings and the EQ master list to validate the listing as complete.

The applicant determined that two commodity groups meet the passive criteria in accordance with NEI 95-10—(1) high-voltage insulators and (2) cables and connections, bus, and electrical portions of electrical and I&C penetration assemblies. The applicant evaluated the identified passive commodity groups to determine whether they are subject to replacement based on a qualified life or specified time period (short lived) or not subject to replacement based on a qualified life or specified time period (long lived). The applicant determined that the other electrical and I&C commodity groups are active and do not require an AMR. The staff reviewed the screening of selected components to verify the correct implementation of the methodology. The staff also reviewed selected electrical screening results, on a sampling basis, to verify proper implementation of the screening process. Based on these audit activities, the staff identified no discrepancies between the methodology documented and the implementation results.

2.1.5.4.3 Conclusion

The staff reviewed the LRA, procedures, electrical drawings, and a sample of the results of the screening methodology. The staff concludes that the applicant's methodology is consistent with the description in the LRA and the applicant's implementing procedures. On the basis of its review of information contained in the LRA, the applicant's implementing procedures, and a review of a sample of electrical screening results, the staff concludes that the applicant's methodology for identifying electrical commodity groups within the scope of license renewal and subject to an AMR is consistent with the requirements of 10 CFR 54.21(a)(1) and, therefore, is acceptable.

2.1.5.5 Screening Methodology Conclusion

On the basis of a review of the LRA, the screening implementing procedures, discussions with the applicant's staff, and a sample review of screening results, the staff concludes that the applicant's screening methodology is consistent with the guidance contained in the SRP-LR and that the applicant identified those passive, long-lived components within the scope of license renewal that are subject to an AMR. The staff concludes that the applicant's methodology is consistent with the requirements of 10 CFR 54.21(a)(1) and, therefore, is acceptable.

2.1.6 Summary of Evaluation Findings

The staff's review of the information presented in LRA Section 2.1, the supporting information in the scoping and screening implementing procedures and reports, the information presented during the scoping and screening methodology audit, and the applicant's responses dated February 13, 2008, to the staff's RAIs formed the basis of the staff's evaluation. The staff verified that the applicant's scoping and screening methodology is consistent with the requirements of 10 CFR 54.4 and 10 CFR 54.21(a)(1). The staff also confirmed that the applicant's description and justification of its scoping and screening methodology are adequate to meet the requirements of 10 CFR 54.21(a)(1). From this review, the staff concludes that the applicant's methodology for identifying systems and structures within the scope of license renewal and SCs requiring an AMR is acceptable.

2.2 Plant-Level Scoping Results

SER Section 2.2A presents plant-level scoping results for IP2; SER Section 2.2B presents plant-level scoping results for IP3.

2.2A IP2 Plant-Level Scoping Results

2.2A.1 Introduction

In LRA Section 2.1, the applicant described the methodology for identifying SSCs within the scope of license renewal and subject to an AMR. The applicant applied the scoping methodology to determine which SSCs must be included within the scope of license renewal. LRA Section 2.2 provides the results of the applicant's review. The staff reviewed the plant-level scoping results to determine whether the applicant had properly identified all systems and structures relied on to mitigate DBEs, as required by 10 CFR 54.4(a)(1); systems and structures whose failure could prevent satisfactory accomplishment of any safety-related functions, as required by 10 CFR 54.4(a)(2); and systems and structures relied on in safety analyses or plant evaluations to perform functions in accordance with 10 CFR 54.4(a)(3).

2.2A.2 Summary of Technical Information in the Application

In LRA Table 2.2-1a-IP2, the applicant listed the plant mechanical systems within the scope of license renewal for IP2. In LRA Table 2.2-1b-IP2, the applicant listed the plant electrical and I&C systems within the scope of license renewal for IP2. In LRA Table 2.2-3, the applicant listed the structures that are within the scope of license renewal for IP2. In LRA Tables 2.2-2-IP2, "Mechanical Systems Not within the Scope of License Renewal," and 2.2-4, the applicant listed the systems and structures that are not within the scope of license renewal. Systems and structures that exist only at one unit are marked in the tables, as appropriate. Based on the DBEs considered in the plant's CLB, other CLB information relating to nonsafety-related systems and structures, and certain regulated events, the applicant identified plant-level systems and structures within the scope of license renewal as defined by 10 CFR 54.4.

2.2A.3 Staff Evaluation

The staff reviewed the scoping and screening methodology and provides its evaluation in SER Section 2.1. To verify that the applicant properly implemented its methodology, the staff's review focused on the implementation results shown in LRA Tables 2.2-1a-IP2, 2.2-1b-IP2, 2.2-2-IP2, 2.2-3, and 2.2-4. In its review, the staff sought to confirm that the applicant had not omitted any plant-level system or structure from the scope of license renewal.

The staff reviewed the systems and structures that the applicant did not identify as within the scope of license renewal to determine whether they perform any intended functions that would require their inclusion within the scope of license renewal. The staff reviewed the applicant's implementation results in accordance with the guidance in SRP-LR Section 2.2, "Plant-Level Scoping Results."

During its review of LRA Section 2.2, the staff identified an area in which it required additional information to complete its review of the applicant's plant-level scoping results. The applicant's responses to the staff's RAIs are discussed below.

In RAI 2.2A-1, dated December 7, 2007, the staff noted that LRA Table 2.2-2-IP2 excludes the hot penetration cooling system from the scope of license renewal and references UFSAR Section 5.1.4.2.2 as the basis for this decision. The staff further noted that UFSAR Section 5.1.4.2.2 provides a local area temperature limit of 250 degrees Fahrenheit (F) and states that air-to-air heat exchangers provide cooling for hot penetrations. The staff noted that cooling of hot containment penetrations minimizes age-related, heat-induced degradation of local concrete surrounding the penetration; therefore, the system may have an intended function, as defined in 10 CFR 54.4(a). The staff requested that the applicant justify the exclusion of the hot penetration cooling system from the scope of license renewal.

In its response, dated January 4, 2008, the applicant stated that the hot penetration cooling system removes heat from penetrations for hot piping systems to limit the temperature of the surrounding concrete during normal plant operation. The applicant further explained that the hot penetration cooling system is not required to function during accident conditions and has no function that meets the requirements of 10 CFR 54.4(a)(1). Additionally, the applicant explained that the hot penetration cooling system is not relied on to perform intended functions in accordance with 10 CFR 54.4(a)(2) or 10 CFR 54.4(a)(3); therefore, it is not within the scope of license renewal. The applicant provided the following evaluation:

> In order to lose significant structural properties, concrete must be held at high temperatures for an extended period of time. The hottest penetrations at IPEC are the MS lines, which normally operate at a temperature of 507°F. The results of a heat transfer analysis indicate that in the improbable case that all cooling air would be lost to the main steam penetration; the surrounding concrete would reach a maximum temperature of 200°F in approximately 100 hours and 280°F in approximately 1000 hours. It is not credible that cooling air would be lost for a significantly long period of time since the failure of the air blower drive motors is alarmed in the control room. Therefore, the failure of the hot penetration cooling system would not adversely impact the concrete in the penetrations.

Based on its review, the staff finds the response to RAI 2.2A-1 acceptable because the applicant adequately explained that the hot penetration cooling system is not safety related and its failure would not adversely affect a safety-related system or structure. The staff confirmed that the hot penetration cooling system is not credited in any accident analyses in the applicant's CLB. The staff's concern described in RAI 2.2A-1 is resolved.

During the NRC's onsite scoping and screening audit, the staff reviewed the applicant's onsite documentation for the potential interaction of SSCs based on the proximity of nonsafety-related SSCs to safety-related SSCs. In RAI 2.1-1, dated January 14, 2008, the staff asked the applicant to provide a technical basis for excluding these systems from scope. In its response, dated February 13, 2008, the applicant provided an evaluation of these systems and amended the LRA to include the IP2 chlorination system within the scope of license renewal in accordance with 10 CFR 54.4(a)(2). Hence, as noted in the above section, the applicant added the IP2 chlorination system to LRA Table 2.3.3-19-A-IP2. Additionally, the applicant added LRA Table 2.3.3-19-44-IP2 to identify the component types subject to an AMR. Based on a review of this response, the staff finds that the applicant has adequately identified systems required to be within the scope of license renewal in accordance with 10 CFR 54.4(a)(2).

2.2A.4 Conclusion

The staff reviewed LRA Section 2.2, the RAI responses, and the UFSAR supporting information to determine whether the applicant failed to identify any systems and structures within the scope of license renewal. The staff found an instance in which the applicant omitted a system that should have been included within the scope of license renewal. The applicant has satisfactorily resolved this issue as discussed in the preceding staff evaluation. Therefore, on the basis of its review, the staff concludes that the applicant has appropriately identified the mechanical systems and structures within the scope of license renewal, as required by 10 CFR 54.4(a).

2.2B IP3 Plant-Level Scoping Results

2.2B.1 Introduction

In LRA Section 2.1, the applicant described the methodology for identifying SSCs within the scope of license renewal and subject to an AMR. The applicant applied the scoping methodology to determine which SSCs must be included within the scope of license renewal. LRA Section 2.2 provides the results of the applicant's review. The staff reviewed the plant-level scoping results to determine whether the applicant had properly identified all systems and structures relied on to mitigate DBEs, as required by 10 CFR 54.4(a)(1); systems and structures whose failure could prevent satisfactory accomplishment of any safety-related functions, as required by 10 CFR 54.4(a)(2); and systems and structures relied on in safety analyses or plant evaluations to perform functions in accordance with 10 CFR 54.4(a)(3).

2.2B.2 Summary of Technical Information in the Application

In LRA Table 2.2-1a-IP3, the applicant listed the plant mechanical systems within the scope of license renewal for IP3. In LRA Table 2.2-1b-IP3, the applicant listed the plant electrical and I&C systems within the scope of license renewal for IP3. In LRA Table 2.2-3, the applicant listed the structures that are within the scope of license renewal for IP3. In LRA Tables 2.2-2-IP3, "Mechanical System Not within the Scope of License Renewal," and 2.2-4, the applicant listed the systems and structures that are not within the scope of license renewal. Systems and structures that exist only at one unit are marked in the tables, as appropriate. Based on the DBEs considered in the plant's CLB, other CLB information relating to nonsafety-related systems and structures, and certain regulated events, the applicant identified plant-level systems and structures within the scope of license renewal as defined by 10 CFR 54.4.

2.2B.3 Staff Evaluation

The staff reviewed the scoping and screening methodology and provides its evaluation in SER Section 2.1. To verify that the applicant properly implemented its methodology, the staff's review focused on the implementation results shown in LRA Tables 2.2-1a-IP3, 2.2-1b-IP3, 2.2-2-IP3, 2.2-3, and 2.2-4. In its review, the staff sought to confirm that no plant-level systems or structures were omitted from the scope of license renewal.

The staff reviewed the systems and structures that the applicant did not identify as within the scope of license renewal to determine whether they perform any intended functions that would require their inclusion within the scope of license renewal. The staff conducted its review of the applicant's implementation in accordance with the guidance in SRP-LR Section 2.2.

During its review of LRA Section 2.2, the staff identified areas in which it required additional information to complete its review of the applicant's plant-level scoping results. The applicant's responses to the staff's RAIs are discussed below.

In RAI 2.2B-1, dated December 7, 2007, the staff noted that LRA Table 2.2-2-IP3 excludes the breathable air system from the scope of license renewal and references UFSAR Section 9.10 as the basis for this decision. The staff further noted that UFSAR Section 9.10 states that the breathable air system is a non-Category I system, except for the penetration into containment, where breathable air is provided inside containment through a spare penetration line. The staff noted that the breathable air system's containment penetration should be within the scope of license renewal in accordance with 10 CFR 54.4(a)(1), and requested that the applicant confirm whether the containment penetration is within the scope of license renewal.

In its response, dated January 4, 2008, the applicant stated that the breathable air containment penetration, designated as "X-X," is within the scope of license renewal and was reviewed as part of the containment penetration system in LRA Section 2.3.2.5. The applicant further explained that the containment penetration for the breathable air system is subject to an AMR.

Based on its review, the staff finds the response to RAI 2.2B-1 acceptable because the applicant adequately explained that containment penetration "X-X" for the breathable air system was evaluated with the containment penetration system. Furthermore, the staff confirmed that the LRA identified the breathable air containment penetration as requiring an AMR. The staff's concern described in RAI 2.2B-1 is resolved.

In RAI 2.2B-2, dated February 13, 2008, the staff noted that nonsafety-related SSCs directly connected to safety-related SSCs must be structurally sound to maintain the pressure boundary integrity of safety class piping. The nonsafety-related piping and supports up to and including the first seismic anchor beyond the safety/nonsafety interface may need to be within scope to ensure that the safety-related portion of the piping will be able to perform its intended function.

LRA Table 2.2-2-IP3 excluded the hydrogen gas system from the scope of license renewal. This system, along with the nitrogen system, provides the volume control tank (VCT) with gas for oxygen scavenging. Since the piping is directly connected to the VCT, the staff questioned whether the system should be considered within the scope of license renewal, in accordance with 10 CFR 54.4(a)(2), because of the potential for physical interaction between the nonsafety- and safety-related equipment. The staff asked the applicant to evaluate placing the hydrogen system or nitrogen system, or both, within scope under 10 CFR 54.4(a)(2) and to evaluate any other interfaces of gas system interaction with safety-related equipment.

In its response, dated March 12, 2008, the applicant stated that the IP3 VCT is within scope and subject to an AMR, in accordance with 10 CFR 54.4(a)(1). The nitrogen system piping (and associated valve components) upstream of check valve 270 are within scope and subject to an AMR, in accordance with 10 CFR 54.4(a)(2), with component types evaluated in the LRA. In addition, the piping and valves connected to check valve 270 have an intended function to maintain integrity to ensure that physical interaction with safety-related components cannot prevent satisfactory accomplishment of a safety function due to structural support. Therefore, the hydrogen system should be within scope, as required by 10 CFR 54.4(a)(2). The applicant revised the LRA to include the hydrogen system. The applicant stated that no additional changes to the LRA were required due to other gas system interaction with safety-related equipment. The staff's concern described in RAI 2.2B-2 is resolved. SER Section 2.3B.3.19 documents the staff's review of the IP3 hydrogen system that the applicant added to the scope of license renewal. SER Section 3.3.2.1 documents the staff's evaluation of AMR results for the IP3 hydrogen system.

2.2B.4 Conclusion

The staff reviewed LRA Section 2.2, the RAI responses, and the UFSAR supporting information to determine whether the applicant failed to identify any systems and structures within the scope of license renewal. The staff identified the omission of the hydrogen system, which the applicant should have included within the scope of license renewal. The applicant has satisfactorily resolved this issue as discussed in the preceding staff evaluation. Therefore, on the basis of its review, the staff concludes that the applicant has appropriately identified the mechanical systems and structures within the scope of license renewal, as required by 10 CFR 54.4(a).

2.3 Scoping and Screening Results: Mechanical Systems

SER Section 2.3A presents the scoping and screening results for IP2 mechanical systems; SER Section 2.3B presents the scoping and screening results for IP3 mechanical systems.

This section documents the staff's review of the applicant's scoping and screening results for mechanical systems. Specifically, this section discusses the following:

- RCS
- engineered safety features
- auxiliary systems
- steam and power conversion systems

In accordance with the requirements of 10 CFR 54.21(a)(1), the applicant must identify and list passive, long-lived mechanical SSCs that are within the scope of license renewal and subject to an AMR. To verify that the applicant properly implemented its methodology, the staff focused its review on the implementation results. This focus allowed the staff to confirm that the applicant had not omitted any mechanical system components that meet the scoping criteria and are subject to an AMR.

The staff's evaluation of the information in the LRA was the same for all mechanical systems. The objective was to determine whether the applicant had identified, as required by 10 CFR 54.4, components and supporting structures for mechanical systems that appear to meet the license renewal scoping criteria. Similarly, the staff evaluated the applicant's screening results to verify that all passive, long-lived components were subject to an AMR, in accordance with 10 CFR 54.21(a)(1).

In its scoping evaluation, the staff reviewed the applicable LRA sections and license renewal drawings, focusing on components that the applicant had not included within the scope of license renewal. The staff reviewed relevant licensing basis documents, including the UFSAR, for each mechanical system to determine whether the applicant had omitted from the scope of license renewal any components with license renewal intended functions, as defined in 10 CFR 54.4(a). The staff also reviewed the licensing basis documents to determine whether the LRA specified all license renewal intended functions in accordance with 10 CFR 54.4(a). The staff requested additional information to resolve any omissions or discrepancies identified.

After its review of the scoping results, the staff evaluated the applicant's screening results. For those SCs with intended functions, the staff sought to determine whether (1) the functions are performed with moving parts or a change in configuration or properties or (2) the SCs are subject to replacement after a qualified life or specified time period, as described in 10 CFR 54.21(a)(1). For SCs that did not meet either of these criteria, the staff sought to confirm that they were subject to an AMR, as required by 10 CFR 54.21(a)(1). The staff requested additional information to resolve any omissions or discrepancies identified.

Two-Tier Scoping Review Process for Balance of Plant Systems

The scope of license renewal as documented in the LRA, includes 144 mechanical systems, among which 96 systems are balance of plant (BOP) systems. These 96 systems include most of the auxiliary systems and all of the steam and power conversion systems. The staff performed a two-tier scoping review for these BOP systems.

In the two-tier scoping review, the staff reviewed the LRA and UFSAR description, focusing on the system intended function, and divided all of the BOP systems into two groups, those that required a simplified Tier 1 review and those that required a more detailed Tier 2 review. The staff selected the systems for a detailed Tier 2 review based on the following screening criteria:

- safety importance and risk significance
  - high safety-significant systems
  - common-cause failure of redundant trains
- operating experience indicating likely passive failures
- previous LRA review experience

Examples of systems that are typically selected for safety importance and risk significance, based on the individual plant examination results, are the component cooling water (CCW) system, the auxiliary feedwater (AFW) system, and the SW system. An example of a system, whose failure could cause failure of redundant trains is a drain system for flood protection. Examples of systems with operating experience that indicates the potential for passive failures include the main steam (MS), feedwater (FW), and SW systems. Examples of systems with omissions identified in previous LRA reviews include the spent fuel cooling system and makeup water sources to safety systems. In addition, the staff ensured that a minimum of 50 percent of the BOP systems received a Tier 2 review.

For systems receiving a simplified Tier 1 review, the staff reviewed the LRA and the UFSAR to determine whether the applicant failed to identify any component types typically found within the scope of license renewal. SER Sections 2.3A.3 and 2.3B.3 identify the IP2 and IP3 BOP systems, respectively, for which the staff conducted a simplified Tier 1 review. For all other BOP systems, the staff performed a detailed Tier 2 review.

For systems receiving a detailed Tier 2 review, the staff reviewed the LRA, UFSAR, and the detailed boundary drawings to determine whether the applicant failed to identify any components within the scope of license renewal and any components subject to an AMR. During its review, the staff evaluated the system functions described in the LRA and UFSAR to verify that the applicant did not omit from the scope of license renewal any components with intended functions, as defined in 10 CFR 54.4(a). The staff then reviewed those components that the applicant identified as within the scope of license renewal to verify that the applicant had not omitted any passive and long-lived components subject to an AMR, in accordance with 10 CFR 54.21(a)(1).

2.3A Scoping and Screening Results: IP2 Mechanical Systems

2.3A.1 Reactor Coolant System

LRA Section 2.3.1 identifies the RCS SCs subject to an AMR for license renewal.

The RCS includes mechanical components in the following subsystems.

- reactor vessel
- reactor vessel internals
- steam generators (SGs)
- RCPs
- pressurizer
- control rod drives
- in-core instrumentation

The applicant described the supporting SCs of the RCS in the following LRA sections:

- 2.3.1.1, "Reactor Vessel"
- 2.3.1.2, "Reactor Vessel Internals"
- 2.3.1.3, "Reactor Coolant Pressure Boundary"
- 2.3.1.4, "Steam Generators"

LRA Section 2.3.1 describes the following RCS subsystems:

Reactor Vessel. The cylindrical reactor vessel has a hemispherical bottom and a flanged and gasketed removable upper head. The upper reactor closure head and the reactor vessel flange are joined by studs. Two metallic O-rings seal the reactor vessel when the reactor closure head is bolted in place. A leak-off connection between the two O-rings monitors leakage across the inner O-ring. The vessel design is in accordance with ASME Code, Section III, "Nuclear Vessels." Coolant enters the reactor vessel through inlet nozzles in a plane just below the vessel flange and above the core, flows downward through the annular space between the vessel wall and the core barrel into a plenum at the bottom of the vessel, reverses direction, and flows up through the core. After mixing in the upper plenum, the mixed coolant stream then flows out of the vessel through exit nozzles on the same plane as the inlet nozzles. The core instrumentation nozzles are on the lower head and the control rod nozzle penetrations are on the upper head.

Reactor Vessel Internals. The reactor vessel internals direct the coolant flow, support the reactor core, and guide the control rods. The reactor vessel contains the core support assembly, upper plenum assembly, fuel assemblies, control cluster assemblies, surveillance specimens, and in-core instrumentation. The lower core support structure, the upper core support structure, and the incore instrumentation support structure are the three major parts of the reactor vessel internals. A one-piece thermal shield, concentric with the reactor core, is located between the core barrel and the reactor vessel. The shield, cooled by the coolant on its downward pass, protects the vessel by attenuating much of the gamma radiation and some of the fast neutrons which escape from the core.

Pressurizer. System pressure is controlled by the pressurizer, which maintains water and steam pressure through the use of electrical heaters and sprays. Steam can either be formed by the heaters or condensed by a pressurizer spray to minimize pressure variations caused by contraction and expansion of the coolant. Control and protective circuits such as the high-pressure trip and code relief valves connected to the top head of the pressurizer protect the RCS against overpressure. The relief valves discharge into the pressurizer relief tank, which condenses and collects the valve effluent. Two power-operated relief valves (PORVs) and three code safety valves protect against pressure surges beyond the pressure-limiting capacity of the pressurizer spray. The PORVs also operate from the overpressure protection system to prevent RCS pressure from exceeding the limits of ASME Code, Section III, Appendix G during low-temperature operation. Steam and water discharge from the power relief and safety valves passes to the pressurizer relief tank partially filled with water at or near ambient containment conditions. The tank normally contains water in a predominantly nitrogen atmosphere. Steam discharged under the water level condenses and cools by mixing with the water. Rupture discs that discharge into the reactor containment protect the tank against a discharge exceeding the design value.

Steam Generators. Each reactor coolant loop has a vertical shell and U-tube steam generator (SG). Reactor coolant enters the inlet side of the channel head at the bottom of the SG through the inlet nozzle, flows through the U-tubes to an outlet channel, and exits the generator through another bottom nozzle. The inlet and outlet channels are separated by a partition. Feedwater to the SG enters just above the top of the U-tubes through a feedwater ring. The water flows downward through an annulus between the tube wrapper and the shell and then upward through the tube bundle where it converts to a steam-water mixture that passes through a primary separator assembly that reduces the water content in the mixture. The separated water combines with the feedwater for another pass through the tube bundle. The remaining higher steam-content mixture rises through additional secondary separators which further reduce the water content.

Reactor Coolant Pumps. Each reactor coolant loop has a vertical single-stage centrifugal pump with a controlled-leakage seal assembly. Reactor coolant pumped by the impeller attached to the bottom of the rotor shaft and drawn up through the impeller discharges through passages in the diffuser and out through a discharge nozzle in the side of the casing. A flywheel at the top of the rotor shaft extends the pump coastdown flow during any loss of power to the pump motor. A portion of the flow from the chemical and volume control system (CVCS) charging pumps is injected into the RCP between the impeller and the controlled-leakage seal. Component cooling water flows to the motor-bearing oil coolers and the thermal barrier cooling coil.

The RCS contains safety-related components relied on to remain functional during and following DBEs. The failure of nonsafety-related SSCs in the RCS potentially could prevent the satisfactory accomplishment of a safety-related function. In addition, the RCS performs functions that support fire protection, PTS, ATWS, and SBO.

Control Rod Drives. The control rod drive system positions the control rods within the core. The reactor uses the Westinghouse magnetic-type control rod drive assemblies on the upper reactor vessel head to insert or withdraw control rods in the core to control generation of nuclear power. The movement of the control rods is accomplished through the sequential operation of three types of magnetic coils. Upon a loss of power to the coils, the released rod cluster control assemblies with full-length absorber rods fall by gravity into the core. Each control rod drive assembly is a hermetically-sealed unit to prevent leakage of reactor coolant. The design of all pressure-containing components meets ASME Code, Section III, Division 1 requirements for Class A vessels.

The control rod drive system contains safety-related components relied upon to remain functional during and following DBEs.

In-Core Instrumentation. The in-core instrumentation system provides information on the neutron flux distribution and fuel assembly outlet temperatures at selected core locations to confirm reactor core design parameters and calculated hot channel factors. The system acquires data and performs no plant operation. The system consists of thermocouples positioned to measure fuel assembly coolant outlet temperature at preselected locations, flux thimbles running the length of selected fuel assemblies to measure the neutron flux distribution within the reactor core by moveable in-core detectors, in-core drives, drive motors, positioning equipment, and instruments. The flux thimbles, seal table, and guide tube form part of the RCPB. The in-core instrumentation system includes the pressure-retaining guide tubes that form parts of the RCPB. For IP2, the RCS and the nuclear instrumentation system include other, nonpressure boundary portions of the in-core instrumentation (listed in LRA Table 2.2-1b-IP2 with the electrical and I&C systems).

The in-core instrumentation system contains safety-related components relied upon to remain functional during and following DBEs.

The RCS Class I piping evaluation boundary extends into portions of systems attached to the RCS. For both units, the RCS AMR includes the Class I components of the systems listed below. The LRA section referenced below includes the non-Class 1 portions of the following systems:

- CVCS (LRA Section 2.3.3.6)
- isolation valve seal water (LRA Section 2.3.2.3)
- primary sampling system (LRA Section 2.3.3.19)
- residual heat removal (RHR) (LRA Section 2.3.2.1)
- safety injection system (LRA Section 2.3.2.4)

IP2 RCS components containing air are evaluated with compressed air systems (LRA Section 2.3.3.4). A small number of IP2 RCS components are evaluated with the primary water makeup systems (LRA Section 2.3.3.7) and the nitrogen systems (LRA Section 2.3.3.5). IP2 RCP lube oil collection system components are part of the IP2 fire protection system, not the RCS. IP2 RCS containment penetration components, which are not part of the RCPB, are evaluated with containment penetrations (LRA Section 2.3.2.5).

Fuel assemblies are not subject to an AMR because they are replaced after a limited number of fuel cycles. The control rods are active components and are not subject to an AMR.

SER Sections 2.3A.1.1–2.3A.1.4 discuss the staff's findings based on its review of LRA Sections 2.3.1.1–2.3.1.4, respectively.

2.3A.1.1 Reactor Vessel

2.3A.1.1.1 Summary of Technical Information in the Application

LRA Section 2.3.1.1 describes the reactor vessel, stating that the evaluation boundary for the reactor vessel encompasses the reactor vessel pressure boundary subcomponents, which include the shell, top and bottom heads, closure head stud assembly, primary nozzles and safe-ends, control rod drive mechanism (CRDM) housing penetrations, bottom-mounted instrumentation flux thimble tube penetrations, guide tubes, and seal table. LRA Section 2.3.1.1 also describes other subcomponents that support the intended functions of the reactor vessel, including the core support pads and core guide lugs, vessel flange, and closure head lifting lugs.

LRA Section 2.3.1 describes the functions of the reactor vessel. LRA Tables 2.3.1-1-IP2 and 2.3.1-1-IP3 identify reactor vessel component types within the scope of license renewal and subject to an AMR, as well as their intended functions.

2.3A.1.1.2 Staff Evaluation

The staff reviewed LRA Section 2.3.1.1 and the UFSARs using the evaluation methodology described in SER Section 2.3 and the guidance in SRP-LR Section 2.3.

During its review, the staff evaluated the system functions described in the LRA and UFSAR to verify that the applicant had not omitted from the scope of license renewal any components with intended functions, as require by 10 CFR 54.4(a). The staff then reviewed those components that the applicant identified as within the scope of license renewal to verify that the applicant had not omitted any passive and long-lived components subject to an AMR, in accordance with 10 CFR 54.21(a)(1).

2.3A.1.1.3 Conclusion

The staff reviewed the LRA and UFSAR to determine whether the applicant failed to identify any SSCs within the scope of license renewal. The staff found no such omissions. In addition, the staff sought to determine whether the applicant failed to identify any components subject to an AMR. The staff found no such omissions. On the basis of its review, the staff concludes that the applicant has adequately identified the reactor vessel components within the scope of license renewal, as required by 10 CFR 54.4(a), and those subject to an AMR, as required by 10 CFR 54.21(a)(1).

2.3A.1.2 Reactor Vessel Internals

2.3A.1.2.1 Summary of Technical Information in the Application

LRA Section 2.3.1.2 describes the reactor vessel internals. For both units, the lower core support structure, the upper core support structure, and the incore instrumentation support structure are the three major parts of the reactor internals.

The lower core support structure is supported at its upper flange from a ledge in the reactor vessel. Within the core barrel are a core baffle and a lower core plate, both of which are attached to the core barrel wall. The lower core support structure provides passageways for the coolant flow. The lower core plate at the bottom of the core below the baffle plates provides support and orientation for the fuel assemblies. Fuel alignment pins (two for each assembly) are also inserted into this plate. Columns are placed between the lower core plate and core support casting to provide stiffness and to transmit the core load to the core support casting. Adequate coolant distribution is obtained through the use of the lower core plate and a diffuser plate.

The support columns establish the spacing between the upper support assembly and the upper core plate and are fastened at top and bottom to these plates and beams.

The rod cluster control assembly guide tube assemblies shield and guide the control rod drive shafts and control rods. They are fastened to the upper support and are guided by pins in the upper core plate for proper orientation and support. The control rod shroud tube, which is attached to the upper support plate and guide tube, provides additional guidance for the control rod drive shafts.

An upper system (thermocouple conduit) is used to convey and support thermocouples penetrating the vessel through the head, and a lower system (flux thimble guide tube) is used to convey and support flux thimbles penetrating the vessel through the bottom. The upper system utilizes the reactor vessel head penetrations. Instrumentation port columns are slip-connected to in-line columns that are in turn fastened to the upper support plate. These port columns protrude through the head penetrations. The thermocouples are carried through these port columns and the upper support plate at positions above their readout locations. The columns of the upper core support system support the thermocouple conduits.

LRA Section 2.3.1 describes the functions of the reactor vessel internals. LRA Tables 2.3.1-2-IP2 and 2.3.1-2-IP3 identify reactor vessel internals component types within the scope of license renewal and subject to an AMR, as well as their intended functions.

2.3A.1.2.2 Staff Evaluation

The staff reviewed LRA Section 2.3.1.2 and the UFSAR using the evaluation methodology described in SER Section 2.3 and the guidance in SRP-LR Section 2.3.

During its review, the staff evaluated the system functions described in the LRA and UFSAR to verify that the applicant had not omitted from the scope of license renewal any components with intended functions, as required by 10 CFR 54.4(a). The staff then reviewed those components that the applicant identified as within the scope of license renewal to verify that the applicant had not omitted any passive and long-lived components subject to an AMR, in accordance with 10 CFR 54.21(a)(1).

During its review, the staff identified an area in which additional information was necessary to complete its review. The applicant responded to the staff's RAI as discussed below.

In RAI 2.3A.1.2-1, dated January 28, 2008, the staff noted that if certain reactor vessel internals failed, they could potentially inhibit core coolability during an accident. Therefore, the staff requested that the applicant clarify whether the sample tubing and sample tubing springs are within the scope of license renewal.

In its response, dated February 27, 2008, the applicant stated that it had evaluated the sample tubing (also known as the irradiation specimen guide) and the sample tubing springs (also known as the specimen plugs). The review included consideration of component functions and the potential impact of component failure on the function of other components. The applicant stated that sample tubing and the sample tubing springs have no license renewal intended function and are not subject to an AMR. Additionally, the applicant stated that it had reviewed Westinghouse Commercial Atomic Power (WCAP)-14577 Rev 1-A, "License Renewal Evaluation: Aging Management for Reactor Internals." Section 3.1 of the staff's SER, which evaluated WCAP-14577 states, "[t]he staff found the list of intended functions to be complete and in accordance with 10 CFR 54.4(a)." Section 2.1.1 of the same SER details the list of functions and states, "Prevent failure of all nonsafety-related systems, structures, and components whose failure could prevent any of these (previously listed) functions." WCAP-14577A, Table 2-2, confirms the applicant's conclusion that no AMR is required for the sample tubing and the sample tubing springs because the components do not have a license renewal function. Therefore, the staff finds the applicant's response acceptable. The staff's concern described in RAI 2.3A.1.2-1 is resolved.

2.3A.1.2.3 Conclusion

The staff reviewed the LRA, UFSAR, and RAI response to determine whether the applicant failed to identify any SSCs within the scope of license renewal. The staff found no such omissions. In addition, the staff sought to determine whether the applicant failed to identify any components subject to an AMR. The staff found no such omissions. On the basis of its review, the staff concludes that the applicant has adequately identified the reactor vessel internals components within the scope of license renewal, as required by 10 CFR 54.4(a), and those subject to an AMR, as required by 10 CFR 54.21(a)(1).

2.3A.1.3 Reactor Coolant Pressure Boundary

2.3A.1.3.1 Summary of Technical Information in the Application

LRA Section 2.3.1.3 describes the RCPB, which includes the pressurizer, RCPs, interconnecting piping and fittings, system valves, component bolting, and piping and valves from connected systems. The RCPB includes multiple components from interconnecting systems, since their safety function is to maintain the RCS pressure boundary integrity. RCPB piping consists of the primary loops to and from the reactor pressure vessel, SG, and RCPs. The main reactor coolant piping and fittings are austenitic stainless steel.

Smaller piping, including the pressurizer surge and spray lines, drains, and connections to other systems, is austenitic stainless steel. Piping connections are welded except for flanged connections at the pressurizer relief tank and at the relief and safety valves. LRA Section 2.3.1.3 provides a listing of the lines comprising the RCPB.

LRA Section 2.3.1 describes the functions of the RCPB.

LRA Tables 2.3.1-3-IP2, 2.3.3-19-30-IP2, 2.3.1-3-IP3, 2.3.3-19-43-IP3, 2.3.3-19-44-IP3, and 2.3.3-19-46-IP3 identify RCPB component types within the scope of license renewal and subject to an AMR, as well as their intended functions.

2.3A.1.3.2 Staff Evaluation

The staff reviewed LRA Section 2.3.1.3, the UFSARs, and license renewal drawings using the evaluation methodology described in SER Section 2.3 and the guidance in SRP-LR Section 2.3.

During its review, the staff evaluated the system functions described in the LRA and UFSAR to verify that the applicant had not omitted from the scope of license renewal any components with intended functions, as required by 10 CFR 54.4(a). The staff then reviewed those components that the applicant identified as within the scope of license renewal to verify that the applicant had not omitted any passive and long-lived components subject to an AMR, in accordance with 10 CFR 54.21(a)(1).

During its review, the staff identified areas in which additional information was necessary to complete its review. The applicant responded to the staff's RAIs as discussed below.

In RAI 2.3A.1.3-1, dated November 9, 2007, the staff noted that License renewal drawing LRA-9321-2738, Sheet 1, depicts the RCPB for IP2. The staff was uncertain as to whether additional drawings depicting the RCPB existed. Therefore, the staff asked the applicant to clarify whether there were any additional sheets depicting the RCPB, and if so, to provide the drawings to the staff.

In its response, dated December 6, 2007, the applicant stated that License renewal drawing LRA-9321-2738 consisted of only one sheet, and the drawing, which identifies the major components in the RCS for IP2, includes the reactor vessel, pressurizer, RCPs, and SGs. The other License renewal drawings listed on page 2.3-20 of the LRA depict the remaining RCPB components for IP2 and IP3 that are in scope and subject to an AMR. License renewal drawing LRA-9321-2738 shows the continuations of lines to the other drawings depicting portions of the RCPB.

Based on its review, the staff finds the applicant's response acceptable because it clarified which drawings depict the components of the RCBP that are in scope and subject to an AMR. The response also confirmed that there were no additional drawings required staff review. Therefore the staff's concern in RAI 2.3A.1.3-1 is resolved.

In RAI 2.3A.1.3-2, dated January 28, 2008, the staff requested that the applicant clarify whether the pressurizer manways are within the scope of license renewal and subject to an AMR. LRA Tables 2.3.1-3-IP2 and 2.3.1-3-IP3 identified the pressurizer manway covers and insert plates as within the scope of license renewal and subject to an AMR.

In its response, dated February 27, 2008, the applicant stated that the pressurizer manway is a ring, integral to the shell of the pressurizer. The manway is part of the pressurizer shell included within the "pressurizer shell and heads" entries in LRA Tables 2.3.1-3-IP2 and 2.3.1-3-IP3. All portions of the manway assembly (i.e., the manway cover, the manway insert plate, and the pressurizer shell including the manway itself) are within the scope of license renewal and subject to AMR. Because the applicant clarified that this component is already within the scope of license renewal, the staff finds the applicant's response acceptable. The staff's concern described in RAI 2.3A.1.3-2 is resolved.

In RAI 2.3A.1.3-3, dated January 28, 2008, the staff requested that the applicant provide additional information and, if necessary, justify the exclusion of the vents associated with the level sensors, as shown on license renewal drawing LRA-208798-0. The applicant did not highlight the level sensor vents in the reactor vessel level indication system as components that are subject to an AMR. The sensor vents appear to provide an RCPB.

In its response, dated February 27, 2008, the applicant stated that the level elements on drawing LRA-208798, LE-1311, LE-1312, LE-1321 and LE-1322, are pressure transmitters. The vents are part of the transmitter body. In accordance with 10 CFR 54.21(a)(1)(i) and NEI 95-10, pressure transmitters are active components that are not subject to an AMR. Normal operational and surveillance activities readily monitor the performance or condition of active components. Because this component is part of an active component and is monitored through normal operational activities, the staff finds the applicant's response acceptable. The staff's concern described in RAI 2.3A.1.3-3 is resolved.

In RAI 2.3.0-2, dated February 13, 2008, the staff noted that on license renewal drawings for the IP2 and IP3 RCP motors, various components of the upper and lower bearing heat exchangers were marked "Not A Long Lived Component," and thus, were not subject to an AMR. Additionally, the staff noted that license renewal drawings of the IP2 and IP3 emergency diesel generator (EDG) jacket water cooling systems also have components marked "Not A Long Lived Component." The staff noted that SRP-LR Section 2.1.3.2.2 describes long-lived SCs as those that are not subject to periodic replacement based on a qualified life or specified time period. Furthermore, this section states that replacement programs may be based on vendor recommendations, plant experience, or any means that establishes a specific replacement frequency under a controlled program.

Because the staff identified that previous LRAs typically have not designated pumps, motors, and heat exchangers as "not long lived" (i.e., these components, or portions thereof, are subject to an AMR), the staff requested the applicant to:

(a)   Identify the component types serviced by the CCW system indicated in the above mentioned drawings that are marked "Not A Long Lived Component."

(b)   Provide a basis for designating these components as "not long lived" to include details on how the "qualified life" of the components was established and describe the program under which aging management activities for the components are performed and any available plant-specific operating experience confirming the effectiveness of management activities.

In its response, dated March 12, 2008, the applicant addressed the staff's concerns for the component types serviced by the CCW system. The applicant stated that it reviewed the documentation specifying the RCP motor upper and lower bearing heat exchangers as short lived and determined that they are actually not subject to periodic replacement. The applicant stated that the RCP motor upper and lower bearing heat exchangers are therefore subject to an AMR. Additionally, in its response, the applicant proposed changes to LRA Section 3.3.2.1.3 and LRA Tables 3.3.2-3-IP2 and 3.3.2-3-IP3 for the CCW system to include the aforementioned heat exchangers with their materials, environments, and aging management programs (AMPs). SER Section 3.3.2.1 documents the staff's review of the AMR line items. SER Sections 2.3A.3.14 and 2.3B.3.14 document the staff's evaluation of the applicant's response for the EDG jacket water cooling system.

Based on its review, the staff finds the applicant's response to RAI 2.3.0-2 acceptable for the RCS because it adequately explained that the RCP motor upper and lower bearing cooler heat exchangers in the CCW system were erroneously designated "Not A Long-Lived Component" and are, therefore, subject to an AMR. Further, in its March 12, 2008, letter, the applicant amended the LRA to include the heat exchangers and their AMP. Therefore, the staff's concern described in RAI 2.3.0-2 is resolved.

2.3A.1.3.3 Conclusion

The staff reviewed the LRA, UFSAR, RAI responses, and drawings to determine whether the applicant failed to identify any SSCs within the scope of license renewal. The staff found no such omissions. In addition, the staff sought to determine whether the applicant failed to identify any components subject to an AMR. The staff found an instance in which the applicant omitted components that should have been subject to an AMR. The applicant has satisfactorily resolved this issue as discussed in the preceding staff evaluation. On the basis of its review, the staff concludes that the applicant has adequately identified the RCPB components within the scope of license renewal, as required by 10 CFR 54.4(a), and those subject to an AMR, as required by 10 CFR 54.21(a)(1).

2.3A.1.4 Steam Generators

2.3A.1.4.1 Summary of Technical Information in the Application

The SGs are designed and manufactured in accordance with ASME Code, Section III. The IP2 SGs were constructed in accordance with the 1980 Edition, through the Winter 1981 Addenda, of the ASME Code. The IP3 SGs were constructed consistent with the 1983 Edition, through the Summer 1984 Addenda, of the ASME Code. The SGs are constructed primarily of carbon (low alloy) steel. The heat transfer tubes are Inconel: Alloy 600 for IP2 and Alloy 690 for IP3. The tubes were thermally treated after tube-forming operations. The interior surfaces of the channel heads and nozzles are clad with austenitic stainless steel, and the tubesheet surfaces in contact with reactor coolant are clad with Inconel. The tube-to-tubesheet joints are welded. The primary nozzles are provided with safe-ends with weld metal overlay.

LRA Section 2.3.1 describes the functions of the SGs. LRA Tables 2.3.1-4-IP2 and 2.3.1-4-IP3 identify SG component types within the scope of license renewal and subject to an AMR, as well as their intended functions.

2.3A.1.4.2 Staff Evaluation

The staff reviewed LRA Section 2.3.1.4, the UFSARs, and license renewal drawings using the evaluation methodology described in SER Section 2.3 and the guidance in SRP-LR Section 2.3.

During its review, the staff evaluated the system functions described in the LRA and UFSAR to verify that the applicant had not omitted from the scope of license renewal any components with

intended functions, as required by 10 CFR 54.4(a). The staff then reviewed those components that the applicant identified as within the scope of license renewal to verify that the applicant had not omitted any passive and long-lived components subject to an AMR in accordance with 10 CFR 54.21(a)(1).

2.3A.1.4.3 Conclusion

The staff reviewed the LRA, UFSAR, and drawings to determine whether the applicant failed to identify any SSCs within the scope of license renewal. The staff found no such omissions. In addition, the staff determined whether the applicant failed to identify any components subject to an AMR. The staff found no such omissions. On the basis of its review, the staff concludes that the applicant has adequately identified the SG components within the scope of license renewal, as required by 10 CFR 54.4(a), and those subject to an AMR, as required by 10 CFR 54.21(a)(1).

2.3A.2 Engineered Safety Features

LRA Section 2.3.2 identifies the engineered safety features SCs subject to an AMR for license renewal.

The applicant described the supporting SCs of the engineered safety features in the following LRA sections:

- 2.3.2.1, "Residual Heat Removal"
- 2.3.2.2, "Containment Spray System"
- 2.3.2.3, "Containment Isolation Support Systems"
- 2.3.2.4, "Safety Injection Systems"
- 2.3.2.5, "Containment Penetrations"

SER Sections 2.3A.2.1–2.3A.2.5 discuss the findings of the staff's review of LRA Sections 2.3.2.1–2.3.2.5, respectively.

2.3A.2.1 IP2 Residual Heat Removal

2.3A.2.1.1 Summary of Technical Information in the Application

LRA Section 2.3.2.1 describes the RHR system, which provides emergency core cooling, as part of the safety injection system, and removes residual heat during later stages of plant cooldown. The RHR system is one of three (RHR, CCW, and spent fuel pit cooling (SFPC)) auxiliary coolant systems. The RHR system consists of two RHR heat exchangers, two seal coolers, two RHR (low-head) pumps, and required piping, valves, and I&C components. The RHR system provides emergency core cooling during the injection phase of a loss-of-coolant accident (LOCA). The RHR heat exchangers, in conjunction with the safety injection recirculation pumps, are used for post-accident heat removal during the LOCA recirculation phase. Outlet flow from the RHR heat exchangers may be directed to the containment spray (CS) headers, to the RCS cold legs, or to the RCS hot legs via the high-head safety injection pumps. The RHR pumps also back up the safety injection system recirculation pumps during a LOCA recirculation phase. In this capacity, the RHR pumps may draw water from the containment sump and deliver it to the RCS cold leg injection lines, to the suction of the high-head safety injection pumps, or to the CS headers. The RHR system removes residual heat during the later stages of plant cooldown and during cold shutdown and refueling operations. After RCS temperature and pressure have been reduced to 350 degrees F and less than 365 pounds per square inch gauge (psig), alignment of the RHR pumps initiates decay heat cooling by taking suction from one reactor hot leg and discharging it through the RHR heat exchangers into the reactor cold legs.

The RHR system contains safety-related components relied on to remain functional during and following DBEs. In addition, the RHR system performs functions that support fire protection and SBO.

In the LRA, ASME Code Class 1 components with the intended function of maintaining the RCPB are reviewed with the RCS (LRA Section 2.3.1). A small number of components are reviewed with the safety injection system in LRA Section 2.3.2.4.

LRA Table 2.3.2-1-IP2 identifies RHR system component types within the scope of license renewal and subject to an AMR, as well as their intended functions.

2.3A.2.1.2 Staff Evaluation

The staff reviewed LRA Section 2.3.2.1, the UFSAR, and license renewal drawings using the evaluation methodology described in SER Section 2.3 and the guidance in SRP-LR Section 2.3.

During its review, the staff evaluated the system functions described in the LRA and UFSAR to verify that the applicant had not omitted from the scope of license renewal any components with intended functions, as required by 10 CFR 54.4(a). The staff then reviewed those components that the applicant identified as within the scope of license renewal to verify that it had not omitted any passive and long-lived components subject to an AMR in accordance with 10 CFR 54.21(a)(1).

2.3A.2.1.3 Conclusion

The staff reviewed the LRA, UFSAR, and drawings to determine whether the applicant failed to identify any SSCs within the scope of license renewal. The staff found no such omissions. In addition, the staff sought to determine whether the applicant failed to identify any components subject to an AMR. The staff found no such omissions. On the basis of its review, the staff concludes that the applicant has adequately identified the RHR system components within the scope of license renewal, as required by 10 CFR 54.4(a), and those subject to an AMR, as required by 10 CFR 54.21(a)(1).

2.3A.2.2 IP2 Containment Spray System

2.3A.2.2.1 Summary of Technical Information in the Application

LRA Section 2.3.2.2 describes the CS system, which cools the containment and removes iodine following an accident. The CS system consists of two trains of pumps, valves, and headers that automatically start and spray refueling water storage tank (RWST) borated water into the containment atmosphere when the system senses high containment pressure following a LOCA or MS line break accident. The spray water enters through nozzles connected to four ring headers in the containment dome. Each spray pump supplies two ring headers. After injection from the RWST is terminated, the system can supply the spray headers with recirculated water from the recirculation sump or the containment sump by a diversion of a portion of the injection flow from the safety injection system. Long-term, post-accident retention of iodine is achieved by four sodium tetraborate baskets in the containment at an elevation (46 feet) that will be flooded under accident conditions, allowing the sodium tetraborate to dissolve into the fluid for pH control. The containment structural evaluation includes the four sodium tetraborate baskets, but they are not described further because they have no license renewal intended function and are therefore not subject to an AMR.

The CS system contains safety-related components relied on to remain functional during and following DBEs.

Containment spray system components that support the RHR system pressure boundary are reviewed in the RHR systems (LRA Section 2.3.2.1). A small number of components are reviewed in the safety injection system (LRA Section 2.3.2.4).

LRA Table 2.3.2-2-IP2 and newly created Table 2.3.3-19-46-IP2 (see evaluation below) identify CS system component types within the scope of license renewal and subject to an AMR, as well as their intended functions.

2.3A.2.2.2 Staff Evaluation

The staff reviewed LRA Section 2.3.2.2, UFSAR Section 6.3, and license renewal drawings using the evaluation methodology described in SER Section 2.3 and the guidance in SRP-LR Section 2.3.

During its review, the staff evaluated the system functions described in the LRA and UFSAR to verify that the applicant had not omitted from the scope of license renewal any components with intended functions, as required by 10 CFR 54.4(a). The staff then reviewed those components that the applicant identified as within the scope of license renewal to verify that the applicant had not omitted any passive and long-lived components subject to an AMR, in accordance with 10 CFR 54.21(a)(1).

During its review of LRA Section 2.3.2.2, the staff identified an area in which additional information was necessary to complete the review of the applicant's scoping and screening results.

During its review of license renewal drawings for the containment spray system, the staff identified portions of piping of the CS system that were not highlighted, indicating that section of piping had no intended functions in accordance with 10 CFR 54.4 (a)(1) or 10 CFR 54.4 (a)(3). LRA Section 2.3.2.2 states that the CS system has no intended function for 10 CFR 54.4(a)(2). However, this section of piping is directly connected to safety-related containment spray piping; therefore, the staff determined that it should be in scope for 10 CFR 54.4(a)(2) for nonsafety-related piping that is structurally attached to safety-related piping. In RAI 2.3A.2.2-1, dated February 13, 2008, the staff asked the applicant to explain this discrepancy. The staff also asked the applicant to indicate any portions of the CS system evaluated for inclusion in the scope of license renewal in accordance with 10 CFR 54.4(a)(2), and to identify any other instances whereby a system was identified as not having any 10 CFR 54.4(a)(2) function but

2-48

did have nonsafety-related components that were not identified as within scope for 10 CFR 54.4(a)(2).

In its response, dated March 12, 2008, the applicant stated that the components identified by the staff do have an intended function to maintain integrity such that no physical interaction with safety-related components could prevent satisfactory accomplishment of a safety function. Hence, the applicant amended the LRA to include the portions of the CS system within the scope of license renewal under the requirements of 10 CFR 54.4(a)(2). The applicant responded to the staff's request by performing a re-evaluation of those safety-related systems that were identified in the LRA as only being in scope for (a)(1) and have no (a)(2) components. The applicant's re-examination identified three instances where a system that performs a safety function was in scope for 10 CFR 54.4(a)(1), but nonsafety-related components were not identified as in scope for 10 CFR 54.4(a)(2). The staff's evaluation of the affected systems is discussed in SER Sections 2.3A.3.3, 2.3B.2.5, and 2.3B.3.3. For the IP2 CS system, in its letter dated March 12, 2008, the applicant amended the LRA to reflect the following changes:

(a)   LRA Table 2.3.3-19-A-IP2 would reflect the CS system as a miscellaneous system within the scope of license renewal for 10 CFR 54.4(a)(2).

(b)   Removal of the CS system from the list of IP2 systems not reviewed for 10 CFR 54.4(a)(2) for spatial interaction.

(c)   Revision of LRA Table 2.3.3-19-B-IP2 to reflect that the CS system now has components subject to an AMR for 10 CFR 54.4(a)(2).

(d)   Creation of a new LRA Table 2.3.3-19-46-IP2 for the five added component types in the CS system for nonsafety-related components potentially affecting a safety function, and subject to an AMR.

(e)   Creation of a new LRA Table 3.3.2-19-46-IP2 for the five added component types, their materials, environments, and AMPs.

Based on its review, the staff finds the applicant's response to RAI 2.3A.2.2-1 for the IP2 CS system acceptable because it adequately explained that the applicant's reevaluation of safety-related systems identified components that should have been within scope under 10 CFR 54.4(a)(2). Additionally, the applicant amended the LRA to include those portions of the CS system identified by the staff as being in scope under 10 CFR 54.4(a)(2). The staff reviewed the applicant's changes to the LRA tables and found that they adequately reflect those components brought into the scope of license renewal, in accordance with 10 CFR 54.4(a)(2), because of their potential for spatial interaction with safety-related components. Therefore, the staff's concern described in RAI 2.3A.2.2-1 for the CS system is resolved. SER Section 3.2.2.1 documents the staff's evaluation of new AMR results for the CS system.

2.3A.2.2.3 Conclusion

The staff reviewed the LRA, UFSAR, RAI responses, and drawings to determine whether the applicant failed to identify any SSCs within the scope of license renewal. The staff found no such omissions. In addition, the staff sought to determine whether the applicant failed to identify any components subject to an AMR. The staff found one instance in which the applicant omitted components that should have been subject to an AMR. The applicant has satisfactorily resolved this issue as discussed in the preceding staff evaluation. On the basis of its review, the staff concludes that the applicant has adequately identified the CS system components within the scope of license renewal, as required by 10 CFR 54.4(a), and those subject to an AMR, as required by 10 CFR 54.21(a)(1).

2.3A.2.3 IP2 Containment Isolation Support Systems

2.3A.2.3.1 Summary of Technical Information in the Application

LRA Section 2.3.2.3 describes the containment isolation support systems, which include the isolation valve seal water systems and the weld channel and penetration pressurization system. The containment isolation support systems consist of piping and valves routed to the various system piping that penetrates the containment. The isolation valve seal water, weld channel, and penetration pressurization systems isolate the containment from the outside environment for various systems with piping penetrating containment. The containment isolation support systems inject fluid or air or gas into system lines between the containment isolation valves that penetrate the containment for pressure boundary integrity against leakage of radioactive fluids to the environment in the event of a LOCA. Individual lines define these barriers of piping and isolation valves systems. Besides satisfying containment isolation criteria, the valving facilitates normal operation and maintenance of the systems for reliable operation of other engineered safeguard systems.

The isolation valve seal water system provides sealing water or gas between the isolation and double-disk isolation valves of containment penetrations located in lines connected to the RCS or exposed to the containment atmosphere during any condition that requires containment isolation. This system limits fission product release from the containment. Although not credited in post-accident dose analyses, the system ensures a containment leak rate in an accident lower than that assumed in the accident analysis and the offsite dose calculations. System components form parts of the containment penetration isolation boundary.

The weld channel and penetration pressurization system provides pressurized gas to all containment penetrations and most liner inner weld seams in the event of a LOCA, so there will be no leakage through these potential paths from the containment to the atmosphere. The weld channel and penetration pressurization system also serves spaces between selected isolation valves. Although not credited in the post-accident dose analyses, the weld channel and penetration pressurization system maintained at a pressure level above the peak accident pressure is designed to keep any postulated leakage in rather than out of the containment. The system supplies regulated clean, dry compressed air from either of the plant's compressed air systems outside the containment to all containment penetrations and most inner liner weld channels. The instrument air system, backed up by the station air system and by a bank of nitrogen cylinders as a standby source of gas pressure, is the primary source of air for this system.

The containment isolation support systems contain safety-related components relied upon to remain functional during and following DBEs.

Isolation valve seal water system components with the intended function of maintaining the RCPB are reviewed in the RCS (LRA Section 2.3.1.3).

LRA Table 2.3.2-3-IP2 identifies containment isolation support systems component types within the scope of license renewal and subject to an AMR, as well as their intended functions.

2.3A.2.3.2 Staff Evaluation

The staff reviewed LRA Section 2.3.2.3, UFSAR Sections 6.5.1, 6.6.2, and 14.3.6.1, and license renewal drawings using the evaluation methodology described in SER Section 2.3 and the guidance in SRP-LR Section 2.3.

During its review, the staff evaluated the system functions described in the LRA and UFSAR to verify that the applicant had not omitted from the scope of license renewal any components with intended functions, as required by 10 CFR 54.4(a). The staff then reviewed those components that the applicant identified as within the scope of license renewal to verify that the applicant had not omitted any passive and long-lived components subject to an AMR, in accordance with 10 CFR 54.21(a)(1).

During its review, the staff identified an area in which additional information was necessary to complete the review of the applicant's scoping and screening results.

In RAI 2.3A.2.3-1, dated November 9, 2007, the staff identified several line-mounted components (valves PRV PCV 1193 through PRV PCV 1200) that were located in lines with a pressure boundary function. However, the applicant had not identified the components themselves as being subject to an AMR. Therefore, the staff requested that the applicant clarify whether these components are subject to an AMR or justify their exclusion.

In its response, dated December 6, 2007, the applicant stated that the valves in question are within the scope of license renewal and subject to an AMR. Furthermore, the applicant noted that LRA Table 2.3.2-3-IP2 identified these valves as component type "valve body," with AMR results summarized in LRA Table 3.2.2-3-IP2. Some of the valves in question have aluminum valve bodies with internal and external environments of gas (internal) and air—indoor (external). The applicant added a line item of "valve body" to LRA Table 3.2.2-3-IP2 to reflect the aluminum material.

Based on its review, the staff finds the applicant's response to RAI 2.3A.2.3-1 acceptable because the applicant clarified that the subject valves are within the scope of license renewal and subject to an AMR and added aluminum valve bodies to the AMR. The staff's concern described in RAI 2.3A.2.3-1 is resolved. SER Section 3.2.2.1 discusses the staff's evaluation of the added AMR for aluminum valve bodies.

By letter dated June 30, 2009, the applicant submitted an annual update to the LRA, identifying changes made to the CLB that materially affect the contents of the LRA. For the containment isolation support system, the applicant identified buried piping in the containment isolation support system that had not been previously identified as being within the scope of license renewal. The piping is part of the air pressure supply that feeds Rack 15 for the steam and feedwater penetrations shown on license renewal drawing LRA-9321-2726-0. The staff reviewed the amendment and finds the addition to the scope to be acceptable. The staff's evaluation of the corresponding AMR results is documented in SER Section 3.2.2.1.

2.3A.2.3.3 Conclusion

The staff reviewed the LRA, UFSAR, RAI responses, and drawings to determine whether the applicant failed to identify any SSCs within the scope of license renewal. The staff found no such omissions. In addition, the staff sought to determine whether the applicant failed to identify any components subject to an AMR. The staff found one instance in which the applicant omitted components that should have been subject to an AMR. The applicant has satisfactorily resolved this issue as discussed in the preceding staff evaluation. On the basis of its review, the staff concludes that the applicant has adequately identified the containment isolation support systems components within the scope of license renewal, as required by 10 CFR 54.4(a), and those subject to an AMR, as required by 10 CFR 54.21(a)(1).

2.3A.2.4 IP2 Safety Injection System

2.3A.2.4.1 Summary of Technical Information in the Application

LRA Section 2.3.2.4 describes the safety injection system, which, in a LOCA, automatically delivers cooling water to the reactor core to limit the fuel clad temperature so the core remains intact and in place with its essential heat transfer geometry preserved. Components comprising the safety injection system code (i.e., the applicant's code for designating systems and boundaries) include the RWST, the three safety injection (high-head) pumps, the accumulators (one for each reactor coolant loop), recirculation pumps and piping, valves, and other components of these subsystems. The three safety injection (high-head) pumps inject RWST borated water into the RCS for core cooling. The safety injection signal automatically opens the required safety injection system isolation valves and starts the safety injection pumps. The injection piping and valves connect the accumulators containing borated water and pressurized with nitrogen to the RCS. Two check valves isolate these tanks from the RCS during normal operation. When RCS pressure falls below accumulator pressure the check valves open, discharging the tank contents into the RCS through the same injection piping used by the safety injection pumps.

After the injection, the recirculation system cools and returns to the RCS any coolant spilled from the break and water collected from the CS. The system recirculation pumps take suction from the recirculation sump in the containment floor and deliver spilled reactor coolant and borated refueling water back to the core through the RHR heat exchangers. For smaller RCS breaks in which recirculated water must be injected against higher pressures for long-term cooling, the system delivers the water from an RHR heat exchanger to the high-head safety injection pump suction and, by this external recirculation route, to the reactor coolant loops. The system also allows either of the RHR pumps to take over the recirculation function.

The safety injection system contains safety-related components relied on to remain functional during and following DBEs. It also contains nonsafety-related components whose failure potentially could prevent the satisfactory accomplishment of a safety-related function. In addition, the safety injection system performs functions that support fire protection.

ASME Code Class 1 components with the intended function of maintaining the RCPB are reviewed in the RCS (LRA Section 2.3.1.3). A small number of components are reviewed in the containment system (LRA Section 2.3.2.2), RHR systems (LRA Section 2.3.2.1), and nitrogen systems (LRA Section 2.3.3.5).

LRA Tables 2.3.2-4-IP2 and 2.3.3-19-37-IP2 identify safety injection system component types within the scope of license renewal and subject to an AMR, as well as their intended functions.

2.3A.2.4.2 Staff Evaluation

The staff reviewed LRA Section 2.3.2.4, the UFSAR, and license renewal drawings using the evaluation methodology described in SER Section 2.3 and the guidance in SRP-LR Section 2.3.

During its review, the staff evaluated the system functions described in the LRA and UFSAR to verify that the applicant had not omitted from the scope of license renewal any components with intended functions, as required by 10 CFR 54.4(a). The staff then reviewed those components that the applicant identified as within the scope of license renewal to verify that the applicant had not omitted any passive and long-lived components subject to an AMR, in accordance with 10 CFR 54.21(a)(1).

2.3A.2.4.3 Conclusion

The staff reviewed the LRA, UFSAR, and drawings to determine whether the applicant failed to identify any SSCs within the scope of license renewal. The staff found no such omissions. In addition, the staff sought to determine whether the applicant failed to identify any components subject to an AMR. The staff found no such omissions. On the basis of its review, the staff concludes that the applicant has adequately identified the safety injection system components within the scope of license renewal, as required by 10 CFR 54.4(a), and those subject to an AMR, as required by 10 CFR 54.21(a)(1).

2.3A.2.5 IP2 Containment Penetrations

2.3A.2.5.1 Summary of Technical Information in the Application

LRA Section 2.3.2.5 describes the following containment penetrations, which are not an independent system but a grouping of containment penetration components that are not evaluated with other systems:

- electrical penetrations
- fuel core component handling system
- hydrogen recombiners

The electrical penetrations pass electrical conductors through the containment boundary. The electrical penetrations system code (i.e., the applicant's code for designating systems and boundaries) is primarily structural and electrical components that are evaluated in the structural and electrical AMRs; however, the system has mechanical components which are evaluated in this section. The penetrations have a pressure connection for continuous pressurization by the weld channel system, which is considered part of the containment isolation boundary.

The fuel core component handling system defuels and refuels the reactor core. The fuel handling system transports and handles fuel safely and effectively. Most system components (e.g., fuel handling bridges) are structural and evaluated with their respective structures. The

2-53

fuel transfer tube and blind flange are fuel core component handling system components that together constitute a containment penetration.

The hydrogen recombiners system, which reduces the hydrogen concentration in the containment volume following a DBA, has two redundant passive autocatalytic recombiners that replaced earlier flame units. The recombiners are passive devices with no moving parts and need no electrical power or any other support system. Recombination is by attraction of oxygen and hydrogen molecules to the surface of a palladium catalyst. The exothermic reaction of the combination generates heat, which causes a convective flow that draws more gases from the containment atmosphere into the unit. Since a recent license amendment (Amendment No. 243), hydrogen recombination is no longer required as a safety function. The system includes containment penetrations from the original flame hydrogen recombiners.

The containment penetrations contain safety-related components relied on to remain functional during and following DBEs.

Containment penetration components evaluated in this section maintain the system pressure boundary inside containment from the first weld from the penetration to the class boundary change outside containment. Components in the Class 1 boundary are reviewed in the RCPB (LRA Section 2.3.1.3). Structural portions of the containment penetrations are evaluated with the containment building (LRA Section 2.4.1). Electrical portions of electrical penetration assemblies are evaluated with electrical components (LRA Section 2.5).

LRA Table 2.3.2-5-IP2 identifies containment penetrations component types within the scope of license renewal and subject to an AMR, as well as their intended functions.

2.3A.2.5.2 Staff Evaluation

The staff reviewed LRA Section 2.3.2.5; UFSAR Sections 5.1.4, 5.1.4.2.1, 6.8, and 9.5.2; and license renewal drawings using the evaluation methodology described in SER Section 2.3 and the guidance in SRP-LR Section 2.3.

During its review, the staff evaluated the system functions described in the LRA and UFSAR to verify that the applicant had not omitted from the scope of license renewal any components with intended functions, as required by 10 CFR 54.4(a). The staff then reviewed those components that the applicant identified as within the scope of license renewal to verify that the applicant had not omitted any passive and long-lived components subject to an AMR, in accordance with 10 CFR 54.21(a)(1).

During its review of LRA Section 2.3A.2.5, the staff identified an area in which additional information was necessary to complete the review of the results of the applicant's scoping and screening effort. The applicant responded to the staff's RAI as discussed below.

In RAI 2.3A.2.5-1, dated November 9, 2007, the staff noted that a drawing referenced for IP2 appeared to be applicable to IP3. Therefore, the staff requested that the applicant confirm the accuracy of the referenced drawings.

In its response, dated December 6, 2007, the applicant stated that an administrative error occurred when transferring the License renewal drawing numbers from the site basis document to the License renewal drawing list. Additionally, the applicant identified the drawings that corresponded to the respective units.

Based on its review, the staff found the applicant's response to RAI 2.3A.2.5-1 acceptable because the applicant identified and corrected an administrative error. Subsequently, the staff reviewed and evaluated the components associated with the containment penetrations on the referenced drawings and found no omissions from the scope of license renewal. The staff's concern described in RAI 2.3A.2.5-1 is resolved.

2.3A.2.5.3 Conclusion

The staff reviewed the LRA, UFSAR, RAI responses, and drawings to determine whether the applicant failed to identify any SSCs within the scope of license renewal. The staff found no such omissions. In addition, the staff sought to determine whether the applicant failed to identify any components subject to an AMR. The staff found no such omissions. On the basis of its review, the staff concludes that the applicant has adequately identified the containment penetrations components within the scope of license renewal, as required by 10 CFR 54.4(a), and those subject to an AMR, as required by 10 CFR 54.21(a)(1).

2.3A.3 Scoping and Screening Results: IP2 Auxiliary Systems

LRA Section 2.3.3 identifies the auxiliary systems SCs subject to an AMR for license renewal.

The applicant described the supporting SCs of the auxiliary systems in the following LRA sections:

- 2.3.3.1, "Spent Fuel Pit Cooling"
- 2.3.3.2, "Service Water"
- 2.3.3.3, "Component Cooling Water"
- 2.3.3.4, "Compressed Air"
- 2.3.3.5, "Nitrogen Systems"
- 2.3.3.6, "Chemical and Volume Control"
- 2.3.3.7, "Primary Water Makeup"
- 2.3.3.8, "Heating, Ventilation and Air Conditioning"
- 2.3.3.9, "Containment Cooling and Filtration"
- 2.3.3.10, "Control Room Heating, Ventilation and Cooling"
- 2.3.3.11, "Fire Protection – Water"
- 2.3.3.12, "Fire Protection – CO_2, Halon, and RCP Oil Collection Systems"
- 2.3.3.13, "Fuel Oil"
- 2.3.3.14, "Emergency Diesel Generators"
- 2.3.3.15, "Security Generators"
- 2.3.3.16, "Appendix R Diesel Generators"
- 2.3.3.17, "City Water"
- 2.3.3.18, "Plant Drains"
- 2.3.3.19, "Miscellaneous Systems In-Scope for (a)(2)"

The applicant developed LRA Section 2.3.3.19 to capture all the systems or portions of systems that are within the scope of license renewal in accordance with 10 CFR 54.4(a)(2). Among the subsections included in LRA Section 2.3.3.19, the staff identified the following auxiliary systems for simplified Tier 1 review:

- chemical feed
- intake structure system
- house service boiler
- main generator
- ignition oil
- integrated liquid waste handling
- nuclear service grade makeup
- boiler blowdown
- secondary sampling
- technical support center diesel
- main turbine

The staff conducted a more detailed Tier 2 review for all remaining auxiliary systems.

Staff Requests for Additional Information

During its review, the staff noted the applicant did not specifically identify components in scope under 10 CFR 54.4(a)(2) on the accompanied drawings. To ensure that the applicant did not omit any components that should be in scope under 10 CFR 54.4(a)(2), the staff asked the applicant to verify that it had included segments of the selected systems in scope under 10 CFR 54.4(a)(2). In the following RAIs, dated February 13, 2008, the staff requested that the applicant confirm its methodology for identifying nonsafety-related portions of systems with a potential to adversely affect safety-related functions by describing the applicable specific portions of system piping that the applicant included within the scope of license renewal, in accordance with 10 CFR 54.4(a)(2):

- RAI 2.3A.3.1-1
- RAI 2.3A.3.2-1
- RAI 2.3A.3.3-1
- RAI 2.3A.3.5-1
- RAI 2.3A.3.13-1
- RAI 2.3A.3.14-2
- RAI 2.3A.3.18-1

In its response, dated March 12, 2008, the applicant stated that all component types identified by the staff on the license renewal drawings in question are within the scope of license renewal in accordance with 10 CFR 54.4(a)(2), and are subject to an AMR.

Based on its review, the staff finds the applicant's response to these RAIs acceptable because the applicant has adequately explained that all component types identified by the staff are within the scope of license renewal in accordance with 10 CFR 54.4(a)(2) and are subject to an AMR. The staff's concern described in these RAIs is resolved.

SER Sections 2.3A.3.1 through 2.3A.3.19 provide the staff's reviews of IP2 systems described in LRA Sections 2.3.3.1 through 2.3.3.19, respectively. The following sections discuss the staff's findings for these systems.

2.3A.3.1 IP2 Spent Fuel Pit Cooling System

2.3A.3.1.1 Summary of Technical Information in the Application

LRA Section 2.3.3.1 describes the SFPC system, which removes residual heat from the spent fuel pit. The SFPC loop has two pumps, a heat exchanger, filter, demineralizer, piping, valves, and instrumentation. One of the pumps draws water from the pit, circulates it through the heat exchanger cooled by CCW, and returns it to the pit. Loop piping is arranged so that any pipeline failure does not drain the spent fuel pit below the top of the stored fuel elements. The spent fuel pit pump suction line, which draws water from the pit, penetrates the spent fuel pit wall above the fuel assemblies. The system also includes the spent fuel pit. Spent fuel storage racks at the bottom of the pit for spent fuel assemblies are the full-length, top-entry type made of stainless steel with Boraflex as a neutron absorber.

The SFPC system contains safety-related components relied upon to remain functional during and following DBEs. It also contains nonsafety-related components whose failure potentially could prevent the satisfactory accomplishment of a safety-related function.

The spent fuel pit (including liner and the spent fuel racks) are included in the evaluation of the fuel storage buildings (LRA Section 2.4.3). The heat exchanger components forming parts of the CCW system pressure boundary are evaluated with the CCW systems (LRA Section 2.3.3.3). A small number of components are evaluated with the primary water makeup systems (LRA Section 2.3.3.7).

LRA Tables 2.3.3-1-IP2 and 2.3.3-19-35-IP2 identify the SFPC system component types within the scope of license renewal and subject to an AMR, as well as their intended functions.

2.3A.3.1.2 Staff Evaluation

The staff reviewed LRA Section 2.3.3.1; UFSAR Sections 9.3.1, 9.5.2.1.5, and 14.2.1; a license renewal drawing; and IP2 Amendment 227, "Credit for Soluble Boron and Burnup in Spent Fuel Pit (TAC No. MB2989)," dated May 29, 2002 (ADAMS Accession No. ML021230367), using the evaluation methodology described in SER Section 2.3 and the guidance in SRP-LR Section 2.3.

During its review, the staff evaluated the system functions described in the LRA and UFSAR to verify that the applicant had not omitted from the scope of license renewal any components with intended functions, as required by 10 CFR 54.4(a). The staff then reviewed those components that the applicant identified as within the scope of license renewal to verify that it had not omitted any passive and long-lived components subject to an AMR, in accordance with 10 CFR 54.21(a)(1).

During its review of LRA Section 2.3.3.1, the staff identified an area in which additional information was necessary to complete the review of the applicant's scoping and screening results. The discussion of the staff's RAIs in SER Section 2.3A.3 details the disposition of RAI 2.3A.3.1-1, dated February 13, 2008.

2.3A.3.1.3 Conclusion

The staff reviewed the LRA, UFSAR, and a drawing to determine whether the applicant failed to identify any SSCs within the scope of license renewal. The staff found no such omissions. In addition, the staff sought to determine whether the applicant failed to identify any components subject to an AMR. The staff found no such omissions. On the basis of its review, the staff concludes that the applicant has adequately identified the SFPC components within the scope of license renewal, as required by 10 CFR 54.4(a), and those subject to an AMR, as required by 10 CFR 54.21(a)(1).

2.3A.3.2 IP2 Service Water System

2.3A.3.2.1 Summary of Technical Information in the Application

LRA Section 2.3.3.2 describes the SW system, which supplies cooling water from the Hudson River to various heat loads in both primary and secondary portions of the plant, in a continuous flow to systems and components necessary for plant safety during either normal operation or abnormal or accident conditions. Sufficient redundancy of active and passive components maintains short- and long-term cooling to vital loads, in accordance with the single-failure criterion. Six identical vertical, centrifugal sump-type pumps at the intake structure supply service water to two independent discharge headers (each is supplied by three pumps). An automatic, self-cleaning, rotary-type strainer in each pump's discharge removes solids. Each header connects to an independent supply line. Either of the two supply lines can supply the essential load, while the other supplies the nonessential load. Essential loads must have an assured supply of cooling water in the event of a loss of offsite power or a LOCA. Nonessential loads are supplied with cooling water by an SW pump started manually, when required, following a LOCA. Nonessential loads include the CCW heat exchangers, circulating water (CW) pump seal injection, turbine building closed cooling water system, hydrogen coolers, stator cooling water heat exchanger, exciter air coolers, and isolated phase bus heat exchangers. The system also provides backup water to clean the traveling screens.

The SW system contains safety-related components relied on to remain functional during and following DBEs. It also contains nonsafety-related components whose failure potentially could prevent the satisfactory accomplishment of a safety-related function. In addition, the SW system performs functions that support fire protection and SBO.

Components that support safe shutdown in the event of a fire in the auxiliary feed pump room are reviewed in LRA Section 2.3.4.5. Components cooling the CCW systems are reviewed in those systems (LRA Section 2.3.3.3). Components cooling the EDG systems are reviewed with those systems (LRA Section 2.3.3.14).

LRA Tables 2.3.3-2-IP2 and 2.3.3-19-39-IP2 identify SW system component types within the scope of license renewal and subject to an AMR, as well as their intended functions.

2.3A.3.2.2 Staff Evaluation

The staff reviewed LRA Section 2.3.3.2, UFSAR Section 9.6.1, and license renewal drawings using the evaluation methodology described in SER Section 2.3 and the guidance in SRP-LR Section 2.3.

During its review, the staff evaluated the system functions described in the LRA and UFSAR to verify that the applicant had not omitted from the scope of license renewal any components with intended functions, as required by 10 CFR 54.4(a). The staff then reviewed those components that the applicant identified as within the scope of license renewal to verify that it had not omitted any passive and long-lived components subject to an AMR, in accordance with 10 CFR 54.21(a)(1).

During its review of LRA Section 2.3.3.2, the staff identified an area in which additional information was necessary to complete the review of the applicant's scoping and screening results. The discussion of the staff's RAIs in SER Section 2.3A.3 details the disposition of RAI 2.3A.3.2-1, dated February 13, 2008.

2.3A.3.2.3 Conclusion

The staff reviewed the LRA, UFSAR, and drawings to determine whether the applicant failed to identify any SSCs within the scope of license renewal. The staff found no such omissions. In addition, the staff sought to determine whether the applicant failed to identify any components subject to an AMR. The staff found no such omissions. On the basis of its review, the staff concludes that the applicant has appropriately identified the SW system components within the scope of license renewal, as required by 10 CFR 54.4(a), and those subject to an AMR, as required by 10 CFR 54.21(a)(1).

2.3A.3.3 IP2 Component Cooling Water System

2.3A.3.3.1 Summary of Technical Information in the Application

LRA Section 2.3.3.3 describes the CCW system, which removes RCS residual and sensible heat via the RHR loop during plant shutdown, cools the letdown flow to the CVCS during power operation, dissipates waste heat from various primary plant components, and cools engineered safeguards and safe-shutdown components. The system includes the pumps, heat exchangers, distribution and return piping and valves, instruments, and controls to cool the following:

- RHR heat exchangers
- RCPs
- non-regenerative heat exchanger
- excess letdown heat exchanger
- CVCS seal water heat exchanger
- sample heat exchangers
- waste gas compressors
- reactor vessel support pads
- RHR pumps
- safety injection pumps
- recirculation pumps
- spent fuel pit heat exchanger
- charging pumps, fluid drive coolers, and crankcase

Some of the CCW-cooled heat exchangers in other systems have no safety function; however, these nonsafety-related heat exchangers form parts of the CCW system pressure boundary.

These heat exchangers are within the scope of license renewal and have an intended function to maintain the pressure boundary but not to transfer heat.

The CCW system was not designed to accommodate a passive failure during initial IP2 construction. The subsequent consideration of a passive failure required commitments for alternate cooling water supplies to safety-related equipment. Connections to primary and city water provide the alternate supplies.

The CCW system contains safety-related components relied on to remain functional during and following DBEs. In addition, the CCW system performs functions that support fire protection and SBO.

A few components within the CCW system support the RHR system pressure boundary and therefore are reviewed with the RHR systems (LRA Section 2.3.2.1). Components cooling the safety injection systems are reviewed with those systems (LRA Section 2.3.2.4). Components cooling the CVCS systems are reviewed with those systems (LRA Section 2.3.3.6).

LRA Table 2.3.3-3-IP2 and newly created Table 2.3.3-19-45-IP2 (see evaluation below) identify CCW system component types within the scope of license renewal and subject to an AMR, as well as their intended functions.

2.3A.3.3.2 Staff Evaluation

The staff reviewed LRA Section 2.3.3.3, UFSAR Sections 6.2.2.3.4 and 9.3, and license renewal drawings using the evaluation methodology described in SER Section 2.3 and the guidance in SRP-LR Section 2.3.

During its review, the staff evaluated the system functions described in the LRA and UFSAR to verify that the applicant had not omitted from the scope of license renewal any components with intended functions, as required by 10 CFR 54.4(a). The staff then reviewed those components that the applicant identified as within the scope of license renewal to verify that it had not omitted any passive and long-lived components subject to an AMR, in accordance with 10 CFR 54.21(a)(1).

During its review, the staff identified an area in which additional information was necessary to complete the review of the applicant's scoping and screening results.

In RAI 2.3A.2.2-1, dated February 13, 2008, the staff asked the applicant to identify instances in which a safety-related system, which has nonsafety-related components, was scoped in per 10 CFR 54.4(a)(1), but those nonsafety-related components were not identified as in scope for 10 CFR 54.4(a)(2).

In its response, dated March 12, 2008, the applicant explained that it identified three instances in which nonsafety-related components were not considered to be within the scope of license renewal under 10 CFR 54.4(a)(2). The applicant further explained that it should have included the CCW systems at IP2 and IP3, as well as the IP3 building vent sampling (BVS) system, among those systems subject to the requirements of 10 CFR 54.4(a)(2). In these instances, the applicant amended the LRA to reflect these changes. For the IP2 CCW system, in its letter dated March 12, 2008, the applicant amended the LRA to reflect the following changes:

(a)   LRA Table 2.3.3-19-A-IP2 would reflect the CCW system as a miscellaneous system within the scope of license renewal pursuant to 10 CFR 54.4(a)(2).

(b)   Removal of the CCW system from the list of IP2 systems not reviewed for spatial interaction, in accordance with 10 CFR 54.4(a)(2).

(c)   Revision of LRA Table 2.3.3-19-B-IP2 to reflect that the CCW system now has components subject to an AMR, pursuant to 10 CFR 54.4(a)(2).

(d)   Creation of a new LRA Table 2.3.3-19-45-IP2 for the five added component types in the CCW system for nonsafety-related components, potentially affecting a safety-related function, and subject to an AMR.

(e)   Creation of a new LRA Table 3.3.2-19-45-IP2 for the five added component types, their materials, environments, and AMPs.

Based on its review, the staff finds the applicant's response to RAI 2.3A.2.2-1 for the IP2 CCW system acceptable because it adequately explained that the applicant's reevaluation of safety-related systems identified some components that should have been within scope for meeting the requirements of 10 CFR 54.4(a)(2). Additionally, the staff finds that the applicant's response amended the LRA to include those portions of the CCW system identified by the staff as being in scope under 10 CFR 54.4(a)(2). The staff reviewed the applicant's addition of new tables to the LRA and found that they adequately reflect those components brought into the scope of license renewal under 10 CFR 54.4(a)(2) because of their potential for spatial interaction with safety-related components. The staff's concern described in RAI 2.3A.2.2-1 for the IP2 CCW system is resolved. SER Section 3.3.2.1 documents the staff's evaluation of new AMR results for the CCW system.

The discussion of the staff's RAIs in SER Section 2.3A.3 details the disposition of RAI 2.3A.3.3-1, dated February 13, 2008.

2.3A.3.3.3 Conclusion

The staff reviewed the LRA, UFSAR, RAI response, and drawings to determine whether the applicant failed to identify any SSCs within the scope of license renewal. The staff found no such omissions. In addition, the staff sought to determine whether the applicant failed to identify any components subject to an AMR. The staff found one instance in which the applicant omitted components that should have been subject to an AMR. The applicant has satisfactorily resolved this issue as discussed in the preceding staff evaluation. On the basis of its review, the staff concludes that the applicant has appropriately identified the CCW system components within the scope of license renewal, as required by 10 CFR 54.4(a), and those subject to an AMR, as required by 10 CFR 54.21(a)(1).

2.3A.3.4 IP2 Compressed Air Systems

2.3A.3.4.1 Summary of Technical Information in the Application

LRA Section 2.3.3.4 describes the compressed air systems, including the instrument air and station air systems. The instrument air system continuously supplies dry, oil-free air from duplicate compressors with duplicate dryers and filters for pneumatic instruments and controls. Indian Point Nuclear Generating Unit 1 (IP1) and IP2 station air systems provide alternate supplies. A connection in the station air system allows a backup supply from portable compressed air equipment. The instrument air system, although designed to meet air capacity requirements, utilizes the higher-capacity IP1 station air compressors as a primary source of supply. Because of the high-capacity output of the IP1 air compressors, they can supply all IP1 and IP2 station and instrument air requirements. The IP2 station air compressor and both IP2 instrument air compressors serve as backups. The system includes the compressors, dryers, filters, receivers, distribution piping and valves, instruments, and controls. Items essential for safe operation and cooldown have air reserves or gas bottles that enable the equipment to function safely until its air supply resumes. The instrument air system includes piping, air bottles, valves, and controls supporting this air reserve function, but excludes the air or gas bottle parts of other systems. The system also may supply air to the post-accident venting system to pressurize containment in support of hydrogen control, but this function is not safety related.

The station air system distributes compressed air to hose connections throughout the plant, primarily for maintenance activities. The station air system also serves as an alternate air supply to the instrument air system. Either an IP2 air compressor or the IP1 compressors and equipment provide station air. The station air system consists of IP1 and IP2 station air equipment, including air compressors, air receivers, filters, dryers, distribution piping, and valves.

The compressed air system contains safety-related components relied on to remain functional during and following DBEs. It also contains nonsafety-related components whose failure potentially could prevent the satisfactory accomplishment of a safety-related function. In addition, the compressed air system performs functions that support fire protection.

Instrument air system components that support safe shutdown in a fire in the auxiliary feed pump room are reviewed in LRA Section 2.3.4.5. Components containing nitrogen are reviewed with the nitrogen systems (LRA Section 2.3.3.5).

LRA Tables 2.3.3-4-IP2, 2.3.3-19-18-IP2, and 2.3.3-19-33-IP2 identify compressed air system component types within the scope of license renewal and subject to an AMR, as well as their intended functions.

2.3A.3.4.2 Staff Evaluation

The staff reviewed LRA Section 2.3.3.4, UFSAR Sections 9.6.4 and 9.6.4.2, and license renewal drawings using the evaluation methodology described in SER Section 2.3 and the guidance in SRP-LR Section 2.3.

During its review, the staff evaluated the system functions described in the LRA and UFSAR to verify that the applicant had not omitted from the scope of license renewal any components with intended functions, as required by 10 CFR 54.4(a). The staff then reviewed those components that the applicant identified as within the scope of license renewal to verify that it had not omitted any passive and long-lived components subject to an AMR, in accordance with 10 CFR 54.21(a)(1).

2.3A.3.4.3 Conclusion

The staff reviewed the LRA, UFSAR, and drawings to determine whether the applicant failed to identify any SSCs within the scope of license renewal. The staff found no such omissions. In addition, the staff sought to determine whether the applicant failed to identify any components subject to an AMR. The staff found no such omissions. On the basis of its review, the staff concludes that the applicant has adequately identified the compressed air system components within the scope of license renewal, as required by 10 CFR 54.4(a), and those subject to an AMR, as required by 10 CFR 54.21(a)(1).

2.3A.3.5 IP2 Nitrogen Systems

2.3A.3.5.1 Summary of Technical Information in the Application

LRA Section 2.3.3.5 describes the gas system, which stores and distributes gases, primarily hydrogen, carbon dioxide (CO_2), and nitrogen, for various uses around the plant. The gas system includes the hydrogen, CO_2, and nitrogen gas subsystems. The system supplies hydrogen to the chemical and VCT for oxygen scavenging of RCS water to support water chemistry control and to the main generator for cooling gas. CO_2 gas purges the main generator of hydrogen to support outage work on the generator. The nitrogen gas subsystem includes the various nitrogen supplies of motive gas to components as a backup to the instrument air supply and for process functions (including cover gas, purge gas, and gas required for operation of level instrumentation). Nitrogen enters containment through several penetrations. For the safe shutdown required by Appendix R to 10 CFR Part 50, nitrogen is necessary for pneumatically actuated components. The nitrogen gas subsystem supplies the atmospheric dump valves, backup nitrogen to AFW system valve actuators, a portable nitrogen bottle that can be carried into containment to operate the auxiliary spray valve, motive gas for the charging pumps suction valve, and pneumatically powered instrumentation. An SBO event requires nitrogen to be supplied to the atmospheric dump valves, the AFW system valve actuators, and pneumatically powered instrumentation.

The gas system contains safety-related components relied on to remain functional during and following DBEs. It also contains nonsafety-related components whose failure potentially could prevent the satisfactory accomplishment of a safety-related function. In addition, the gas system performs functions that support fire protection and SBO.

Gas system component parts of containment penetrations are reviewed with the containment penetrations (LRA Section 2.3.2.5). A small number of components are reviewed with the compressed air systems (LRA Section 2.3.3.4), the city water system (LRA Section 2.3.3.17), the plant drains (LRA Section 2.3.3.18), and the AFW systems (LRA Section 2.3.4.3).

LRA Tables 2.3.3-5-IP2 and 2.3.3-19-14-IP2 identify gas system component types within the scope of license renewal and subject to an AMR, as well as their intended functions.

2.3A.3.5.2 Staff Evaluation

The staff reviewed LRA Section 2.3.3.5, UFSAR Sections 4.3.4.2, 7.2.1.5, 9.2, 10.2.2, and 10.2.6.3, and license renewal drawings using the evaluation methodology described in SER Section 2.3 and the guidance in SRP-LR Section 2.3.

During its review, the staff evaluated the system functions described in the LRA and UFSAR to verify that the applicant had not omitted from the scope of license renewal any components with intended functions, as required by 10 CFR 54.4(a). The staff then reviewed those components that the applicant identified as within the scope of license renewal to verify that it had not omitted any passive and long-lived components subject to an AMR, in accordance with 10 CFR 54.21(a)(1).

During its review of LRA Section 2.3.3.5, the staff identified an area in which additional information was necessary to complete the review of the applicant's scoping and screening results. The discussion of the staff's RAIs in SER Section 2.3A.3 details the disposition of RAI 2.3A.3.5-1, dated February 13, 2008.

2.3A.3.5.3 Conclusion

The staff reviewed the LRA, UFSAR, RAI response, and drawings to determine whether the applicant failed to identify any SSCs within the scope of license renewal. The staff found no such omissions. In addition, the staff sought to determine whether the applicant failed to identify any components subject to an AMR. The staff found no such omissions. On the basis of its review, the staff concludes that the applicant has appropriately identified the nitrogen system components within the scope of license renewal, as required by 10 CFR 54.4(a), and those subject to an AMR, as required by 10 CFR 54.21(a)(1).

2.3A.3.6 IP2 Chemical and Volume Control System

2.3A.3.6.1 Summary of Technical Information in the Application

LRA Section 2.3.3.6 describes the CVCS, which controls RCS inventory (amounts of makeup and letdown) and chemistry (RCS boron concentration and other chemical additions). The system cleans up reactor coolant by degasification and purification, injects seal water to the RCPs, depressurizes the RCS via a pressurizer auxiliary spray flowpath, and injects control poison in the form of a boric acid solution from the boric acid storage tanks.

During normal plant operation, reactor coolant letdown flows through the shell side of the regenerative heat exchanger, which reduces its temperature by transferring heat to the charging fluid. The coolant then flows through a letdown orifice, which regulates flow and reduces coolant pressure. The cooled, low-pressure water leaves the reactor containment and enters the primary auxiliary building (PAB). After passing through the non-regenerative heat exchanger and one of the mixed-bed demineralizers, the fluid flows through the reactor coolant filter and enters the VCT.

The coolant flows from the VCT to three positive-displacement, variable-speed charging pumps, which raise the pressure to a level above that in the RCS. The high-pressure water flows from the PAB to the reactor containment along two parallel paths—one returning directly to the RCS through the tube side of the regenerative heat exchanger to the RCS cold leg, and the other injecting water into the RCP seals through seal injection filters. The RCP seal water returns to the CVCS through a seal water filter and heat exchanger back to the VCT.

The RWST and the boric acid storage tank can supply borated water to the charging system. The RWST is available to the charging pumps for injection of borated water. The boric acid system has boric acid transfer pumps, a boric acid filter, and storage tanks to maintain a large inventory of concentrated boric acid solution.

The CVCS contains safety-related components relied on to remain functional during and following DBEs. It also contains nonsafety-related components whose failure potentially could prevent the satisfactory accomplishment of a safety-related function. In addition, the CVCS performs functions that support fire protection, ATWS, and SBO.

CVCS components that maintain the RCS pressure boundary are reviewed with the RCS pressure boundary (LRA Section 2.3.1.3). Some system components are reviewed with the primary water makeup systems (LRA Section 2.3.3.7).

LRA Tables 2.3.3-6-IP2 and 2.3.3-19-5-IP2 identify CVCS component types within the scope of license renewal and subject to an AMR, as well as their intended functions.

2.3A.3.6.2 Staff Evaluation

The staff reviewed LRA Section 2.3.3.6, UFSAR Section 9.2.2, and license renewal drawings using the evaluation methodology described in SER Section 2.3 and the guidance in SRP-LR Section 2.3.

During its review, the staff evaluated the system functions described in the LRA and UFSAR to verify that the applicant had not omitted from the scope of license renewal any components with intended functions, as required by 10 CFR 54.4(a). The staff then reviewed those components that the applicant identified as within the scope of license renewal to verify that it had not omitted any passive and long-lived components subject to an AMR, in accordance with 10 CFR 54.21(a)(1).

2.3A.3.6.3 Conclusion

The staff reviewed the LRA, UFSAR, and drawings to determine whether the applicant failed to identify any SSCs within the scope of license renewal. The staff found no such omissions. In addition, the staff sought to determine whether the applicant failed to identify any components subject to an AMR. The staff found no such omissions. On the basis of its review, the staff concludes that the applicant has adequately identified the CVCS components within the scope of license renewal, as required by 10 CFR 54.4(a), and those subject to an AMR, as required by 10 CFR 54.21(a)(1).

2.3A.3.7 IP2 Primary Water System

2.3A.3.7.1 Summary of Technical Information in the Application

LRA Section 2.3.3.7 describes the primary water makeup system, which supplies makeup water to primary plant systems as required to support normal plant operation (e.g., tanks, piping, valves, pumps) The system includes containment penetration. The primary water makeup system can supply backup cooling water to safety-related components in a passive failure of the CCW system.

The primary water makeup system contains safety-related components relied on to remain functional during and following DBEs. It also contains nonsafety-related components whose failure potentially could prevent the satisfactory accomplishment of a safety-related function.

LRA Tables 2.3.3-7-IP2 and 2.3.3-19-29-IP2 identify primary water makeup system component types within the scope of license renewal and subject to an AMR, as well as their intended functions.

2.3A.3.7.2 Staff Evaluation

The staff reviewed LRA Section 2.3.3.7, the UFSAR, and license renewal drawings using the evaluation methodology described in SER Section 2.3 and the guidance in SRP-LR Section 2.3.

During its review, the staff evaluated the system functions described in the LRA and UFSAR to verify that the applicant had not omitted from the scope of license renewal any components with intended functions, as required by 10 CFR 54.4(a). The staff then reviewed those components that the applicant identified as within the scope of license renewal to verify that it had not omitted any passive and long-lived components subject to an AMR, in accordance with 10 CFR 54.21(a)(1).

2.3A.3.7.3 Conclusion

The staff reviewed the LRA, UFSAR, and drawings to determine whether the applicant failed to identify any SSCs within the scope of license renewal. The staff found no such omissions. In addition, the staff sought to determine whether the applicant failed to identify any components subject to an AMR. The staff found no such omissions. On the basis of its review, the staff concludes that the applicant has adequately identified the primary water makeup system components within the scope of license renewal, as required by 10 CFR 54.4(a), and those subject to an AMR, as required by 10 CFR 54.21(a)(1).

2.3A.3.8 IP2 Heating, Ventilation and Air Conditioning Systems

2.3A.3.8.1 Summary of Technical Information in the Application

LRA Section 2.3.3.8 describes the heating, ventilation, and air conditioning (HVAC) systems that maintain the area environment for personnel and equipment.

The HVAC systems include various ventilation subsystems serving various areas of the plant. With the exception of the containment cooling and filtration system and a few components in

2-66

the operation of other mechanical systems, the HVAC system encompasses all IP2 ventilation systems and components and some from IP1. The main HVAC systems supporting plant operation include the following systems:

- containment purge supply and exhaust
- containment pressure relief
- containment iodine removal
- control rod drive mechanism (CRDM) cooling
- PAB ventilation
- fuel storage building ventilation
- cable spreading room/electrical tunnel ventilation
- 480 volt (V) switchgear room ventilation
- battery room exhaust
- EDG building ventilation
- auxiliary feed pump room ventilation
- diesel fire pump house ventilation
- electric fire pump room ventilation
- plant vent
- shield wall area enclosure heating and ventilation system
- SBO/Appendix R diesel generator ventilation
- portable HVAC credited in Appendix R
- security diesel room ventilation
- turbine hall ventilation
- technical support center ventilation
- administration building ventilation

LRA Section 2.3.3.9 addresses containment cooling and filtration, and LRA Section 2.3.3.10 addresses control room HVAC.

The containment purge supply and exhaust system supplies fresh air to purge the containment for personnel access. The system consists of a makeup air unit to supply fresh air, a filtration unit to filter the air released from containment, supply and exhaust ductwork, containment penetration piping, and valves. The system need not be in operation during DBAs or any regulated events. The system has two penetrations with safety-related piping and valves that support the containment isolation function. The pressure boundary function of system portions are also necessary to prevent air from being drawn into the shared fan housing for the containment purge and PAB exhaust fans.

The containment pressure relief system accommodates normal pressure changes in the containment during reactor power operation. This system consists of a filtration unit, fan, pressure relief ductwork, containment penetration piping, and valves. The system need not be in operation during DBAs or any regulated events. The system has a penetration with safety-related piping and valves that support the containment isolation function.

The containment iodine removal system consists of two auxiliary particulate and charcoal filter units in the containment, primarily used for pre-access cleanup. During power operation, the containment air particulate and gas monitor indications help determine whether to use either or both of these units. These units, wholly contained within containment, are not safety related or required during DBAs or regulated events.

The CRDM cooling system maintains the control rod drive operating coil stacks below their maximum allowable temperature during normal operation. Four fans take suction from the control rod drive shroud and discharge into the containment atmosphere. This equipment is not required to function during accident conditions or in response to regulated events.

The PAB ventilation system ventilates the waste hold-up tank pit and enclosed spaces in the PAB. The waste hold-up tanks in the waste hold-up tank pit are the central collection points for liquid radioactive waste. The PAB houses equipment and components required for normal plant operation, as well as accident mitigation. The PAB heating and ventilation system maintains an operating environment for personnel and equipment during normal operating and post-accident conditions with supply and exhaust fans with ductwork and dampers. None of the applicant's dose consequence analyses credit filtration. The PAB ventilation system is in use during normal operating conditions (plant start-up, power operation, and normal shutdown). This system must also operate during DBAs and for safe shutdown following a fire.

The fuel storage building heating and ventilation system heats and ventilates that building, minimizes leakage of unfiltered air from the building during fuel-handling operations, and filters building exhaust. The system has two fresh air tempering units with supply fans and heaters, exhaust roughing, high-efficiency particulate air (HEPA) and carbon filters, an exhaust fan, motor-operated dampers, and ducts. The applicant originally credited the system in the fuel-handling accident; however, the analysis described in UFSAR Section 14.2.1.1, which uses the alternate source term, no longer assumes operation of the ventilation system or any holdup of the radionuclides released from the spent fuel pit. Consequently, the system has no safety functions.

The cable spreading room/electrical tunnel exhaust system ventilates the 33-foot elevation of the control building. The system consists of two exhaust fans mounted above the tunnel in a plenum. Intake louvers on the north and south walls draw air into the cable- spreading room. The system maintains an operating environment for personnel and equipment during normal operating and post-accident conditions and is required for cooling during DBAs, as well as regulated events.

The 480-V electrical switchgear room ventilation system ventilates that room at the 15-foot elevation of the control building, using three fans mounted in the north wall. The fans take suction from the switchgear room and discharge outside. A fixed louver with fire damper allows air to flow into the room. The system maintains an operating environment for personnel and equipment during normal operating and post-accident conditions and is required for cooling during DBAs, as well as regulated events.

Battery rooms in the control and superheater buildings have exhaust fans to prevent long-term buildup of hydrogen during normal operation when the batteries charge. These exhaust fans need not function during DBAs or regulated events.

The EDG building ventilation system has exhaust fans, exhaust dampers, and intake louvers. These HVAC components are required to support diesel operation during DBAs, as well as regulated events such as the Appendix R safe shutdown.

The heating and ventilation system of the auxiliary boiler feed pump building, which is in use during normal operating conditions, consists of several exhaust fans for cooling. A roll-up door can be opened for cooling during emergency operation of the AFW system. Following a fire, portable blowers can ventilate this area; therefore, the applicant stated that operation of the auxiliary boiler feed pump building heating and ventilation system is not required during DBAs or regulated events.

The diesel fire pump house ventilation system cools the structure housing the diesel fire pump. This structure is cooled by louvers, and the diesel itself is cooled by fire water. These HVAC components are required to support fire system operation credited in Appendix R evaluations.

The electric fire pumps are located in two rooms in the IP1 turbine building cooled by exhaust fans and dampers that cool the electric fire pumps. These HVAC components are required to support fire system operation credited in Appendix R evaluations.

The plant vent system, which provides a flowpath for the exhaust to the atmosphere, includes the plant vent duct and some vent flow monitoring instrumentation. The offsite dose analyses does not credit the plant vent as the release point but, because of its proximity to the control room air intake, the control room dose calculations do consider the plant vent to be the release point.

The IP2 shield wall area enclosure heating and ventilation system heats and ventilates the shield wall area enclosure. Components and piping primarily associated with the MS and FW systems are located in the main enclosure. The shield wall area enclosure heating and ventilation system is in use during normal operating conditions, such as plant start-up, power operation, and normal shutdown. The operation of this equipment is not required during DBAs or regulated events.

IP2 installed a new SBO and Appendix R diesel generator credited with supplying backup power to the plant to assist in safe shutdown following a fire or an SBO. Its associated ventilation equipment is required for its function. The IP2 SBO/Appendix R diesel generator ventilation system utilizes louvers, an exhaust fan, and outlet ductwork. The fan will operate when the diesel operates.

The Appendix R safe-shutdown report indicates that, for a fire in certain plant areas, portable blowers and flexible ductwork can ventilate the safe-shutdown equipment, and are therefore required by Appendix R. Power can be supplied by portable generators.

The IP2 security diesel generator is credited for emergency lighting for some areas to support safe shutdown following a fire. The ventilation equipment that cools this diesel consists of dampers, ductwork, and an engine-driven blower that ventilates the room when the engine operates. This ventilation is required for the operation of the security diesel credited with providing power for lighting, as required by Appendix R.

Using fixed and adjustable louvers and awning sashes, the turbine building ventilation system draws in air exhausted by power roof ventilators and wall exhaust fans. This cooling is not required during DBAs and regulated events.

The technical support center ventilation system maintains environmental conditions in the center. The system, which includes fans, dampers, filters, and cooling equipment, performs no safety-related functions during accident conditions and is not required for any regulated events.

The administration building ventilation system, which heats, ventilates, and provides air conditioning to administration building personnel and equipment, is not required during DBAs or regulated events.

The HVAC system contains safety-related components relied on to remain functional during and following DBEs. It also contains nonsafety-related components whose failure could prevent the satisfactory accomplishment of a safety-related function. In addition, the HVAC system performs functions that support fire protection and SBO.

Some HVAC components are reviewed with the compressed air systems (LRA Section 2.3.3.4) or with the containment cooling and filtration systems (LRA Section 2.3.3.9).

LRA Tables 2.3.3-8-IP2 and 2.3.3-19-17-IP2 identify HVAC system component types within the scope of license renewal and subject to an AMR, as well as their intended functions.

2.3A.3.8.2 Staff Evaluation

The staff reviewed LRA Section 2.3.3.8; UFSAR Sections 5.3.2, 9.8, and 9.10; and a license renewal drawing using the evaluation methodology described in SER Section 2.3 and the guidance in SRP-LR Section 2.3.

During its review, the staff evaluated the system functions described in the LRA and UFSAR to verify that the applicant had not omitted from the scope of license renewal any components with intended functions, as required by 10 CFR 54.4(a). The staff then reviewed those components the applicant identified as within the scope of license renewal to verify that it had not omitted any passive and long-lived components subject to an AMR, in accordance with 10 CFR 54.21(a)(1).

2.3A.3.8.3 Conclusion

The staff reviewed the LRA, UFSAR, and a drawing to determine whether the applicant failed to identify any SSCs within the scope of license renewal. The staff found no such omissions. In addition, the staff sought to determine whether the applicant failed to identify any components subject to an AMR. The staff found no such omissions. On the basis of its review, the staff concludes that the applicant has adequately identified the HVAC system components within the scope of license renewal, as required by 10 CFR 54.4(a), and those subject to an AMR, as required by 10 CFR 54.21(a)(1).

2.3A.3.9 IP2 Containment Cooling and Filtration System

2.3A.3.9.1 Summary of Technical Information in the Application

LRA Section 2.3.3.9 describes the containment cooling and filtration system. The IP2 containment cooling and filtration system cools the containment. Air-handling units, discharging into a common header ductwork distribution system, achieve air recirculation cooling during normal operation and ensure adequate flow of cooled air throughout the containment. Each air-handling unit consists of equipment arranged so that, during normal and accident operation, air flows through the unit in the following sequence: cooling coils, moisture separators (demisters), centrifugal fan with direct-drive motor, and distribution header. The system rejects heat to SW system cooling coils in normal operation, emergency operation, and safe-shutdown cooling following a fire.

The containment cooling and filtration system contains safety-related components relied on to remain functional during and following DBEs. In addition, the containment cooling and filtration system performs functions that support fire protection and SBO.

LRA Table 2.3.3-9-IP2 identifies containment cooling and filtration system component types within the scope of license renewal and subject to an AMR, as well as their intended functions.

2.3A.3.9.2 Staff Evaluation

The staff reviewed LRA Section 2.3.3.9, UFSAR Sections 5.3.2.2 and 6.4.2, and license renewal drawings using the evaluation methodology described in SER Section 2.3 and the guidance in SRP-LR Section 2.3.

During its review, the staff evaluated the system functions described in the LRA and UFSAR to verify that the applicant had not omitted from the scope of license renewal any components with intended functions, as required by 10 CFR 54.4(a). The staff then reviewed those components the applicant identified as within the scope of license renewal to verify that it had not omitted any passive and long-lived components subject to an AMR, in accordance with 10 CFR 54.21(a)(1).

2.3A.3.9.3 Conclusion

The staff reviewed the LRA, UFSAR, and drawings to determine whether the applicant failed to identify any SSCs within the scope of license renewal. The staff found no such omissions. In addition, the staff sought to determine whether the applicant failed to identify any components subject to an AMR. The staff found no such omissions. On the basis of its review, the staff concludes that the applicant has adequately identified the containment cooling and filtration system components within the scope of license renewal, as required by 10 CFR 54.4(a), and those subject to an AMR, as required by 10 CFR 54.21(a)(1).

2.3A.3.10 IP2 Control Room Heating, Ventilation and Cooling System

2.3A.3.10.1 Summary of Technical Information in the Application

LRA Section 2.3.3.10 describes the control room ventilation system, which maintains the central control room in a safe, habitable environment during normal operation and under accident conditions. The system has an air-conditioning unit with fan, steam heating coil, roughing filter to recirculate air inside the control room, a backup fan in parallel with the air-conditioning unit, and a filter unit consisting of HEPA filters, charcoal filters, post-filters, and booster fans to permit filtration of incoming air for a slight positive pressure in the control room during accident conditions. System ducts, dampers, and controls allow three system operating modes: Mode 1 (normal operation) with outside air makeup, Mode 2 (safety injection or high radiation) with outside filtered air, and Mode 3 (toxic gas or smoke) with all outside air isolated. Control room dose analyses credit the operation of this system, including the filtration of incoming air. IP1 and IP2 share a central control room. The IP1 control room ventilation equipment is modified for recirculation mode only.

The central control room system contains safety-related components relied upon to remain functional during and following DBEs. It also contains nonsafety-related components whose failure potentially could prevent the satisfactory accomplishment of a safety-related function. In addition, the central control room system performs functions to maintain the central control room in a safe, habitable environment during an Appendix R event and SBO.

LRA Tables 2.3.3-10-IP2 and 2.3.3-19-17-IP2 identify central control room HVAC system component types within the scope of license renewal and subject to an AMR, as well as their intended functions.

2.3A.3.10.2 Staff Evaluation

The staff reviewed LRA Section 2.3.3.10, UFSAR Section 9.9, and license renewal drawings using the evaluation methodology described in SER Section 2.3 and the guidance in SRP-LR Section 2.3.

During its review, the staff evaluated the system functions described in the LRA and UFSAR to verify that the applicant had not omitted from the scope of license renewal any components with intended functions, as required by 10 CFR 54.4(a). The staff then reviewed those components the applicant identified as within the scope of license renewal to verify that it had not omitted any passive and long-lived components subject to an AMR, in accordance with 10 CFR 54.21(a)(1).

2.3A.3.10.3 Conclusion

The staff reviewed the LRA, UFSAR, and drawings to determine whether the applicant failed to identify any SSCs within the scope of license renewal. The staff found no such omissions. In addition, the staff sought to determine whether the applicant failed to identify any components subject to an AMR. The staff found no such omissions. On the basis of its review, the staff concludes that the applicant has adequately identified the central control room HVAC system components within the scope of license renewal, as required by 10 CFR 54.4(a), and those subject to an AMR, as required by 10 CFR 54.21(a)(1).

2.3A.3.11 IP2 Fire Protection – Water

2.3A.3.11.1 Summary of Technical Information in the Application

LRA Section 2.3.3.11 describes the fire protection system, which provides fire protection for the station through the use of water, dry chemicals, foam, detection and alarm systems, and rated fire barriers, doors, and dampers. Passive mechanical components in the fire protection system include many firefighting subsystem components and features, such as piping, fire dampers, valves, hydrants, portable fire extinguishers, and two fire water storage tanks. Also included under this system code (i.e., the applicant's code for designating systems and boundaries) are the IP1 fire pumps and some associated IP1 fire protection components, such as hydrants, valves, fire extinguishers, and strainers. Plant drain components in the fire protection system are passive fire protection features required to ensure adequate protection of safety-related equipment from water damage in areas containing fixed suppression systems.

The fire protection—water system draws water from two storage tanks, a 1.5-million-gallon tank supplied by the city water distribution system for fire protection purposes and a 300,000-gallon fire water storage tank of city water as a redundant supply for the water-based fire protection systems. The pumping facilities consist of two electric fire pumps taking suction from the site's city water main. Two small electric pumps also maintain pressure for the fire water system. A diesel fire pump for redundant pumping capabilities normally takes suction from the 300,000-gallon fire water storage tank. The pumping facilities meet flow and pressure requirements for water-based fire protection systems. The fire protection water distribution system consists of outdoor underground piping, indoor distribution piping, isolation valves, strainers, hose stations, and outdoor hydrants.

The water-based fire suppression systems include the wet pipe sprinkler systems, preaction sprinkler systems, deluge water spray systems, foam water spray systems, and hydrants and hose stations.

According to the LRA, the fire protection—water system has no intended function under 10 CFR 54.4(a)(1). The scoping and screening methodology identified the following fire water system intended functions, in accordance with 10 CFR 54.4(a)(2):

- Maintain integrity of nonsafety-related components such that no physical interaction with safety-related components could prevent satisfactory accomplishment of a safety function.

- Provide a backup source of makeup water to the spent fuel pit.

The scoping and screening methodology also identified the following the fire water system intended functions, in accordance with 10 CFR 54.4(a)(3):

- Provide fixed automatic and manual fire suppression (including hydrants, hose stations and portable extinguishers) to extinguish fires in vital areas of the plant (10 CFR 50.48).

- Ensure adequate protection of safety-related equipment from water damage in areas susceptible to flooding (10 CFR 50.48).

- Ensure that drain systems in areas containing combustible materials prevent the spreading of fires into other areas of the plant (10 CFR 50.48).

LRA Section 2.3.3.12 evaluates the fire protection—CO_2, Halon 1301, and RCP oil collection systems.

The drain portion of the system is evaluated with plant drains (LRA Section 2.3.3.18). The fuel oil subsystem components are evaluated with fuel oil systems (LRA Section 2.3.3.13). A small number of components are evaluated with city water systems (LRA Section 2.3.3.17).

The applicant evaluated those nonsafety-related components that were not evaluated with other systems and whose failure could prevent satisfactory accomplishment of safety functions with miscellaneous systems within the 10 CFR 54.4(a)(2) scope of license renewal (LRA Section 2.3.3.19).

LRA Tables 2.3.3-11-IP2 and 2.3.3-19-11-IP2 identify fire protection—water system component types within the scope of license renewal and subject to an AMR, as well as their intended functions.

2.3A.3.11.2 Staff Evaluation

The staff reviewed LRA Section 2.3.3.11, UFSAR Section 9.6.2, and license renewal drawings using the evaluation methodology described in SER Section 2.3 and the guidance in SRP-LR, Section 2.3. During its review, the staff evaluated the system functions described in the LRA and UFSAR to verify that the applicant had not omitted from the scope of license renewal any components with intended functions, as required by 10 CFR 54.4(a). The staff then reviewed those components that the applicant identified as within the scope of license renewal to verify that it had not omitted any passive and long-lived components subject to an AMR in accordance with 10 CFR 54.21(a)(1).

The staff also reviewed the following IP2 fire protection CLB documents listed in the IP2 Operating License Condition 2.K: NRC fire protection SERs for IP2, dated November 30, 1977; February 3, 1978; January 31, 1979; October 31, 1980; August 22, 1983; March 30, 1984; October 16, 1984; September 16, 1985; November 13, 1985; March 4, 1987; January 12, 1989; and March 26, 1996.

The staff also reviewed IP2 commitments made in response to the requirements of 10 CFR 50.48 (i.e., an approved fire protection program), using its commitment responses to Branch Technical Position (BTP) Auxiliary and Power Conversion Systems Branch (APCSB) 9.5-1, "Guidelines for Fire Protection for Nuclear Power Plants," dated May 1, 1976, and Appendix A to BTP APCSB 9.5-1, dated August 23, 1976.

During its review of LRA Section 2.3.3.11, the staff identified areas in which additional information was necessary to complete its review of the applicant's scoping and screening results. The applicant responded to the staff's RAIs as discussed below.

In RAI 2.3A.3.11-1, dated October 24, 2007, the staff questioned why the license renewal drawings identified certain fire protection system components as not subject to an AMR. Specifically, License renewal drawing LRA-227551-0 shows the following fire protection system components as not subject to an AMR (i.e., they are not highlighted in green):

- maintenance and outage building
- PAB and boric acid building charcoal filter deluge system

License renewal drawing LRA-227552-0 shows the following fire protection system components as not subject to an AMR (i.e., they are not highlighted in green):

- No. 11 fire pump room
- fuel oil tank/water meter house
- ignition oil tank and pump room deluge system
- main and auxiliary transformer deluge system

License renewal drawing LRA-227553-0 shows the following fire protection system components as not subject to an AMR (i.e., they are not highlighted in green):

- staircase Nos. 2, 3, 4, 5 and 6
- turbine oil piping system

License renewal drawing LRA-227554-0 shows the following fire protection system component as not subject to an AMR (i.e., it is not highlighted in green):

- staircase Nos. 1, 8, and 9

License renewal drawing LRA-9321-4006-0 shows the following fire protection system components as not subject to an AMR (i.e., they are not highlighted in green):

- fire hydrants
- fire hose connections
- fire hose stations

In the RAI, the staff requested that the applicant verify whether the above components are within the scope of license renewal, in accordance with 10 CFR 54.4(a), and subject to an AMR, in accordance with 10 CFR 54.21(a)(1). The staff requested that the applicant justify excluding these components from the scope of license renewal and an AMR.

In its response, dated November 16, 2007, the applicant provided scoping and screening results for the fire protection system components in question in license renewal drawing LRA-227551-0. For the maintenance and outage building, the applicant stated the following:

> The maintenance and outage building adjacent to the fuel storage building of IP2 houses offices and facilities for maintenance personnel. The maintenance and outage building fire protection components are not required for 10 CFR 50.48 as the building does not house and is not in proximity to safety-related equipment, nor does it contain equipment required for safe-shutdown. The maintenance and outage building fire protection components are not described in the January 31, 1979, fire protection SER.

Based on its review, the staff finds the applicant's response acceptable because the maintenance and outage building does not have a license renewal intended function. The maintenance and outage building does not require fire protection in accordance with the provisions of 10 CFR 50.48; therefore, the associated fire protection components are not within the scope of license renewal.

For the PAB charcoal filter deluge system, the applicant stated the following:

> Drawing LRA-227551-0 detail E shows piping and solenoid valves downstream of FP-587 for the PAB charcoal filter deluge system. These portions of the system were inadvertently not highlighted on the drawing as subject to an AMR for license renewal. The PAB charcoal filter deluge system is in-scope and

2-75

subject to an AMR. Applicable component types are included in LRA Table 2.3.3-11-IP2 with the AMR results in LRA Table 3.3.2-11-IP2.

Based on its review, the staff finds the applicant's response acceptable because it indicated that the PAB charcoal filter deluge system is within the scope of license renewal and subject to an AMR.

For the boric acid building charcoal filter deluge system, the applicant stated the following:

> The boric acid building charcoal filter deluge system is in-scope and subject to an AMR as shown on drawing LRA-227551-0 in detail E and detail F. Applicable component types are included in LRA Table 2.3.3-11-IP2 with the AMR results in LRA Table 3.3.2-11-IP2.

Based on its review, the staff finds the applicant's response acceptable because it indicated that the boric acid building charcoal filter deluge system is within the scope of license renewal and subject to an AMR.

In its response, dated November 16, 2007, the applicant provided scoping and screening results for the fire protection system components in license renewal drawing LRA-227552-0. For the No.11 Fire Pump Room, the applicant stated the following:

> The portion of the fire protection system labeled on drawing LRA-227552-0 as No.11 fire pump room includes systems for gas turbine No. 1 Transformer, expanded portion of the maintenance area, the L&P Transformer, the bulk H2 storage (screenwell house) for Unit 1, and the maintenance material processing area. This portion of the system is not required to meet 10 CFR 50.48 requirements for the following reasons. Deluge valve FP-294 feeds the line that is blind-flanged to the gas turbine No. 1 Transformer which is retired in place. A fire in this area cannot adversely impact safety-related equipment. Deluge valve FP-1008 feeds the expanded portion of the maintenance area which houses no safety-related equipment. A fire in this area cannot affect areas containing safety-related equipment. Deluge valve FP-242 supplies spray system No. 1 which protects the L&P Transformer which is retired in place. A fire in this area cannot adversely impact safety-related equipment. Deluge valve FP-261 supplies the line for spray system No. 4 to the bulk H2 storage (screenwell house) for Unit 1, and deluge valve FP-890 supplies the line for the maintenance material processing area.
>
> These areas do not contain safety-related equipment, and a fire in the areas cannot affect areas containing safety-related equipment. None of these fire protection systems are described in the January 31, 1979, fire protection SER.

Based on its review, the staff finds the applicant's response acceptable because the portion of the fire protection system identified does not have a license renewal intended function and is not subject to an AMR, in accordance with 10 CFR 54.4(a) or 10 CFR 54.21(a)(1), respectively.

For the fuel oil tank/water meter house, the applicant stated the following:

> As shown on drawing LRA-227552-0 detail J (fuel oil tank/water meter house), hydrants No. 18 and No. 19 provide fire protection coverage for fuel oil storage tanks. The fuel oil storage tanks are associated with house service boiler and ignition oil tanks. These fuel oil tanks have no intended function for license renewal. They are not required to meet 10 CFR 50.48 requirements since a fire in this portion of the yard cannot affect safety-related or safe-shutdown equipment. In addition, this equipment is not described in the January 31, 1979, fire protection SER. Fire protection components associated with the water meter house (piping and valves) are in-scope and subject to an AMR and are shown on drawing LRA-192505. These components (piping and valves) are part of the city water system discussed in Section 2.3.3.17 of the LRA.

Based on its review, the staff finds the applicant's response acceptable. The fuel oil tank does not have a license renewal intended function and is, therefore, excluded from the scope of license renewal and is not subject to an AMR. The staff notes that the water meter house fire protection components are within the scope of license renewal, subject to an AMR, and shown on drawing LRA-192505.

For the ignition oil tank and pump room deluge system, the applicant stated the following:

> The ignition oil tank and pump rooms are in the Indian Point Nuclear Generating Unit 1 (IP1) super-heater building (not adjacent to IP2 areas containing safety-related equipment). These rooms do not contain safety-related equipment or systems required for safe-shutdown. Three-hour rated walls, penetrations, and doors will prevent a fire in the ignition oil tank and pump room from spreading to safety-related areas associated with IP2. The ignition oil tank and pump room deluge system is not required to meet 10 CFR 50.48 and is not described in the January 31, 1979, fire protection SER.

Based on its review, the staff finds the applicant's response acceptable because the ignition oil tank and pump room deluge system is not required by 10 CFR 50.48 and is, therefore, outside the scope of license renewal.

For the main and auxiliary transformer deluge system, the applicant stated the following:

> The main and auxiliary transformer deluge systems and their associated components for the oil filled transformers adjacent to the control building were initially determined to have no license renewal intended function. They were considered required only to protect the transformers to satisfy requirements of the plant insurance carrier. However, the spray systems provide for defense in depth in addition to installed 3-hour rated fire barriers and are now considered in-scope and subject to an AMR for license renewal. Applicable component types that are subject to an AMR are included in LRA Table 2.3.3-11-IP2 with the AMR results in LRA Table 3.3.2-11-IP2.

Based on its review, the staff finds the applicant's response acceptable because it clarifies that (1) the main and auxiliary transformer deluge systems and their associated components have no license renewal intended function and (2) the spray systems provide for defense in depth, in addition to the installed 3-hour-rated fire barriers, and are considered in scope and subject to an AMR for license renewal. The staff's concern is resolved.

In its response, dated November 16, 2007, the applicant provided scoping and screening results for the fire protection system components in license renewal drawing LRA-27553-0. For staircase Nos. 2, 3, 4, 5 and 6, the applicant stated the following:

> Staircase No. 2 is located in the IP1 service building adjacent to the IP1 turbine building. The service building for IP1 houses administrative offices. Staircases No. 5 and No. 6 are located in the IP1 super-heater building at the south exterior wall. None of these areas are in proximity to areas containing safety-related equipment. Fires in the areas of Staircases No. 2, 5, and 6 are prevented from spreading to nearby safety-related areas (IP2 control building) by three-hour rated walls, penetrations, and doors. Fire protection equipment in Staircases No. 2, 5, and 6 are not required for 10 CFR 50.48 and are not described in the January 31, 1979, fire protection SER.

> Fire protection equipment for Staircase No. 4 at Elevation 53', located in the control building, is in-scope and subject to an AMR as shown on drawing LRA-227553-0 at detail WW.

> The supply to the radwaste/HP offices downstream of valve FP-363 and components downstream of normally closed valve FP-155 are not required for 10 CFR 50.48 because these areas do not contain safety-related equipment nor can a fire in the radwaste/HP offices impact areas containing safety-related equipment.

> Fire protection for Staircase No. 3 at Elevation 15', 33', and 53' in the control building is in-scope and subject to an AMR as shown on drawing LRA-227553-0 at detail W.

> The supply to the technical support building (TSC) downstream of valve FP-865 is not required for 10 CFR 50.48 because this area does not contain safety-related equipment nor can a fire in the technical support building impact areas containing safety-related equipment.

Based on its review, the staff finds the applicant's response acceptable because it clarifies that (1) fire protection systems in staircase Nos. 2, 5, and 6 are not required to comply with the requirements of 10 CFR 50.48, (2) fire protection system equipment for staircase No. 4 is within the scope of license renewal and subject to an AMR, as shown on license renewal drawing LRA-227553-0, and (3) the fire protection system for staircase No. 3 also is in scope and subject to an AMR, as shown on license renewal drawing LRA-227553.

For the turbine oil piping system, the applicant stated the following:

> Turbine oil piping sprinkler system components downstream of valve FP-65 provide coverage for the file room, one stop shop building, and the work control center building, none of which contains, or can impact areas containing,

safety-related equipment. The turbine oil piping sprinkler system is therefore not required for compliance with 10 CFR 50.48. However, hose reel FP-66 and associated piping are in-scope and subject to an AMR for license renewal as shown on drawing LRA-227553-0 at detail X.

Based on its review, the staff finds the applicant's response acceptable because it clarifies the portion of the turbine oil piping sprinkler system components that are not required under 10 CFR 50.48. These components are not within the scope of license renewal because the areas that they cover do not contain safety-related equipment and a fire in these locations cannot impact areas containing safety-related equipment.

In its response, dated November 16, 2007, the applicant provided scoping and screening results for the fire protection system components in license renewal drawing LRA-227554-0. For staircase Nos. 1, 8, and 9, the applicant stated the following:

> Staircase No. 1 is located in the IP1 nuclear service building and Staircases No. 8 and 9 are located in the IP1 nuclear service chemical system building. The nuclear service building is adjacent to the IP1 containment building and houses no safety-related equipment. The nuclear service chemical system building is adjacent to the IP1 containment building and houses no safety-related equipment. These buildings are not in proximity to areas containing safety-related equipment. Fires in the areas of Staircases No. 1, 8, and 9 are prevented from spreading to safety-related areas (control building) by three-hour rated walls, penetrations, and doors. Fire protection equipment in Staircases Nos. 1, 8, and 9 is not required for 10 CFR 50.48. These staircases are associated with IP1 and the associated fire protection system components are no longer required for compliance with 10 CFR 50.48 since the IP1 operating license was revoked in June 1980 as stated in the October 31, 1980, supplement to the January 31, 1979, fire protection SER.

Based on its review, the staff finds the applicant's response acceptable because it clarifies that staircase Nos. 1, 8, and 9 are associated with IP1, and the associated fire protection system components are no longer required for compliance with 10 CFR 50.48.

In its response, dated November 16, 2007, the applicant provided scoping and screening results for the fire protection system components depicted on license renewal drawing LRA-9321-4006-0 that are in question. For fire hydrants, the applicant stated the following:

> Hydrants for the IP2 screenwell structure (Hydrants 21 and 22), main transformer yard (Hydrant 25), emergency diesel generators building (Hydrant 27), primary auxiliary building (Hydrants 26, 28, and 29), and auxiliary feed pump building (Hydrant 24) are required for 10 CFR 50.48. These hydrants are highlighted on LRA drawing LRA-9321-4006-0.

> Hydrants that are not highlighted are those for the IP1 screenwell house (Hydrants 11 and 12), IP1 fuel oil tank farm (Hydrants 17 and 18), east of IP1 fuel handling building (Hydrant 16), station security building (Hydrant 15), and southeast of the IP1 containment building (Hydrants 13 and 14). These hydrants

are not required for 10 CFR 50.48. The IP1 screenwell house does contain equipment for safe-shutdown in the event of fire in another area. Fires are not assumed to occur in multiple fire zones, so a fire in the screenwell house is not a concern. The other areas listed do not present a significant fire hazard to areas containing equipment used for safe-shutdown.

Based on its review, the staff finds the applicant's response acceptable because it clarifies that hydrants 21, 22, 24, 25, 26, 27, 28, and 29 are required by 10 CFR 50.48. The applicant has highlighted these hydrants on license renewal drawing LRA-9321-4006-0. In addition, hydrants 11, 12, 13, 14, 15, 16, 17, and 18 are not highlighted because they are associated with IP1. The IP1 hydrants are no longer required for compliance with 10 CFR 50.48.

For fire hose connections, the applicant stated the following:

> Fire hose connections that are not highlighted on drawing LRA-9321-4006-0 coordinates (B2) are located at the IP1 screenwell house dock and are not required for 10 CFR 50.48. The hose connection at the IP1 screenwell house is isolated with a blank flange.

Based on its review, the staff finds the applicant's response acceptable because it clarifies that fire hose connections that are not highlighted on license renewal drawing LRA-9321-4006-0 are associated with the IP1 screenwell house dock and are no longer required for compliance with 10 CFR 50.48, since the IP1 operating license was revoked in June 1980.

For fire hose stations, the applicant stated the following:

> The fire hose stations that are not highlighted on drawing LRA-9321-4006-0 are in the IP1 fuel handling building and are not required for 10 CFR 50.48. This area does not contain equipment used for safe-shutdown and is an area that does not present a significant fire hazard to areas containing equipment used for safe-shutdown. Fire hose stations associated with IP1 and the associated fire protection system components are no longer required for compliance with 10 CFR 50.48 since the IP1 operating license was revoked in June 1980 as stated in the October 31, 1980, supplement to the January 31, 1979, fire protection SER.

Based on its review, the staff finds the applicant's response acceptable because it clarifies that fire hose stations that are not highlighted on license renewal drawing LRA-9321-4006-0 are in the IP1 fuel-handling building and are no longer required by 10 CFR 50.48, since the IP1 operating license was revoked in June 1980.

At the request of the staff, the applicant clarified its statements made in its November 16, 2007 response to RAI 2.3A.3.11-1 regarding IP1 fire protection components that were stated to be no longer required for compliance with 10 CFR 50.48. By letter dated August 6, 2009, the applicant clarified that IP1 fire protection components identified in its response dated November 16, 2007 that are specifically used only to support IP1 do not have an intended function for IP2 or IP3. Since they are not required to demonstrate compliance with 10 CFR 50.48 for IP2 or IP3, the applicant determined that they are not within the scope of license renewal. Entergy further stated that the IP1 components are credited in the IP1 fire protection program which meets the requirements in 10 CFR 50.48(f). The staff notes that 10 CFR 50.48(f) applies to reactors that have permanently ceased operations, and does not apply to IP2 or IP3.

The staff finds the applicant's response acceptable because it clarified that the IP1 components that support IP1 fire protection program are not needed to support the operation of IP2 or IP3, and therefore, they are not within the scope of license renewal.

Based on its review, the staff finds the applicant's response to RAI 2.3A.3.11-1, as clarified, acceptable. The staff's concern described in RAI 2.3A.3.11-1 is resolved.

In RAI 2.3A.3.11-2, dated October 24, 2007, the staff stated that LRA Tables 2.3.3-11-IP2 and 2.3.3-11-IP3 exclude several types of fire protection components that are discussed in the fire protection SERs or UFSAR or both and which also appear on the license renewal drawings as subject to an AMR (i.e., they are highlighted in green). These components include the following:

- hose connections
- hose racks
- yard hose houses
- interior fire hose stations
- pipe fittings
- pipe supports
- couplings
- threaded connections
- restricting orifices
- interface flanges
- chamber housings
- heat-actuated devices
- tank heaters
- thermowells
- water motor alarms
- expansion joint
- filter housing
- gear box housing
- heat exchanger (bonnet)
- heat exchanger (shell)
- heat exchanger (tube)
- heater housing
- diesel-driven fire pump engine's muffler
- orifice
- sight glass
- strainer housing
- turbocharger housing
- flexible hose
- latch door pull box
- pneumatic actuators
- actuator housing
- dikes for oil spill confinement
- buried underground fuel oil tanks for EDGs
- expansion tank

- fire water main loop valves
- post-indicator valves
- jacket cooling water keep-warm pump and heater
- lubricating oil collection system components for each RCP
- lubricating oil cooler
- auxiliary lubricating oil makeup tank
- rocker lubricating oil pump
- floor drains and curbs for fire-fighting water
- backflow prevention devices
- flame retardant coating for cables
- fire retardant coating for structural steel supporting walls and ceilings

The staff requested that the applicant verify whether LRA Tables 2.3.3-11-IP2 and 2.3.3-11-IP3 should include the components listed above. If they are excluded from the scope of license renewal and not subject to an AMR, the staff requested that the applicant justify their exclusion.

In its response, dated November 16, 2007, the applicant provided the results of scoping and screening for the listed fire protection system component types as follows:

Hose connections—As stated in LRA Section 2.0 Page 2.0-1, the component type "piping" includes pipe, pipe fittings (such as elbows and reducers), flow elements, orifices, and thermowells. Hose connections are pipe fittings subject to an AMR as indicated in LRA Tables 2.3.3-11-IP2 and 2.3.3-11-IP3 under the component type "piping," with the AMR results provided in LRA Tables 3.3.2-11-IP2 and 3.3.2-11-IP3.

Hose racks—Hose racks subject to an AMR are included in the structural AMR as component type "fire hose reels." This item is included in LRA Table 2.4-4, with the AMR results provided in LRA Table 3.5.2-4.

Yard hose houses—Yard hose houses (small buildings over hydrants which contain fire hose and fire fighting equipment) are not subject to an AMR. Failure of a yard hose house would not prevent fire suppression capability of the associated hydrant.

Interior fire hose stations—Interior fire hose stations are subject to an AMR. They are included in LRA Table 2.4-4 under component type "fire hose reels," with the AMR results provided in LRA Table 3.5.2-4.

Pipe fittings—As stated in LRA Section 2.0 on Page 2.0-1, the component type piping" may include pipe, pipe fittings (such as elbows and reducers), flow elements, orifices, and thermowells. Pipe fittings are subject to an AMR and included in LRA Tables 2.3.3-11-IP2 and 2.3.3-11-IP3 under the component type "piping" with the AMR results in LRA Tables 3.3.2-11-IP2 and 3.3.2-11-IP3.

Pipe supports—Pipe supports are subject to an AMR and are included in the structural AMR as component type "component and piping supports." This item is included in LRA Table 2.4-4, with the AMR results provided in LRA Table 3.5.2-4.

Couplings—As stated in LRA Section 2.0 Page 2.0-1, the component type "piping" may include pipe, pipe fittings (such as elbows and reducers), flow elements, orifices, and thermowells. Couplings are subject to an AMR and included in LRA Tables 2.3.3-11-IP2 and 2.3.3-11-IP3 under the component type "piping," with the AMR results provided in LRA Tables 3.3.2-11-IP2 and 3.3.2-11-IP3.

Threaded connections—As stated in LRA Section 2.0 Page 2.0-1, the component type "piping" may include pipe, pipe fittings (such as elbows and reducers), flow elements, orifices, and thermowells. Threaded connections are considered pipe fittings and are included in LRA Tables 2.3.3-11-IP2 and 2.3.3-11-IP3 under the component type "piping," with the AMR results provided in LRA Tables 3.3.2-11-IP2 and 3.3.2-11-IP3.

Restricting orifices—As stated in LRA Section 2.0 Page 2.0-1, the component type "piping" may include pipe, pipe fittings (such as elbows and reducers), flow elements, orifices, and thermowells. Restricting orifices in the fire protection water systems are included in the "piping" line item in LRA Tables 2.3.3-11-IP2 and 2.3.3-11-IP3, with the AMR results provided in LRA Tables 3.3.2-11-IP2 and 3.3.2-11-IP3.

Interface flanges—As stated in LRA Section 2.0 Page 2.0-1, the component type "piping" may include pipe, pipe fittings (such as elbows and reducers), flow elements, orifices, and thermowells. Interface flanges are subject to an AMR and included in LRA Tables 2.3.3-11-IP2 and 2.3.3-11-IP3 under the component type "piping," with the AMR results provided in LRA Tables 3.3.2-11-IP2 and 3.3.2-11-IP3.

Chamber housings—Deluge valves for IP2 and IP3 include a retard chamber, piping, and valves whose purposes are to prevent false alarms due to system pressure surges and to provide a flow path to the water gong alarm during system actuation. Since failure of these components of the deluge valve would not prevent fire suppression capability for the sprinkler system, they are not subject to an AMR.

Heat-actuated devices—Heat actuated devices are active components not subject to an AMR.

Tank heaters—Tank heaters are active components not subject to an AMR.

Thermowells—Thermowells are included in Tables 2.3.3-11-IP2 and 2.3.3-11-IP3, with the AMR results provided in LRA Tables 3.3.2-11-IP2 and 3.3.2-11-IP3.

Water motor alarms—Water motor alarms are local bells mechanically driven by water flow. Water motor alarms are active components not subject to an AMR.

Expansion joint—Expansion joint is a component type in the fire pump diesel exhaust system and is included in Tables 2.3.3-11-IP2 and 2.3.3-11-IP3, with the AMR results provided in LRA Tables 3.3.2-11-IP2 and 3.3.2-11-IP3.

Filter housing—Filter housing is only associated with IP3 components shown on drawing LRA-9321-40903-0. Filter housing is a component type shown in Table 2.3.3-11-IP3, with the AMR results provided in LRA Table 3.3.2-11-IP3.

Gear box housing—Gear box housings are part of the vendor supplied fire pump diesel engine assembly which is an active component not subject to an AMR.

Heat exchanger (bonnet)—There is no heat exchanger (bonnet) associated with the fire protection systems.

Heat exchanger (shell)—There is no heat exchanger (shell) associated with the fire protection systems.

Heat exchanger (tube)—There is no heat exchanger associated with the fire water systems. The IP3 CO2 system includes a heat exchanger consisting of a coil (tube) in air, which is addressed in LRA Table 2.3.3.12-IP3 as component type heat exchanger (tube), with the AMR results provided in LRA Table 3.3.2-12-IP3.

Heater housing—Heater housings are included in Tables 2.3.3-11-IP2 and 2.3.3 11-IP3, with the AMR results provided in LRA Tables 3.3.2-11-IP2 and 3.3.2-11-IP3.

Diesel driven fire pump engine muffler—The diesel driven fire pump engine muffler is component type "silencer" included in Tables 2.3.3-11-IP2 and 2.3.3-11-IP3, with the AMR results provided in LRA Tables 3.3.2-11-IP2 and 3.3.2-11-IP3.

Orifice—As stated in LRA Section 2.0 Page 2.0-1, the component type "piping" may include pipe, pipe fittings (such as elbows and reducers), flow elements, orifices, and thermowells. Orifices in the fire protection water systems are included in LRA Tables 2.3.3-11-IP2 and 2.3.3-11-IP3 under the component type "piping," with the AMR results provided in LRA Tables 3.3.2-11-IP2 and 3.3.2-11-IP3.

Sight glass—Sight glasses are not a component type in the fire protection systems subject to an AMR.

Strainer housing—Strainer housings are included in Tables 2.3.3.11-IP2 and 2.3.3-11-IP3, with the AMR results provided in LRA Tables 3.3.2-11-IP2 and 3.3.2-11-IP3.

Turbocharger housing—Turbocharger housing is a part of the fire pump diesel engine assembly, which is an active component not subject to an AMR.

Flexible hose—Flexible hoses are replaced at specified intervals and are therefore not subject to an AMR per 10 CFR 54.21(a)(1)(ii).

Latch door pull box—Latch door pull boxes are active electro-mechanical devices not subject to an AMR.

Pneumatic actuators—Pneumatic actuators are active components not subject to an AMR.

Actuator housing—The actuator housing is part of the valve actuator which is an active assembly with no pressure boundary function; therefore, it is not subject to an AMR.

Dikes for oil spill confinement—There are no dikes for oil spill confinement within the scope of license renewal for fire protection.

Buried underground fuel oil tanks for emergency diesel generators—Buried underground Fuel oil tanks for the emergency diesel generators are addressed in LRA Section 2.3.3.13, "Fuel Oil."

Expansion tank—Expansion tank is not a component in the fire water system.

Fire water main loop valves—Fire water main loop valves are included in component type "valve body" and are included in Tables 2.3.3.11-IP2 and 2.3.3.11-IP3, with the AMR results provided in LRA Tables 3.3.2-11-IP2 and 3.3.2-11-IP3.

Post-indicator valves—Post-indicator valves are included in component type "valve body" and are included in Tables 2.3.3.11-IP2 and 2.3.3.11-IP3, with the AMR results provided in LRA Tables 3.3.2-11-IP2 and 3.3.2-11-IP3.

Jacket cooling water keep-warm pump and heater—The jacket cooling water keep-warm pump and heater are parts of the diesel engine assembly, which is an active assembly not subject to an AMR.

Lubricating oil collection system components for each reactor coolant pump—The lubricating oil collection system components for each reactor coolant pump are subject to an AMR and are addressed in LRA Section 2.3.3.12 and Tables 2.3.2-12-IP2 and 2.3.2-12-IP3, with the AMR results provided in LRA Tables 3.3.2-12-IP2 and 3.3.2-12-IP3.

Lubricating oil cooler—The lubricating oil cooler is a part of the fire pump diesel engine assembly, which is an active assembly not subject to an AMR.

Auxiliary lubricating oil makeup tank—The auxiliary lubricating oil makeup tank is not a component in the fire protection systems.

Rocker lubricating oil pump—The rocker lubricating oil pump is a part of the fire pump diesel engine assembly, which is an active component and not subject to an AMR.

Floor drains and curbs for fire-fighting water—Floor drains for fire-fighting water are addressed in LRA Section 2.3.3.18, "Plant Drains" and Tables 2.3.3-18-IP2 and 2.3.3-18-IP3 under component type "piping," with the AMR results provided in LRA Tables 3.3.2-18-IP2 and 3.3.2-18-IP3. Curbs are included in the structural AMR under component types "floor slabs, interior walls and ceilings" (for concrete). They are included in LRA Table 2.4-3, with the AMR results provided in LRA Table 3.5.2-3.

Backflow prevention devices—Backflow prevention devices are addressed in LRA Section 2.3.3.18 and Tables 2.3.3-18-IP2 and 2.3.3-18-IP3 under the component type "valve body," with the AMR results provided in LRA Tables 3.3.2-11-IP2 and 3.3.2-11-IP3.

Flame retardant coating for cables—Flame retardant coatings for cables are subject to an AMR and are included in the category of bulk commodities evaluated in the structural AMR. Flame retardant coatings are a subcomponent of component types "fire barrier penetration seal" and "fire stop." These component types are included in LRA Table 2.4-4, with the AMR results provided in LRA Table 3.5.2-4.

Fire retardant coating for structural steel supporting walls and ceilings—Fire retardant coating for structural steel supporting walls and ceilings are subject to an AMR and are included in the structural AMR as component type "fire proofing." This line item is included in LRA Table 2.4-4, with the AMR results provided in LRA Table 3.5.2-4.

In reviewing its response to the RAI, the staff found that the applicant had addressed and resolved each item in the RAI, as discussed in the following paragraphs. Although the description of the "piping" line item provided in LRA Tables 2.3.3-11-IP2 and 2.3.3-11-IP3 does not list these components specifically, the applicant stated that it considers this line item to include the hose connections, pipe fittings, couplings, threaded connections, restricting orifices, interface flanges, and orifices. LRA Tables 3.3.2-11-IP2 and 3.3.2-11-IP3 provide the AMR results for these components. In addition, the applicant addressed floor drains in LRA Section 2.3.3.18, "Plant Drains," and Tables 2.3.3-18-IP2 and 2.3.3-18-IP3 under component type "piping," with AMR results provided in LRA Tables 3.3.2-18-IP2 and 3.3.2-18-IP3. The structural AMR includes curbs under component type "floor slabs, interior walls and ceiling" (for concrete) in LRA Table 2.4-3, with AMR results provided in LRA Table 3.5.2-3. Further, the applicant considers that some components in LRA Table 2.4-4, with AMR results in LRA Table 3.5.2-4, include certain components identified in the RAI. Specifically, the applicant indicated that (1) hose racks and interior fire hose stations are considered "fire hose reels," (2) pipe supports are considered "piping supports," (3) flame retardant coating for cables is considered a subcomponent of component types "fire barrier penetration seal" and "fire stop," and (4) fire retardant coating for structural steel supporting walls and ceilings is considered "fire proofing." The staff finds this portion of the applicant's response to RAI 2.3A.3.11-2 acceptable because it confirms that the components in question are within the scope of license renewal and subject to an AMR. In addition, the response also directed the staff to the AMR results in the LRA.

In its response, the applicant also confirmed that thermowells and expansion joints are a component type within the fire pump diesel exhaust system; the diesel-driven fire pump engine muffler is included in component type "silencer"; and the fire water main loop valves, post-indicator valves, and backflow prevention devices are included in component type "valve body" in LRA Tables 2.3.3-11-IP2 and 2.3.3-11-IP3, with the AMR results provided in LRA Tables 3.3.2-11-IP2 and 3.3.2-11-IP3. Filter housing is only associated with IP3 components shown on license renewal drawing LRA-9321-40903-0 and included in LRA Table 2.3.3-11-IP3, with the AMR results provided in LRA Table 3.3.2-11-IP3. LRA Section 2.3.3.13, "Fuel Oil," addresses buried underground fuel oil tanks for EDGs. Lubricating oil collection system components for each RCP are addressed in LRA Section 2.3.3.12 and Tables 2.3.3-12-IP2 and 2.3.3-12-IP3, with the AMR results provided in LRA Tables 3.3.2-12-IP2 and 3.3.2-12-IP3. The staff finds this portion of the applicant's response to RAI 2.3A.3.11-2 acceptable because it confirms that the components in question are within the scope of license renewal and subject to an AMR. Furthermore, the response directed the staff to the AMR results in the LRA.

The staff found that the applicant did not include the following components in the line item descriptions in the LRA: (1) heat-actuated devices, (2) tank heaters, (3) water motor alarm, (4) gear box housings, (5) turbocharger housing, (6) latch door pull box, (7) pneumatic actuators, (8) actuator housings, (9) jacket cooling water keep-warm pump and heater, (10) lubricating oil cooler, and (11) rocker lubricating oil pump. Because these components are active, they are not subject to an AMR.

The following components are not part of the fire protection systems (water) in IP2 and IP3: (1) heat exchanger (bonnet), (2) heat exchanger (shell), (3) heat exchanger (tube); (4) sight glass expansion tanks, (5) auxiliary lubricating oil makeup tanks, and (6) dikes for oil spill confinement. Since these components are not used in the fire protection system—water at IP2 and IP3, the staff finds that the applicant appropriately omitted them from the scope of license renewal.

Although they are within the scope of license renewal, flexible hoses are replaced at specified intervals. Therefore, the staff finds that flexible hoses are not subject to an AMR, in accordance with 10 CFR 54.21(a)(1)(ii).

The applicant determined that yard hose houses are not subject to an AMR because their failure will not result in a failure of the fire suppression function of the associated fire hydrant. Similarly, the applicant determined that chamber housings are not subject to an AMR because their failure will not result in a failure of the fire suppression function of the sprinkler system. The yard hose houses and chamber housings are passive, long-lived components that were identified as within the scope of license renewal. Therefore, the staff considers these components to be subject to an AMR, in accordance with 10 CFR 54.21(a)(1). This was identified as Open Item 2.3A.3.11-1.

Based on its review, the staff found the applicant's response to RAI 2.3A.3.11-2 partially acceptable because it resolved the staff's concerns regarding scoping and screening of fire protection system components listed in the RAI, with the exception of (a) yard hose houses and (b) chamber housings.

By the letter January 27, 2009, the applicant stated that yard hose houses for IP-2 are not a building; they are a metal cabinet storage location containing fire hoses and supporting tools (spanner, gated wyes and nozzles). The hose contained therein is subject to periodic inspection, testing and replacement in accordance with NFPA standards. Yard hose houses provide no function that supports 10 CFR 50.48 requirements: therefore, they are not in the scope of license renewal.

Chamber housings are small surge suppression volumes that function to minimize false actuation alarms due to system pressure surges. The chambers receive water from a small bypass line upon actuation of a deluge fire suppression system. When the chamber fills, water flow continues through the chamber to a drain line. Due to the limited amount of water flowing to the chamber housings, neither normal operation nor failure of the chamber housing would prevent satisfactory operation of the fire suppression system. In addition, the chamber housings shown on IP2 drawings are associated with deluge valves that do not perform a function that is credited for compliance with 10 CFR 50.48. The fire suppression systems with chamber housings serve maintenance areas and a file room in the technical support center.

The applicant clarified that yard fire hydrants are housed in small sheds; and chamber housings are small surge suppression volumes that function to minimize false actuation alarms due to system pressure surges. The staff determined that failure of these components, which is a second level support system, need not be considered in determining the SCs within the scope of the license renewal under 10 CFR 54.4(a)(3). The staff concludes that the above components were correctly excluded from the scope of license renewal. The staff's concern identified in Open Item 2.3A.3.11-1 has been resolved. Therefore, Open Item 2.3A.3.11-1 is closed.

2.3A.3.11.3 Conclusion

The staff reviewed the LRA, UFSAR, RAI responses, and drawings to determine whether the applicant failed to identify any SSCs within the scope of license renewal. The staff found no such omissions. In addition, the staff sought to determine whether the applicant failed to identify any components subject to an AMR. On the basis of its review, the staff concludes that the applicant has adequately identified the fire protection - water system components within the scope of license renewal, as required by 10 CFR 54.4(a), and those subject to an AMR, as required by 10 CFR 54.21(a)(1).

2.3A.3.12 IP2 Fire Protection—Carbon Dioxide, Halon, and RCP Oil Collection Systems

2.3A.3.12.1 Summary of Technical Information in the Application

LRA Section 2.3.3.12 describes the fire protection—CO_2, Halon 1301 and RCP oil collection system, which consists of fixed fire suppression systems utilizing Halon 1301 as well as oil leakage collection for the RCPs. IP2 does not have a CO_2 fire suppression system within the scope of license renewal. The Halon 1301 systems consist of gas storage tanks and the necessary piping, valves, and instrumentation. The RCP oil collection system consists of drain pans, collection tanks, and the necessary piping, valves, and instrumentation to collect any leakage of the RCP lube oil system.

A fixed Halon fire suppression system meets 10 CFR 50.48 requirements for the cable spreading room as a total-flooding, manually-actuated system divided into four zones of discharge nozzles. The RCP oil collection system can collect lube oil from all potential pressurized and unpressurized RCP lube oil system leakage sites and drain it to a vented closed tank that can hold the required lube oil system inventory.

The fire protection Halon and RCP oil collection systems have no intended function under 10 CFR 54.4(a)(1). The scoping and screening methodology identified the following RCP oil collection system intended function, in accordance with 10 CFR 54.4(a)(2): "Maintain integrity of nonsafety-related components such that no physical interaction with safety-related components could prevent satisfactory accomplishment of a safety function."

The scoping and screening methodology also identified the following Halon and RCP oil collection systems' intended functions, in accordance with 10 CFR 54.4(a)(3):

- Provide fixed automatic and manual fire suppression to extinguish fires in vital areas of the plant (10 CFR 50.48).

- Provide each RCP an oil collection system that is designed to contain and direct the oil to remote storage containers in the event of an oil leak.

LRA Tables 2.3.3-12-IP2 and 2.3.3-19-11-IP2 identify fire protection—CO_2, Halon 1301 and RCP oil collection system component types within the scope of license renewal and subject to an AMR, as well as their intended functions.

2.3A.3.12.2 Staff Evaluation

The staff reviewed LRA Section 2.3.3.12, UFSAR Section 9.6.2, and license renewal drawings using the evaluation methodology described in SER Section 2.3 and the guidance in SRP-LR Section 2.3.

During its review, the staff evaluated the system functions described in the LRA and UFSAR to verify that the applicant had not omitted from the scope of license renewal any components with intended functions, as required by 10 CFR 54.4(a). The staff then reviewed those components that the applicant identified as within the scope of license renewal to verify that it had not omitted any passive and long-lived components subject to an AMR, in accordance with 10 CFR 54.21(a)(1).

The staff also reviewed NRC fire protection SERs for IP2, dated November 30, 1977; February 3, 1978; January 31, 1979; October 31, 1980; August 22, 1983; March 30, 1984; October 16, 1984; September 16, 1985; November 13, 1985; March 4, 1987; January 12, 1989; and March 26, 1996.

The staff also reviewed IP2 commitments made in response to the requirements of 10 CFR 50.48 (i.e., an approved fire protection program), using its commitment documents associated with BTP APCSB 9.5-1 and Appendix A to BTP APCSB 9.5-1.

During its review of LRA Section 2.3.3.12, the staff identified areas in which additional information was necessary to complete its review of the applicant's scoping and screening results. The applicant responded to the staff's RAIs as discussed below.

In RAI 2.3A.3.12-1, dated October 24, 2007, the staff questioned why LRA Table 2.3.3-12-IP2 and 2.3.3-12-IP3 excluded several types of CO_2 and Halon 1301 fire suppression system components discussed in the fire protection SERs or the UFSAR or both and which are identified in the License renewal drawing as subject to an AMR (i.e., they are highlighted in brown). These components include the following:

- strainer housing
- pipe fittings
- pipe supports
- couplings
- odorizer
- threaded connections
- flexible hose
- latch door pull box
- pneumatic actuators

The staff requested that the applicant verify whether LRA Tables 2.3.3-12-IP2 and 2.3.3-12-IP3 should include the components listed above. If they are excluded from the scope of license renewal and not subject to an AMR, the staff requested that the applicant justify their exclusion.

In its response, dated November 16, 2007, the applicant provided the following:

Strainer housings—Based on a review of LRA drawings D-8775-002-0, D-8775-004-0, D-8775-005-0 and 9321-24403-0, there are no strainer housings in the Halon systems.

Pipe fittings—As stated in LRA Section 2.0 Page 2.0-1, the term "piping" may include pipe, pipe fittings (such as elbows and reducers), flow elements, orifices, and thermowells. Pipe fittings are subject to an AMR and included in LRA Tables 2.3.3-12-IP2 and 2.3.3-12-IP3, with AMR results provided in Tables 3.3.2-12-IP2 and 3.3.2-12-IP3, under the component type "piping."

Pipe supports—Pipe supports are subject to an AMR and are included in the structural AMR as shown in LRA Table 2.4-4, under "component and piping supports."

Couplings—As stated in LRA Section 2.0, Page 2.0-1, the term "piping" may include pipe, pipe fittings (such as elbows and reducers), flow elements, orifices, and thermowells. Couplings are considered to be pipe fittings, subject to an AMR and included in the "piping" line item in LRA Tables 2.3.3-12-IP2 and 2.3.3-12-IP3, with AMR results provided in Tables 3.3.2-12-IP2 and 3.3.2-12-IP3.

Odorizer—As stated in LRA Section 2.0, Page 2.0-1, the term "piping" may include pipe, pipe fittings (such as elbows and reducers), flow elements, orifices, and thermowells. Odorizer housings are subject to an AMR and are included in component type "piping" in LRA Tables 2.3.3-12-IP2 and 2.3.3-12-IP3, with AMR results provided in Tables 3.3.2-12-IP2 and 3.3.2-12-IP3. The internals of the odorizer are active (short-lived components) subcomponents and not subject to an AMR.

Threaded connections—As stated in LRA Section 2.0 Page 2.0-1, the term "piping" may include pipe, pipe fittings (such as elbows and reducers), flow elements, orifices, and thermowells. Threaded connections are pipe fittings subject to an AMR and included in the "piping" line item in LRA Tables 2.3.3-12-IP2 and 2.3.3-12-IP3, with AMR results provided in Tables 3.3.2-12-IP2 and 3.3.2-12-IP3.

Flexible hose—There are no flexible hoses utilized in the in-scope Halon systems. LRA drawing D-8775-005-0 is based on a vendor drawing that indicates flex hoses at the gas cylinders. Flexible hoses are not used in the IP2 and IP3 configuration. Flexible hoses are utilized in the RCP oil collection system for IP2 and IP3 as indicated in Tables 2.3.3-12-IP2 and 2.3.3-12-IP3, with AMR results provided in Tables 3.3.2-12-IP2 and 3.3.2-12-IP3. These hoses are stainless steel hoses that are not replaced on a specified frequency.

Latch door pull box—Latch door pull boxes are active electro-mechanical devices and not subject to an AMR.

Pneumatic actuators—Pneumatic actuators (in the form of gas operated pilot valves) are utilized in the in-scope Halon 1301 systems. Actuation is by means of active electrical devices which actuate pilot valves utilizing gas pressure as the motive force. The pilot valves and process valves are included under the component type "valve body" and are subject to an AMR.

Based on its review, the staff finds the applicant's response to RAI 2.3A.3.12-1 acceptable. Although the description of the "piping" line item provided in LRA Tables 2.3.3-11-IP2 and 2.3.3-11-IP3 does not list these components specifically, the applicant stated that it considers pipe fittings, pipe supports, couplings, odorizer, and threaded connections to be included in LRA Tables 2.3.3-12-IP2 and 2.3.3-12-IP3 under the component type "piping" with the AMR results provided in LRA Tables 3.3.2-12-IP2 and 3.3.2-12-IP3.

The applicant has included pneumatic actuators in LRA Tables 2.3.3-12-IP2 and 2.3.3-12-IP3 under the component type "valve body," with the AMR results provided in LRA Tables 3.3.2-12-IP2 and 3.3.2-12-IP3. In similar license renewal reviews, components excluded from the list of components subject to an AMR and from the associated definition of a line item term, such as the "piping" line item, are often modified to include components that were not previously named, either in the component list or in the definition, for completeness. Because the applicant considers the line items specified to include these components, the staff finds that these components have been appropriately included within the scope of license renewal and identified as being subject to an AMR.

Further, the applicant noted that some components in LRA Table 2.4-4 are presented in LRA Table 3.5.2-4. Specifically, the applicant indicated that (1) hose racks and interior fire hose stations are considered "fire hose reels," (2) pipe supports are considered "piping supports,"

(3) flame retardant coating is considered a subcomponent of "fire barrier penetration seal" and "fire stop," and (4) fire retardant coating for structural steel supporting walls and ceilings is considered "fire proofing."

Also, the applicant confirmed that the Halon 1301 systems do not utilize flexible hoses, and these systems do not include strainer housings.

The staff found that the line item descriptions in the LRA do not include latch door pull boxes. The staff accepts the applicant's explanation that latch door pull boxes are active components and, therefore, not subject to an AMR. The staff's concern described in RAI 2.3A.3.12-1 is resolved.

2.3A.3.12.3 Conclusion

The staff reviewed the LRA, UFSAR, RAI responses, and drawings to determine whether the applicant failed to identify any SSCs within the scope of license renewal. The staff found no such omissions. In addition, the staff sought to determine whether the applicant failed to identify any components subject to an AMR. The staff found no such omissions. On the basis of its review, the staff concludes that the applicant has adequately identified the fire protection Halon 1301, and RCP oil collection systems components within the scope of license renewal, as required by 10 CFR 54.4(a), and those subject to an AMR, as required by 10 CFR 54.21(a)(1).

2.3A.3.13 IP2 Fuel Oil Systems

2.3A.3.13.1 Summary of Technical Information in the Application

LRA Section 2.3.3.13 describes the fuel oil systems for IP2 and IP3 EDGs, the IP2 security diesel generator, IP2 and IP3 Appendix R diesel generators, and IP2 and IP3 fire protection diesel-driven fire pumps.

The IP2 fuel oil system code (i.e., the applicant's code for designating systems and boundaries) which includes the 1-million-gallon IP1 fuel oil tank and many of its associated components, but not the safety-related EDG fuel oil components, and has no safety-related components. The fuel oil system includes components that supply the bulk fuel oil to site components, including the house heating boiler and the bulk fuel oil supply to IP3. The IP1 fuel oil tank and its piping are not required to support fire diesel or EDG operation. These components have separate fuel oil tanks.

The fuel oil section includes the gas turbine system description because the only intended function of the gas turbine system for license renewal is performed by its fuel oil subsystem. The fuel supply for gas turbines in the IP2 gas turbine system supplements fuel oil storage for the IP2 and IP3 EDGs. This shared fuel storage consists of two onsite 30,000-gallon fuel oil tanks and a 200,000-gallon storage tank at the Buchanan Substation site. A 29,000-gallon minimum from these storage tanks is dedicated for EDG use. The tanks are not connected directly to the EDG fuel oil storage tanks, but trucking facilities can transfer oil within 1 day's notice.

Each diesel fuel oil storage and transfer system supplying fuel to the EDGs has its own fuel oil day tank, as well as an underground storage tank. The day tanks are within the diesel generator buildings. An engine-driven fuel oil pump feeds the fuel from the day tank to supply the engine. The day tank fills automatically during engine operation from its dedicated underground storage tank adjacent to the diesel generator building. Each underground storage tank has a motor-driven pump to transfer fuel to the day tank.

Independent diesel fuel oil storage and transfer systems supply fuel to the IP2 and IP3 fire protection diesel engines. The IP2 fuel oil storage tank, pump, and components are in the IP2 diesel fire pump house.

An independent diesel fuel oil storage and transfer system supplies fuel to the IP2 security diesel generator, which has its own fuel oil day tank within the security access building diesel generator room as well as an independent underground storage tank adjacent to that building.

An independent diesel fuel oil storage and transfer system supplies fuel to the IP2 SBO/Appendix R diesel generator from the gas turbine fuel oil storage tanks and transfer pumps in the oil room. The SBO/Appendix R diesel generator has its own day tank, which supplies fuel to the engine. The day tank fills automatically during engine operation from the storage tanks by the transfer pumps.

The fuel oil system and subsystems contain safety-related components relied on to remain functional during and following DBEs. They also contain nonsafety-related components whose failure potentially could prevent the satisfactory accomplishment of a safety-related function. In addition, the fuel oil system and subsystems perform functions that support fire protection and SBO.

LRA Tables 2.3.3-13-IP2 and 2.3.3-19-10-IP2 identify fuel oil system and subsystems component types within the scope of license renewal and subject to an AMR as well as their intended functions.

2.3A.3.13.2 Staff Evaluation

The staff reviewed LRA Section 2.3.3.13, UFSAR Sections 8.1, 8.2, and 8.2.3, and license renewal drawings using the evaluation methodology described in SER Section 2.3 and the guidance in SRP-LR Section 2.3.

During its review, the staff evaluated the system functions described in the LRA and UFSAR to verify that the applicant had not omitted from the scope of license renewal any components with intended functions, as required by 10 CFR 54.4(a). The staff then reviewed those components that the applicant identified as within the scope of license renewal to verify that it had not omitted any passive and long-lived components subject to an AMR, in accordance with 10 CFR 54.21(a)(1).

During its review of LRA Section 2.3.3.13, the staff identified an area in which additional information was necessary to complete the review of the applicant's scoping and screening results. For discussion and disposition of RAI 2.3A.3.13-1, dated February 13, 2008, see SER Section 2.3A.3 in the discussion of "Staff's RAIs."

2.3A.3.13.3 Conclusion

The staff reviewed the LRA, UFSAR, RAI response, and drawings to determine whether the applicant failed to identify any SSCs within the scope of license renewal. The staff found no such omissions. In addition, the staff sought to determine whether the applicant failed to identify any components subject to an AMR. The staff found no such omissions. On the basis of its review, the staff concludes that the applicant has appropriately identified the fuel oil system components within the scope of license renewal, as required by 10 CFR 54.4(a), and those subject to an AMR, as required by 10 CFR 54.21(a)(1).

2.3A.3.14 IP2 Emergency Diesel Generator System

2.3A.3.14.1 Summary of Technical Information in the Application

LRA Section 2.3.3.14 describes the EDG system, which supplies emergency shutdown power upon loss of all other alternating current auxiliary power and consists of three EDG sets, each with a diesel engine coupled to a 480-V generator. Each emergency diesel includes two redundant air motors for automatic starting, an air storage tank and compressor system, its own starting air subsystem, fuel oil subsystem, intake air subsystem, exhaust subsystem, lube oil subsystem, and jacket water cooling subsystem. The EDG system also includes ventilation equipment for the diesel generator building.

The EDG system contains safety-related components relied on to remain functional during and following DBEs. It also contains nonsafety-related components whose failure potentially could prevent the satisfactory accomplishment of a safety-related function. In addition, the EDG system performs functions that support fire protection.

Some of the valves in this system are parts of the SW system pressure boundary reviewed with the SW system (LRA Section 2.3.3.2). The fuel oil subsystem components are reviewed with fuel oil (LRA Section 2.3.3.13). A small number of components are reviewed with the city water system (LRA Section 2.3.3.17).

LRA Tables 2.3.3-14-IP2 and 2.3.3-19-9-IP2 identify EDG system component types within the scope of license renewal and subject to an AMR, as well as their intended functions.

2.3A.3.14.2 Staff Evaluation

The staff reviewed LRA Section 2.3.3.14, UFSAR Section 8.2.3, and license renewal drawings using the evaluation methodology described in SER Section 2.3 and the guidance in SRP-LR Section 2.3.

During its review, the staff evaluated the system functions described in the LRA and UFSAR to verify that the applicant had not omitted from the scope of license renewal any components with intended functions, as required by 10 CFR 54.4(a). The staff then reviewed those components that the applicant identified as within the scope of license renewal to verify that the applicant had not omitted any passive and long-lived components subject to an AMR, in accordance with 10 CFR 54.21(a)(1).

During its review of LRA Section 2.3.3.14, the staff identified areas in which additional information was necessary to complete the review of the applicant's scoping and screening results. The discussion of the staff's RAIs in SER Section 2.3A.3 details the disposition of RAI 2.3A.3.14-2, dated February 13, 2008. The applicant responded to additional staff RAIs as discussed below.

In RAI 2.3A.3.14-1, dated December 7, 2007, the staff noted that a license renewal drawing for the IP2 jacket water to the EDGs identifies that the jacket water pumps for diesel engine Nos. 21, 22, and 23, respectively, are not subject to an AMR, in accordance with 10 CFR 54.21(a), because they are not long-lived components. The staff stated that SRP-LR, Table 2.3-2, "Examples of Mechanical Components Screening and Basis for Disposition," provides examples of passive, long-lived components, such as diesel engine jacket water skid-mounted equipment. To complete its review, the staff requested that the applicant confirm that the jacket water pumps are short-lived components and describe its method for periodic replacement.

In its response, dated January 4, 2008, the applicant stated that the IP2 EDG maintenance procedures specify that the jacket water pumps in question are scheduled for replacement every 16 years, in accordance with station maintenance procedures. Therefore, they are not subject to an AMR.

Based on its review, the staff finds the applicant's response to RAI 2.3A.3.14-1 acceptable because it adequately explained that the practice of replacing the jacket water pumps meets the intent of 10 CFR 54.21(a)(1)(ii) for short-lived components and that the maintenance procedures ensure the pumps' periodic replacement. Therefore, the staff agrees that the jacket water pumps are not subject to an AMR. The staff's concern described in RAI 2.3A.3.14-1 is resolved.

In RAI 2.3A.3.14-2, dated December 7, 2007, the staff noted that license renewal drawings for the IP2 and IP3 EDG jacket water cooling systems and EDG fuel oil systems identify multiple "flexible conn [connections]" as not being long-lived components; therefore, they are not subject to an AMR. In LRA Section 2.1.2.1.3, "Mechanical System Drawings," the applicant stated, "Flexible elastomer hoses/expansion joints that are periodically replaced (not long-lived) and therefore not subject to aging management review are indicated as such on the drawings." Screening guidance provided in Table 2.1-3 of the SRP-LR describes items considered to be consumables as short lived and subject to periodic replacement. The staff requested that the applicant describe the programs that manage the replacement activities for these flexible connections.

In its response, dated January 4, 2008, the applicant stated that EDG maintenance procedures specify that the flexible connections in the EDG jacket water and fuel oil systems are components that are periodically replaced. The applicant further explained that, in accordance with 10 CFR 54.21(a)(1)(ii), components that are subject to periodic replacement based on a specified time period are not subject to an AMR.

Based on its review, the staff finds the applicant's response to RAI 2.3A.3.14-2 acceptable because it adequately explained that flexible connections are periodically replaced, as directed by EDG maintenance procedures. Therefore, these connections are not subject to an AMR, in accordance with 10 CFR 54.21(a)(1)(ii). The staff agrees that the flexible connections designated as not long lived are not subject to an AMR. The staff's concern described in RAI 2.3A.3.14-2 is resolved.

In RAI 2.3.0-2, dated February 13, 2008, the staff noted that the license renewal drawings for the IP2 and IP3 EDG jacket water cooling systems have components marked "Not a Long Lived Component." The staff noted that SRP-LR Section 2.1.3.2.2 describes long-lived SCs as those that are not subject to periodic replacement based on a qualified life or specified time period. Furthermore, the LRA states that replacement programs may be based on vendor recommendations, plant experience, or any means that establish a specific replacement frequency under a controlled program.

Previous LRAs typically have not designated pumps, motors, and heat exchangers as not long lived. Therefore, the staff requested that the applicant do the following:

(a)   Identify the component types serviced by the CCW that are shown as "Not a Long Lived Component."

(b)   Provide a basis for designating these components as not long lived, including details as to how the "qualified life" of the components was established, a description of the program under which aging management activities for the components are performed, and any available plant-specific operating experience confirming the effectiveness of management activities.

In its response, dated March 12, 2008, the applicant identified the components designated as not long lived in the EDG jacket cooling water as flexible connections and pump casings. The applicant explained that these components are replaced on an established frequency, in accordance with vendor recommendations. The applicant stated that the plant-specific operating experience did not identify any instances of EDG jacket cooling water flexible connection or pump failures, thus confirming the effectiveness of the replacement activities.

Based on its review, the staff finds the applicant's response to RAI 2.3.0-2 for the EDG system acceptable because it adequately provides the basis for the applicant's designation of the EDG jacket cooling water flexible connections and pump casings as short-lived components, in accordance with the guidance found in SRP-LR Section 2.1.3.2.2. The staff's concern described in RAI 2.3.0-2 is resolved.

2.3A.3.14.3 Conclusion

The staff reviewed the LRA, UFSAR, RAI responses, and drawings to determine whether the applicant failed to identify any SSCs within the scope of license renewal. The staff found no such omissions. In addition, the staff sought to determine whether the applicant failed to identify any components subject to an AMR. The staff found no such omissions. On the basis of its review, the staff concludes that the applicant has appropriately identified the EDG system components within the scope of license renewal, as required by 10 CFR 54.4(a), and those subject to an AMR, as required by 10 CFR 54.21(a)(1).

2.3A.3.15 IP2 Security Generator System

2.3A.3.15.1 Summary of Technical Information in the Application

LRA Section 2.3.3.15 describes the security system, which provides plant security equipment, most of which is not mechanical. The security diesel generates back-up electrical power to security equipment, including lighting of the operator access and egress routes for Appendix R events.

The security system performs functions that support fire protection.

The fuel oil subsystem components are reviewed with fuel oil (LRA Section 2.3.3.13).

LRA Table 2.3.3-15-IP2 identifies security system component types within the scope of license renewal and subject to an AMR, as well as their intended functions.

2.3A.3.15.2 Staff Evaluation

The staff reviewed LRA Section 2.3.3.15 using the evaluation methodology described in SER Section 2.3 and the guidance in SRP-LR Section 2.3.

During its review, the staff evaluated the system functions described in the LRA to verify that the applicant had not omitted from the scope of license renewal any components with intended functions, as required by 10 CFR 54.4(a). The staff then reviewed those components that the applicant identified as within the scope of license renewal to verify that it had not omitted any passive and long-lived components subject to an AMR, in accordance with 10 CFR 54.21(a)(1).

2.3A.3.15.3 Conclusion

The staff reviewed the LRA to determine whether the applicant failed to identify any SSCs within the scope of license renewal. The staff found no such omissions. In addition, the staff sought to determine whether the applicant failed to identify any components subject to an AMR. The staff found no such omissions. On the basis of its review, the staff concludes that the applicant has adequately identified the security generator system components within the scope of license renewal, as required by 10 CFR 54.4(a), and those subject to an AMR, as required by 10 CFR 54.21(a)(1).

2.3A.3.16 IP2 Appendix R Diesel Generator System

2.3A.3.16.1 Summary of Technical Information in the Application

By letter dated April 30, 2008, the applicant amended its LRA to reflect the installation of the IP2 SBO/Appendix R diesel generator.

LRA Section 2.3.3.16, as amended, describes the SBO/Appendix R diesel generator system, which supplies power to selected equipment and power supplies relied on in Appendix R and SBO events. With sufficient power for safe-shutdown loads, the SBO/Appendix R diesel generator is the source of alternate alternating current power for IP2, as required by 10 CFR 50.63. The SBO/Appendix R diesel generator provides power during Appendix R and

SBO events. The IPA for license renewal included the SBO/Appendix R diesel generator within the scope of license renewal.

The SBO/Appendix R diesel is located inside the IP1 turbine building. The SBO/Appendix R diesel generator is designed to operate upon a complete loss of power. The SBO/Appendix R diesel generator includes batteries, a battery charger, jacket water heater and cooler, turbochargers, aftercoolers, aftercooler coolers, jacket water pump, lube oil cooler, lube oil pump, and necessary filters and piping. The SBO/Appendix R diesel generator can supply safe-shutdown loads through the 6.9-kilovolt (kV) distribution system and the emergency 480-V buses and motor control centers or the turbine building switchgear and motor control centers.

The SBO/Appendix R diesel generator system performs functions that support fire protection and SBO.

Fuel oil supply components are evaluated with fuel oil (LRA Section 2.3.3.13). Appendix R diesel generator system ventilation is evaluated with the HVAC systems (LRA Section 2.3.3.8).

LRA Table 2.3.3-16-IP2 identifies SBO/Appendix R diesel generator system component types within the scope of license renewal and subject to an AMR, as well as their intended functions.

2.3A.3.16.2 Staff Evaluation

The staff reviewed LRA Section 2.3.3.16 and license renewal drawings using the evaluation methodology described in SER Section 2.3 and the guidance in SRP-LR Section 2.3.

During its review, the staff evaluated the system functions described in the LRA to verify that the applicant had not omitted from the scope of license renewal any components with intended functions, as required by 10 CFR 54.4(a). The staff then reviewed those components that the applicant identified as within the scope of license renewal to verify that it had not omitted any passive and long-lived components subject to an AMR, in accordance with 10 CFR 54.21(a)(1).

The staff reviewed the amended LRA, changes to the UFSAR, and two license renewal drawings to determine whether the applicant failed to identify any SSCs typically found within the scope of license renewal. SER Sections 3.3.2.1 and 3.3A.2.3.13 document the staff's evaluation of the amended AMR results for the SBO/Appendix R diesel generator system.

2.3A.3.16.3 Conclusion

The staff reviewed the LRA, the UFSAR, the LRA amendment, and drawings to determine whether the applicant failed to identify any SSCs within the scope of license renewal. The staff found no such omissions. In addition, the staff sought to determine whether the applicant failed to identify any components subject to an AMR. The staff found no such omissions. On the basis of its review, the staff concludes that the applicant has adequately identified the SBO/Appendix R diesel generator system components within the scope of license renewal, as required by 10 CFR 54.4(a), and those subject to an AMR, as required by 10 CFR 54.21(a)(1).

2.3A.3.17 IP2 City Water

2.3A.3.17.1 Summary of Technical Information in the Application

LRA Section 2.3.3.17 describes the city water system, which supplies water to various plant components. Originally installed for IP1, the city water system now functions for all three units. The city water system code includes IP1 and IP2 components. Water for the system comes from the Village of Buchanan. Within the boundary of the plant system are the supply piping from the water main, pressure-regulating valves, strainers, water meters, and backflow preventers. After metering, the water flows to a manifold which directs it to either the plant or the 1.5-million-gallon city water storage tank. The plant uses city water to supply fire protection systems, the SBO/Appendix R diesel generator, sanitary and drinking facilities (e.g., emergency showers, eye wash stations, humidifiers, hose connections, sinks, water coolers, water heaters, and lavatories), radiation monitors for purging, and various equipment for makeup or cooling; to supply backup to the AFW pumps; and to serve other emergency purposes. The system is also a CCW backup for bearing and seal water cooling for the charging, safety injection, and RHR pumps.

The city water system contains safety-related components relied on to remain functional during and following DBEs. It also contains nonsafety-related components whose failure potentially could prevent the satisfactory accomplishment of a safety-related function. In addition, the city water system performs functions that support fire protection and SBO.

LRA Section 2.3.4.5 reviews components that support safe shutdown in the event of an auxiliary feed pump room fire.

LRA Tables 2.3.3-17-IP2 and 2.3.3-19-7-IP2 identify city water system component types within the scope of license renewal and subject to an AMR, as well as their intended functions.

2.3A.3.17.2 Staff Evaluation

The staff reviewed LRA Section 2.3.3.17, UFSAR Sections 9.6.3 and 10.2.6.3, and license renewal drawings using the evaluation methodology described in SER Section 2.3 and the guidance in SRP-LR Section 2.3.

During its review, the staff evaluated the system functions described in the LRA and UFSAR to verify that the applicant had not omitted from the scope of license renewal any components with intended functions, as required by 10 CFR 54.4(a). The staff then reviewed those components that the applicant identified as within the scope of license renewal to verify that the applicant had not omitted any passive and long-lived components subject to an AMR, in accordance with 10 CFR 54.21(a)(1).

During its review of LRA Section 2.3.3.17, the staff identified areas in which additional information was necessary to complete the review of the applicant's scoping and screening results. The applicant responded to the staff's RAIs as discussed below.

In RAI 2.3A.3.17-1, dated December 7, 2007, the staff noted that the applicant highlighted a small portion of 2-inch city water line No. 35 on a license renewal drawing in purple, indicating that it is within the scope of license renewal and subject to an AMR for the city water system.

The staff also noted that the identified piping, which is shown in a detail, makes no reference to a continuation drawing, and the detailed area of the drawing references another drawing, which the applicant did not include in the LRA.

The staff questioned whether this piping section contains additional components that should be included within the scope of license renewal. Therefore, the staff requested that the applicant either explain why the LRA did not list the parent drawing for the detailed area under license renewal drawings for the city water system or provide the parent drawing and any other continuation drawings that contain components within the scope of license renewal.

In its response, dated January 4, 2008, the applicant stated that the parent drawing is an equipment general arrangement drawing that includes all of the components shown on the drawing which included the detail. The applicant further stated that no additional components are shown on the parent drawing, and the section of 2-inch city water line No. 35 in the detail is continued on another license renewal drawing that shows the "ghost image" of a valve connecting to the 2-inch city water line No. 35.

Based on its review, the staff finds the applicant's response to RAI 2.3A.3.17-1 acceptable because it adequately explained that the parent drawing is an equipment general arrangement drawing that includes all of the components shown in the detail on the license renewal drawing. The staff finds no need for the applicant to bring the additional components for the city water system within the scope of license renewal. Although not identified on the license renewal drawing, the applicant explained that the continuation of the 2-inch city water line No. 35 appears on another license renewal drawing, which had been provided to the staff with the LRA. The staff's concern described in RAI 2.3A.3.17-1 is resolved.

In RAI 2.3A.3.17-2, dated December 7, 2007, the staff stated that a license renewal drawing shows piping highlighted in purple, indicating that the piping for the city water system is within the scope of license renewal and subject to an AMR.

The staff noted that, at four fire protection system valves on the drawing, the system designation changes from the city water system to the fire protection system. Additionally, the staff noted that, at one other fire protection system valve on the drawing, the system designation changes from the city water system to the AFW system. Although the identified system designation changes, the highlighting remains purple. The staff noted that the components indicated should be subject to an AMR under the scope of the city water system. The staff requested that the applicant explain how the color coding applies to the multiple systems identified above.

In its response, dated January 4, 2008, the applicant stated that the fire protection system, which is highlighted in green, is a high-pressure water system that serves structures and strategically located hydrants. The applicant further explained that the city water system, which is highlighted in purple, is a low-pressure system that provides backup to the high-pressure system and includes the low-pressure hydrants. The applicant explained that components in both systems are used for fire protection and that, when performing scoping and screening of components for license renewal, the applicant included components that are part of the low-pressure city water system flowpath and required to accomplish city water system functions in the city water system, regardless of their component identification or the system designator shown on the drawing. Further, the applicant explained that it included components that are part of the high-pressure fire protection system flowpath and required to accomplish fire protection system functions in the fire protection system, regardless of their component identification or the system designator shown on the drawing. The applicant also stated that the system designators shown on the license renewal drawings do not define system boundaries, thus ensuring that all components required to accomplish system functions are included within the scope of license renewal. The applicant included the fire protection valves as part of the city water system with a pressure boundary intended function because they are fed by the low-pressure city water system and are required to accomplish the city water system functions identified in LRA Section 2.3.3.17.

Based on its review, the staff finds the applicant's response to RAI 2.3A.3.17-2 acceptable because it adequately explained that system designators shown on the license renewal drawings do not define system boundaries. The applicant included the fire protection valves as part of the city water system with a pressure boundary intended function because they are fed by the low-pressure city water system and are needed to accomplish the city water system functions related to fire protection. The staff's concern described in RAI 2.3A.3.17-2 is resolved.

In RAI 2.3A.3.17-3, dated December 7, 2007, the staff stated that, in the upper left corner of a license renewal drawing for the city water system, two 6-inch pipe lines are shown with a fire protection designation highlighted in purple, indicating that they are within the scope of license renewal and subject to an AMR. The staff requested that the applicant explain why the two fire protection lines are highlighted in purple (indicating that they are part of the city water system for license renewal) instead of green (indicating that they are part of the fire protection—water system).

In its response, dated January 4, 2008, the applicant stated that the fire protection system, which is highlighted in green, is a high-pressure water system that serves structures and strategically located hydrants. The city water system, which is highlighted in purple, is a low-pressure system that provides backup to the high-pressure system and includes the low-pressure hydrants. The applicant further explained that it included components that are part of the high-pressure fire protection system flowpath and required to accomplish fire protection system functions in the fire protection system, regardless of their component identification or the system designator shown on the drawing. The applicant also explained that the system designators shown on the license renewal drawings do not define system boundaries.

Based on its review, the staff finds the applicant's response to RAI 2.3A.3.17-3 acceptable because it adequately explained that the fire protection lines are within the scope of license renewal as part of the city water system, and they have a pressure boundary intended function. The fire protection lines are fed by the low-pressure city water system and are needed to accomplish the city water system functions related to fire protection. Therefore, the staff's concern described in RAI 2.3A.3.17-3 is resolved.

In RAI 2.3A.3.17-4, dated December 7, 2007, the staff stated that a license renewal drawing shows a short piece of piping for the city water system highlighted in purple, indicating that it is within the scope of license renewal and subject to an AMR. The staff noted that this short piece of city water system piping refers to two drawings for upstream piping. Since this short piece of city water system piping is within the scope of license renewal and continues onto the two upstream drawings, these two drawings should also have city water system piping within the scope of license renewal. However, the staff noted that the applicant did not list these two drawings in the LRA as license renewal drawings for the IP2 and IP3 city water system. The staff requested that the applicant explain why it had not listed the two referenced drawings in the LRA under license renewal drawings for the city water system.

In its response, dated January 4, 2008, the applicant stated that the two referenced drawings are not system flow diagrams, but equipment general arrangement drawings, which were not clear enough to use as license renewal drawings. The applicant further explained that it reviewed these two drawings to confirm that all components shown on them that are required to accomplish city water system functions were included within the scope of license renewal and subject to an AMR. The applicant concluded that the only components shown on these drawings are piping and 11 valves, all of which are within the scope of license renewal and subject to an AMR.

Based on its review, the staff finds the applicant's response to RAI 2.3A.3.17-4 acceptable because it adequately explained that the two drawings are equipment general arrangement drawings for which a review was performed to confirm that all components shown are required for city water system functions and were included within the scope of license renewal and subject to an AMR. The staff understands that the only components shown on these two drawings are piping and 11 valves, which are already within the scope of license renewal and subject to an AMR. The staff's concern described in RAI 2.3A.3.17-4 is resolved.

In RAI 2.3A.3.17-5, dated December 7, 2007, the staff noted that, in the LRA, the applicant stated that the IP2 city water system has the intended function under 10 CFR 54.4(a)(3) of providing a supply of water to fire protection system components, including the fire pumps, fire hydrants, hose reel stations inside containment, fire water tank, and various sprinkler and deluge systems. The staff also noted that a license renewal drawing shows piping for the city water system highlighted in blue, indicating that it is within the scope of license renewal and subject to an AMR. The piping continues onto three additional drawings for downstream piping, which are not listed in the LRA. The staff noted that this additional piping and associated components are necessary for the city water system to accomplish its intended function to supply water from the IP2 city water system to the hose reel stations inside containment. The staff was concerned that the additional drawings might show city water system components that were not identified in the LRA. The staff requested that the applicant provide the three drawings and any other drawings, as necessary, showing the LRA scope of the IP2 city water system.

In its response, dated January 4, 2008, the applicant stated that the three drawings referenced are not system flow diagrams, but equipment general arrangement drawings, which were not clear enough to use as license renewal drawings. The applicant explained that it reviewed these three drawings to confirm that all of the components shown on them that are required to accomplish city water system functions were included within scope of license renewal and subject to an AMR. The only components shown on these drawings are piping and 19 valves, which are within the scope of license renewal and subject to an AMR.

Based on its review, the staff finds the applicant's response to RAI 2.3A.3.17-5 acceptable because it adequately explained that the three drawings are equipment general arrangement drawings for which a review was performed to confirm that all the components shown on them that are required for city water system functions were included within the scope of license renewal and subject to an AMR. The staff understands that the only components shown on these three drawings are piping and 19 valves, which are already within the scope of license renewal and subject to an AMR. The staff's concern described in RAI 2.3A.3.17-5 is resolved.

In RAI 2.3A.3.17-6, dated December 7, 2007, the staff stated that a license renewal drawing for the city water system showed a fire hydrant highlighted in purple, indicating that it is within the scope of license renewal and subject to an AMR because it supports an intended function, in accordance with 10 CFR 54.4(a). The staff noted that the LRA component table for the city water system does not include the component type "hydrant." The staff stated that 10 CFR 54.21(a)(1) requires that the applicant identify and list those components subject to an AMR. The staff requested that the applicant identify where it evaluated the hydrants in the IP2 city water system for aging management.

In its response, dated January 4, 2008, the applicant stated that the site component database identifies the hydrants in the IP2 city water system as valves, and this designation was maintained during the AMR process. The applicant further explained that it included the hydrants in component type "valve body" in LRA Table 2.3.3-17-IP2, with AMR results provided in LRA Table 3.3.2-17-IP2.

Based on its review, the staff finds the response to RAI 2.3A.3.17-6 acceptable because it adequately explained that the applicant's site component database identifies the hydrants in the IP2 city water system as valves. The staff noted that the applicant has included hydrants in the component type "valve body" in the LRA component table and AMR results table for the IP2 city water system. The staff's concern described in RAI 2.3A.3.17-6 is resolved.

2.3A.3.17.3 Conclusion

The staff reviewed the LRA, UFSAR, RAI responses, and drawings to determine whether the applicant failed to identify any SSCs within the scope of license renewal. The staff found no such omissions. In addition, the staff sought to determine whether the applicant failed to identify any components subject to an AMR. The staff found no such omissions. On the basis of its review, the staff concludes that the applicant has appropriately identified the city water system components within the scope of license renewal, as required by 10 CFR 54.4(a), and those subject to an AMR, as required by 10 CFR 54.21(a)(1).

2.3A.3.18 IP2 Plant Drains

2.3A.3.18.1 Summary of Technical Information in the Application

LRA Section 2.3.3.18 describes the plant drains, which are passive fire protection features required for adequate protection of safety-related equipment from water damage in areas with fixed-suppression systems. Plant drain components also prevent drain systems in areas with combustible materials from spreading fires into other areas of the plant. Some plant drains protect safety-related equipment from flooding effects.

Various systems include plant drain components, but, for the purposes of this evaluation, all plant drain components are grouped. SRP-LR Section 2.1.3.1 allows the grouping of similar components from various plant systems into a single, consolidated evaluation.

Both the fire protection and waste disposal systems include plant drain components. The waste disposal system collects and processes all potentially radioactive primary plant wastes, both gaseous and liquid, for removal from the site. The system collects, compresses, stores, samples, and releases gaseous waste from the primary and auxiliary systems. Gases vented to the vent header flow to the waste gas compressor suction header. One of the two compressors operates continuously, while the second unit serves as backup for peak load conditions. From the compressors, gas flows to one of the four large gas decay tanks. The header arrangement at the inlet allows the operator to fill the tank, reuse the gas, or discharge it to the environment. Six additional small gas decay tanks can be used during degassing of the reactor coolant before a cold shutdown. The system collects and processes liquid wastes throughout the plant from equipment, radioactive chemical laboratory, decontamination, demineralizer regeneration, and floor drains. Waste liquids drain by gravity to the waste holdup tank, the collection point for liquid wastes, or to the sump tank, the containment, or the PAB sumps, from which they are pumped to the waste holdup tank. The system sends the liquid waste holdup tank contents to the IP1 waste collection system and collects and transfers liquid drained from the RCS directly to the CVCS for processing.

The system includes the vent header, waste gas compressors, large and small waste gas decay tanks, waste gas analyzer, pumps, collection tanks, station drainage piping, floor drains, instruments and controls, piping, valves, several containment penetrations and accompanying isolation components, and piping, valves, instruments and controls to monitor condensation from the containment fan cooler units.

The plant drains system contains safety-related components relied on to remain functional during and following DBEs. The system also contains nonsafety-related components whose failure potentially could prevent the satisfactory accomplishment of a safety-related function. In addition, the plant drains system performs functions that support fire protection.

A small number of waste disposal system components are reviewed with the CS system (LRA Section 2.3.2.2), the safety injection system (LRA Section 2.3.2.4), the city water system (LRA Section 2.3.3.17), the primary water makeup system (LRA Section 2.3.3.7), the CCW systems (LRA Section 2.3.3.3), and the RCS pressure boundary (LRA Section 2.3.1.3).

LRA Tables 2.3.3-18-IP2 and 2.3.3-19-42-IP2 identify plant drains system component types within the scope of license renewal and subject to an AMR, as well as their intended functions.

2.3A.3.18.2 Staff Evaluation

The staff reviewed LRA Section 2.3.3.18, UFSAR Section 11.1, and a license renewal drawing using the evaluation methodology described in SER Section 2.3 and the guidance in SRP-LR Section 2.3.

During its review, the staff evaluated the system functions described in the LRA and UFSAR to verify that the applicant had not omitted from the scope of license renewal any components with intended functions, as required by 10 CFR 54.4(a). The staff then reviewed those components that the applicant identified as within the scope of license renewal to verify that the applicant had not omitted any passive and long-lived components subject to an AMR, in accordance with 10 CFR 54.21(a)(1).

During its review of LRA Section 2.3.3.18, the staff identified an area in which additional information was necessary to complete its review of the applicant's scoping and screening results. The discussion of the staff's RAIs in SER Section 2.3A.3 details the disposition of RAI 2.3A.3.18-1, dated February 13, 2008.

2.3A.3.18.3 Conclusion

The staff reviewed the LRA, UFSAR, RAI response, and a drawing to determine whether the applicant failed to identify any SSCs within the scope of license renewal. The staff found no such omissions. In addition, the staff sought to determine whether the applicant failed to identify any components subject to an AMR. The staff found no such omissions. On the basis of its review, the staff concludes that the applicant has appropriately identified the plant drains system components within the scope of license renewal, as required by 10 CFR 54.4(a), and those subject to an AMR, as required by 10 CFR 54.21(a)(1).

2.3A.3.19 IP2 Miscellaneous Systems in Scope for 10 CFR 54.4(a)(2)

2.3A.3.19.1 Summary of Technical Information in the Application

In LRA Section 2.3.3.19, "Miscellaneous Systems In Scope for (a)(2)," the applicant described those systems that it included within the scope of license renewal with the potential for physical interaction with safety-related components, as required by 10 CFR 54.4(a)(2), and described the components in these systems subject to an AMR. LRA Table 2.3.3-19-A-IP2 lists all of these systems, as well as the LRA section in which the applicant evaluated these systems. LRA Section 2.3.3.19 describes in detail those systems without correlating LRA sections, which include the following:

- boiler blowdown
- chemical feed
- house service boiler
- intake structure system
- ignition oil
- integrated liquid waste handling
- main generator
- main turbine
- miscellaneous
- nuclear service grade makeup
- post-accident containment air sample
- post-accident containment vent (retired in place)
- primary sampling
- radiation monitoring
- secondary sampling
- technical support center diesel
- chlorination (added by applicant by letter dated February 13, 2008)

Also in LRA Section 2.3.3.19, the applicant identified the following IP2 systems that were not reviewed for 10 CFR 54.4(a)(2) for spatial interaction because the applicant included all of the system's passive mechanical components as (a)(1), functional (a)(2), or (a)(3):

- AFW
- containment cooling and filtration
- CCW
- control rod drive
- CS system
- electrical penetrations
- fuel and core component handling
- in-core instrumentation
- isolation valve seal water

The following briefly describes the IP2 systems included within the scope of license renewal based only on the criterion of 10 CFR 54.4(a)(2) and subject to an AMR.

Chemical Feed. The chemical feed system provides the means to add chemicals to secondary water systems for proper water chemistry control. LRA Table 2.3.3-19-3-IP2 identifies the chemical feed system component types within the scope of license renewal and subject to an AMR, as well as their intended functions.

Intake Structure System. The intake structure system provides coarse filtering of the Hudson River water supplied to the CW system and the SW system. The system also includes mechanical components associated with the chlorine and hypochlorite addition subsystems. LRA Table 2.3.3-19-8-IP2 identifies the intake structure system component types within the scope of license renewal and subject to an AMR, as well as their intended functions.

Main Generator. The main generator system produces the primary electrical output of the unit. The system includes the main generator, its supporting auxiliaries, and components in the stator cooling water and hydrogen seal oil systems. LRA Table 2.3.3-19-15-IP2 identifies the main generator system component types within the scope of license renewal and subject to an AMR, as well as their intended functions.

House Service Boiler. The house service boiler system produces steam for plant heating via the auxiliary steam system. The system includes the house service boilers and components in the fuel oil, FW, and condensate collection systems. LRA Table 2.3.3-19-16-IP2 identifies the house service boiler system component types within the scope of license renewal and subject to an AMR, as well as their intended functions.

Ignition Oil. The ignition oil system supplies ignition oil to the house service boilers. Most of the ignition oil components are associated with the house service boiler system. LRA Table 2.3.3-19-20-IP2 identifies the ignition oil system component types within the scope of license renewal and subject to an AMR, as well as their intended functions.

Integrated Liquid Waste Handling. The integrated liquid waste handling system processes liquid waste collected by the waste disposal system. LRA Table 2.3.3-19-21-IP2 identifies the integrated liquid waste handling system component types within the scope of license renewal and subject to an AMR, as well as their intended functions.

Miscellaneous. The applicant created the miscellaneous system for the purpose of license renewal to group together various structural, electrical, and mechanical components that were not described elsewhere. LRA Table 2.3.3-19-24-IP2 identifies the miscellaneous system component types within the scope of license renewal and subject to an AMR, as well as their intended functions.

Nuclear Service Grade Makeup. The nuclear service grade makeup system supplies water to various service systems. The system includes components of the IP1 water treatment facility. LRA Table 2.3.3-19-25-IP2 identifies the nuclear service grade makeup system component types within the scope of license renewal and subject to an AMR, as well as their intended functions.

Post-Accident Containment Air Sample. The post-accident containment air sample system provides a means to monitor hydrogen concentration inside containment following an accident. Based upon a recent license amendment (License Amendment No. 243), hydrogen monitoring is no longer required as a safety function; however, the system contains component in scope of license renewal under 10 CFR 54.4(a)(2). LRA Table 2.3.3-19-26-IP2 identifies the post-accident containment air sample system component types within the scope of license renewal and subject to an AMR as well as their intended functions.

Post-Accident Containment Vent. The post-accident containment vent system backs up the hydrogen recombiner to reduce post-LOCA hydrogen concentration in containment atmosphere. Based upon a recent license amendment (License Amendment No. 243), the hydrogen recombiner is no longer required as a safety function; however, the system contains component in scope of license renewal under 10 CFR 54.4(a)(2). LRA Table 2.3.3-19-27-IP2 identifies post-accident containment vent system component types within the scope of license renewal and subject to an AMR as well as their intended functions.

Primary Sampling. The primary sampling system performs high-radiation sampling and in-line monitoring and laboratory analysis of representative samples under normal or post-accident conditions. Some of the primary sampling system components support and are reviewed with other systems (e.g., RHR (LRA Section 2.3.2.1), safety injection (LRA Section 2.3.2.4), and containment penetrations (LRA Section 2.3.2.5)). LRA Table 2.3.3-19-28-IP2 identifies primary sampling system component types within the scope of license renewal and subject to an AMR, as well as their intended functions.

Radiation Monitoring. The radiation monitoring system warns of any radiation health hazard and any plant malfunction that might cause health hazards or plant damage. Some of the radiation monitoring system components support, and are reviewed with, other systems (e.g., SW system (LRA Section 2.3.3.2) and containment penetrations (LRA Section 2.3.2.5)). LRA Table 2.3.3-19-31-IP2 identifies radiation monitoring system component types within the scope of license renewal and subject to an AMR, as well as their intended functions.

Boiler Blowdown. The boiler blowdown purification system collects and stores or processes blowdown from an SG with a primary-to-secondary leak. LRA Table 2.3.3-19-34-IP2 identifies boiler blowdown system component types within the scope of license renewal and subject to an AMR, as well as their intended functions.

Secondary Sampling. The secondary sampling system continuously samples and analyzes plant secondary systems. The system has components necessary to collect and transport samples to the sampling stations located in the turbine building. LRA Table 2.3.3-19-38-IP2 identifies secondary sampling system component types within the scope of license renewal and subject to an AMR, as well as their intended functions.

Technical Support Center Diesel. The technical support center diesel system backs up the power supply to the technical support center. The technical support center diesel system includes the diesel generator, fuel oil supply, and supporting instruments and controls. LRA Table 2.3.3-19-40-IP2 identifies technical support center diesel system component types within the scope of license renewal and subject to an AMR, as well as their intended functions.

Main Turbine. The main turbine system receives steam from the SGs, converts a portion of the steam thermal energy to electricity from the main generator, and supplies extraction steam for FW heating. LRA Table 2.3.3-19-41-IP2 identifies main turbine system component types within the scope of license renewal and subject to an AMR, as well as their intended functions.

Chlorination (added by applicant by letter dated February 13, 2008). The chlorination system provides sodium hypochlorite to the intake bays to limit microorganism fouling in these bays and in the systems that use raw water. LRA Table 2.3.3-19-44-IP2 identifies chlorination system component types within the scope of license renewal and subject to an AMR, as well as their intended functions. SER Sections 3.3.2.1 and 3.3A.2.3.36 document the staff's review of the AMR results.

2.3A.3.19.2 Staff Evaluation

The staff reviewed LRA Section 2.3.3.19 and the following UFSAR (IP2) or safety analysis report (SAR) (IP1) Sections that were associated with these systems:

• boiler blowdown[1]	UFSAR Section 10.2.1.5
• chemical feed[1]	UFSAR Section 10.2.6.4
• main generator[1]	UFSAR Section 8
• house service boiler[1]	UFSAR Section 9.6.5
• integrated liquid waste handling[1]	SAR Section 3.7.3 and UFSAR Section 11.1.2.1
• main turbine[1]	UFSAR Section 10.2.2
• nuclear service grade makeup[2]	SAR Section 3.7.2
• post-accident containment air sample[2]	UFSAR Section 6.8.2.3
• post-accident containment vent (retired in place)[2]	UFSAR Section 6.8.2.2
• primary sampling[2]	UFSAR Section 9.4
• radiation monitoring[2]	UFSAR Section 11.2.3
• secondary sampling[1]	UFSAR Section 9.4
• miscellaneous[2]	UFSAR Sections 5.1.9, 5.1.11, and 5.2.2
• intake structure system[1]	—
• ignition oil[1]	—
• technical support center diesel[1]	—
• chlorination[1]	—

[1] The staff conducted a simplified Tier 1 system review of these systems, as described in SER Section 2.3.

[2] The staff conducted a detailed Tier 2 system review of these systems, as described in SER Section 2.3.

For those systems receiving a simplified Tier 1 evaluation, the staff reviewed the applicable LRA sections and UFSAR or SAR sections (if applicable) using the evaluation methodology described in SER Section 2.3 and the guidance in SRP-LR Section 2.3. For those systems receiving a detailed Tier 2 evaluation, the staff reviewed the applicable LRA sections, UFSAR or SAR sections (if applicable), and license renewal drawings (system components are shown on other associated system drawings). Based on information provided in the UFSAR or SAR and LRA, the staff evaluated the system functions described in LRA Section 2.3.3.19 to verify that the applicant had not omitted from the scope of license renewal any components with intended functions, as required by 10 CFR 54.4(a). The staff then reviewed those components that the applicant identified as within the scope of license renewal to verify that it had not omitted any passive and long-lived components subject to an AMR, in accordance with 10 CFR 54.21(a)(1).

The staff reviewed the list of IP2 systems the applicant identified in LRA Section 2.3.3.19 as not having any components in scope for 10 CFR 54.4(a)(2) for spatial interaction because they were already included in scope under 10 CFR 54.4(a)(1), functional (a)(2), or (a)(3). In RAI 2.3A.2.2-1, dated February 13, 2008, the staff requested the applicant to explain why piping segments directly attached to IP2 CS system (a)(1) piping were not highlighted on boundary drawings to show included in scope for license renewal. The staff's review of the applicant's response, dated March 12, 2008, is documented in SER Section 2.3A.2.2.2.

In RAI 2.1-1, dated January 14, 2008, the staff asked the applicant to provide a technical basis for excluding nonsafety-related systems located in proximity to safety-related systems from the scope of license renewal. In its response, dated February 13, 2008, the applicant provided an evaluation of these systems and amended the LRA to include the IP2 chlorination system within the scope of license renewal under 10 CFR 54.4(a)(2). Additionally, the applicant added LRA Table 2.3.3-19-44-IP2 to identify the component types subject to an AMR.

During its review, the staff noted the applicant did not specifically identify components within the scope of license renewal under 10 CFR 54.4(a)(2) on the license renewal drawings. To determine that the applicant did not omit any components from scope under 10 CFR 54.4(a)(2), the staff used a sampling approach, as recommended in SRP-LR Section 2.3.3.1. In multiple RAIs, dated February 13, 2008, the staff asked the applicant to verify that various segments of selected systems were included in scope under 10 CFR 54.4(a)(2). This sampling approach enabled the staff to confirm that the applicant had properly implemented its methodology for identifying nonsafety-related portions of systems with a potential to adversely affect safety-related functions, in accordance with 10 CFR 54.4(a)(2).

In its response, dated March 12, 2008, the applicant stated that all components identified by the staff on the license renewal drawings are within the scope of license renewal, in accordance with 10 CFR 54.4(a)(2), and subject to an AMR. Based on a review of the applicant's response, the staff finds that the applicant has adequately identified the components required to be within the scope of license renewal, in accordance with 10 CFR 54.4(a)(2), as well as those subject to an AMR.

2.3A.3.19.3 Conclusion

For each system described above, the staff reviewed LRA Section 2.3.3.19, the applicable UFSAR or SAR section, and license renewal drawings to determine whether the applicant failed to identify any SSCs within the scope of license renewal. In addition, the staff sought to determine whether the applicant failed to identify any components subject to an AMR. The staff found instances in which the applicant omitted systems and components that should have been included within the scope of license renewal. The applicant has satisfactorily resolved these issues as discussed in the preceding staff evaluation. On the basis of its review, the staff finds that, for all of the systems identified in LRA Section 2.3.3.19, the applicant has appropriately identified the components within the scope of license renewal as required by 10 CFR 54.4(a), and those subject to an AMR, as required by 10 CFR 54.21(a)(1).

2.3A.4 Scoping and Screening Results: IP2 Steam and Power Conversion Systems

LRA Section 2.3.4 identifies the steam and power conversion systems' SCs subject to an AMR for license renewal.

The applicant described the supporting SCs of the steam and power conversion systems in the following LRA sections:

- 2.3.4.1, "Main Steam"
- 2.3.4.2, "Main Feedwater"
- 2.3.4.3, "Auxiliary Feedwater"
- 2.3.4.4, "Steam Generator Blowdown"
- 2.3.4.5, "IP2 AFW Pump Room Fire Event"
- 2.3.4.6, "Condensate"

SER Sections 2.3A.4.1–2.3A.4.6, respectively, describe the staff's review of the IP2 systems described in LRA Sections 2.3.4.1–2.3.4.6. The staff's findings for these systems are discussed below.

2.3A.4.1 IP2 Main Steam System

2.3A.4.1.1 Summary of Technical Information in the Application

LRA Section 2.3.4.1 describes the MS system, which conducts steam from the four SGs inside the containment structure to the turbine generator unit in the turbine generator building. The system has four MS pipes, one from each SG to the turbine stop and control valves, connected near the turbine. Each steam pipe has a main steam isolation valve (MSIV) and a non-return valve outside the containment. There are five code safety valves and one PORV on each MS line outside the reactor containment and upstream of the isolation and non-return valves. A flow venturi upstream of the isolation valve measures steam flow. Steam pressure is also measured upstream of the isolation valve. The MS system supplies steam to the main boiler FW pump turbines and the AFW pump turbine. The system includes the main boiler FW pump turbines and the turbine steam bypass and low-pressure steam dump systems, which channel excess steam flow to the condenser. The steam generator blowdown (SGBD) flowpath also includes MS system components.

The MS system contains safety-related components relied on to remain functional during and following DBEs. It also contains nonsafety-related components whose failure potentially could prevent the satisfactory accomplishment of a safety-related function. In addition, the MS system performs functions that support fire protection and SBO.

Main steam components in the SGBD flowpath are reviewed with the SGBD system (LRA Section 2.3.4.4). Components supporting the AFW system are reviewed with the AFW system (LRA Section 2.3.4.3). Components that support safe shutdown in a fire in the auxiliary feed pump room are evaluated in AFW pump room fire event (LRA Section 2.3.4.5). A small number of components are reviewed with the compressed air systems (LRA Section 2.3.3.4).

LRA Tables 2.3.4-1-IP2 and 2.3.3-19-23-IP2 identify MS system component types within the scope of license renewal and subject to an AMR, as well as their intended functions.

2.3A.4.1.2 Staff Evaluation

The staff reviewed LRA Section 2.3.4.1, UFSAR Section 10.2, and license renewal drawings using the evaluation methodology described in SER Section 2.3 and the guidance in SRP-LR Section 2.3.

During its review, the staff evaluated the system functions described in the LRA and UFSAR to verify that the applicant had not omitted from the scope of license renewal any components with intended functions, as required by 10 CFR 54.4(a). The staff then reviewed those components that the applicant identified as within the scope of license renewal to verify that it had not omitted any passive and long-lived components subject to an AMR, in accordance with 10 CFR 54.21(a)(1).

During its review of LRA Section 2.3.4.1, the staff identified an area in which additional information was necessary to complete the review of the applicant's scoping and screening results. The applicant responded to the staff's RAI as discussed below.

In RAI 2.3A.4.1-1, dated December 7, 2007, the staff noted that license renewal drawings for the IP2 MS system show the following valves within the scope of license renewal and subject to an AMR: PCV-1134, PCV-1135, PCV-1136, PCV-1137, MS-1-21, MS-1-22, MS-1-23, MS-1-24, PCV-1120, PCV-1121, PCV-1122, PCV-1123, PCV-1124, PCV-1125, PCV-1126, PCV-1127, PCV-1128, PCV-1129, PCV-1130, and PCV-1131. The staff also noted that these valves are air operated and have associated air cylinders and air tubing that the applicant excluded from the scope of license renewal. Since some of these valves appear to rely on pressurized air (pneumatic operation) to change position and fulfill their intended function, the staff requested that the applicant explain why the instrument air system, its tubing, and associated solenoid-operated valves (SOVs) to the valves in question are not within the scope of license renewal, in accordance with 10 CFR 54.4(a).

In its response, dated January 4, 2008, the applicant stated that the air operators are active components; therefore, they are not subject to an AMR, in accordance with 10 CFR 54.21(a)(1)(i) and NEI 95-10, Appendix B. The applicant explained that the SOVs and air tubing associated with air-operated valves in the MS system are within the scope of license renewal, but are not subject to an AMR because the majority of the air-operated valves shown on the MS license renewal drawings to be within the scope of license renewal fail to their required position for accident mitigation. As such, these valves do not require pressurized air to fulfill their intended function, and pressure boundary of the air tubing is not necessary. The applicant stated that an exception is the atmospheric dump valves and MSIVs, which close on loss of air but are credited with being reopened, as necessary, in an accident scenario using standby nitrogen in bottles or compressed air stored in accumulators. The applicant explained that components used to reopen the MS system valves are subject to an AMR.

Based on its review, the staff finds the applicant's response to RAI 2.3A.4.1-1 acceptable because it explained that, for most of the air-operated valves, a failure of the air supply system will not result in a loss of the intended function because the MS valves fail to their safe positions. This explanation is consistent with Section 5.2.3.1 of NEI 95-10, Revision 6, which governs fail-safe components. For those air-operated valves that rely on an air supply system (i.e., those MS system valves that do not fail to their safe position), the applicant included the passive pneumatic components (accumulator tanks, tubing, and valves) of those air-operated valves within the scope of license renewal, in accordance with 10 CFR 54.4(a); these components are subject to an AMR, in accordance with 10 CFR 54.21(a)(1). The staff's concern described in RAI 2.3A.4.1-1 is resolved.

2.3A.4.1.3 Conclusion

The staff reviewed the LRA, UFSAR, RAI response, and drawings to determine whether the applicant failed to identify any SSCs within the scope of license renewal. The staff found no such omissions. In addition, the staff sought to determine whether the applicant failed to identify any components subject to an AMR. The staff found no such omissions. On the basis of its review, the staff concludes that the applicant has appropriately identified the MS system components within the scope of license renewal, as required by 10 CFR 54.4(a), and those subject to an AMR, as required by 10 CFR 54.21(a)(1).

2.3A.4.2 IP2 Main Feedwater System

2.3A.4.2.1 Summary of Technical Information in the Application

LRA Section 2.3.4.2 describes the main FW system, which has two half-size, steam-driven main FW pumps that increase condensate pressure for delivery through the final stage of FW heating and the FW regulating valves to the SGs. The FW system includes the high-pressure FW heaters, the SGs, the piping and valves from the outlet of the main feed pumps through the heaters to the SGs, and the main feed pump turbine drip tank drain pumps. The main feed pumps are part of the condensate system, and the main feed pump turbines are part of the MS system.

The main FW system contains safety-related components relied on to remain functional during and following DBEs. The failure of nonsafety-related SSCs in the main FW system could prevent the satisfactory accomplishment of a safety-related function. In addition, the main FW system performs functions that support fire protection.

The SGs and secondary-side instrumentation piping and valves are reviewed with the SGs (LRA Section 2.3.1.4). Components that support safe shutdown in the auxiliary feed pump room fire are evaluated in LRA Section 2.3.4.5. System components containing air are reviewed with the compressed air systems (LRA Section 2.3.3.4).

LRA Tables 2.3.4-2-IP2 and 2.3.3-19-12-IP2 identify main FW system component types within the scope of license renewal and subject to an AMR, as well as their intended functions.

2.3A.4.2.2 Staff Evaluation

The staff reviewed LRA Section 2.3.4.2, UFSAR Section 10.2.6, and a license renewal drawing using the evaluation methodology described in SER Section 2.3 and the guidance in SRP-LR Section 2.3.

The staff evaluated the system functions described in the LRA and UFSAR to verify that the applicant had not omitted from the scope of license renewal any components with intended functions, as required by 10 CFR 54.4(a). The staff then reviewed those components that the applicant identified as within the scope of license renewal to verify that the applicant had not omitted any passive and long-lived components subject to an AMR, in accordance with 10 CFR 54.21(a)(1).

During its review of LRA Section 2.3.4.2, the staff identified an area in which additional information was necessary to complete the review of the applicant's scoping and screening results. The applicant responded to the staff's RAI as discussed below.

In RAI 2.3A.4.2-1, dated December 7, 2007, the staff noted that license renewal drawings identify that valves FCV-417-L, FCV-417, FCV-427-L, FCV-427, FCV-437-L, FCV-437, FCV-447-L, FCV-447, BF2-21, and BF2-22 for the IP2 main FW system are within the system evaluation boundary but are not highlighted, indicating that they are not subject to an AMR. The staff asked the applicant to explain the valves' exclusion from an AMR.

In its response, dated January 4, 2008, the applicant explained that these FW system valves are located upstream of the containment isolation check valves in nonsafety-related piping but are classified as safety-related because of their active function to provide FW isolation. The applicant also stated that these valves "have no passive intended function for 54.4(a)(1) or (a)(3) because their failure would accomplish the safety function of isolating feedwater flow to the SGs." The applicant further stated that these valves perform their function with moving parts; therefore, in accordance with 10 CFR 54.21(a)(1)(i), they are not subject to an AMR and are not highlighted on the license renewal drawing. However, the applicant did indicate that the valves in question are within the scope of license renewal for meeting the requirements of 10 CFR 54.4(a)(2) because of their potential for spatial interaction with safety-related equipment and are, therefore, subject to an AMR.

The staff did not agree with the applicant's rationale that the valves do not have a passive intended function in accordance with 10 CFR 54.4(a)(1). The staff discussed the applicant's view during a telephone call on March 7, 2008. The applicant subsequently amended its RAI response by letter dated March 24, 2008, and reiterated that the FW system valves are safety-related; however, although not highlighted, the applicant stated that these valves and the remainder of the FW system components on the associated license renewal drawing are in scope and subject to an AMR based upon meeting the requirements of 10 CFR 54.4(a)(2) because of their potential for spatial interaction with safety-related equipment.

Based on its review, the staff finds the applicant's amended response to RAI 2.3B.4.2-1 acceptable because the applicant confirmed that the valves in question are within the scope of license renewal pursuant to 10 CFR 54.4(a) and subject to an AMR pursuant to 10 CFR 54.21(a)(1). Although the staff does not agree with the applicant's basis for determining how the valve bodies are subject to an AMR, the staff's concern is resolved because the AMR was performed, and the AMR results were provided in LRA Table 3.3.2-19-12-IP2. The staff's concern described in RAI 2.3A.4.2-1 is resolved.

In RAI 2.3A.4.2-2, dated December 30, 2007, the staff noted that UFSAR Section 14.1.10, Excessive Heat Removal Due To Feedwater System Malfunctions, states that accidental full opening of a feedwater control valve causes excessive feedwater flow, resulting in a transient is similar to, but less severe than, the hypothetical steamline break transient described in UFSAR Section 14.2.5. Therefore, the excessive feedwater flow failure is bounded by the steamline break analysis. In the steamline break analysis, in the event of the failure of the main feedwater control valve, the applicant takes credit the main feedwater stop valves, BFD-5's, to close within 120 seconds. In its revised response to RAI 2.3A.4.2-1, dated March 24, 2008, the applicant stated that the feedwater control valves and the remainder of the feedwater system components on the associated license renewal drawing are within scope of license renewal based upon meeting the requirements of 10 CFR 54.4(a)(2), having the potential for spatial interaction with safety-related equipment, and are subject to an AMR.

Based the applicant's UFSAR, the main feedwater stop valves (BFD-5's) have an intended function that meets the criteria of 10 CFR 54.4(a)(1); however, these valves are neither included within the "system intended function boundary," nor are they highlighted on the license renewal drawings for having a intended function in accordance with 10 CFR 54.4(a)(1). By letter dated December 30, 2008, the staff requested the applicant to justify the exclusion of the main feedwater stop valves (BFD-5's), from scope of license renewal in accordance with 10 CFR 54.4(a)(1). This issue was also identified as Open Item 2.3.4.2-1.

By letter dated January 27, 2009, the applicant stated that based upon a review of the qualifications of the main feedwater stop valves, the BFD-5's are classified as nonsafety-related. Further, the applicant explained that the valves are classified nonsafety-related in the site component database and are located outside the Class I boundary [as corrected by letter dated March 13, 2009] on license renewal drawing LRA-9321-2019-0. As indicated in the IP2 UFSAR, these valves provide a backup isolation function for feedwater in the event of such accidents as a feedwater or steamline break. Credit for nonsafety-related components as a backup to safety-related components in mitigating breaks in seismically-qualified steam line piping is consistent with regulatory guidance provided in Section 15.1.5, "Steam System Piping Failures Inside and Outside of Containment (PWR)," of the Standard Review Plan (NUREG-0800) and is also consistent with the allowance for feedwater regulating and bypass valves to be nonsafety-related, as discussed in NUREG-0138, "Staff Discussion of Fifteen Technical Issues Listed in Attachment to November 3, 1976 Memorandum from Director, NRR to NRR Staff." The applicant concluded that, consistent with the CLB, regulatory guidance, and NUREG-0138, the BFD-5 valves are classified as nonsafety-related, and as such, meet the criteria to be included in scope for license renewal under 10 CFR 54.4(a)(2).

Based on the information provided by the applicant, the staff finds applicant's response to RAI 2.3A.4.2-2 acceptable because the BFD-5 isolation valves are nonsafety-related components, and the valves are included in the scope for license renewal under 10 CFR 54.4(a)(2). Therefore, the staff's concern described in RAI 2.3A.4.2-2 is resolved. As a result, Open Item 2.3.4.2-1 is closed.

2.3A.4.2.3 Conclusion

The staff reviewed the LRA, UFSAR, RAI responses, and a drawing to determine whether the applicant failed to identify any SSCs within the scope of license renewal. The staff found no such omissions. In addition, the staff sought to determine whether the applicant failed to identify any components subject to an AMR. The staff found no such omissions. The staff concludes that the applicant has appropriately identified the main FW system components within the scope of license renewal, as required by 10 CFR 54.4(a), and those subject to an AMR, as required by 10 CFR 54.21(a)(1).

2.3A.4.3 IP2 Auxiliary Feedwater System

2.3A.4.3.1 Summary of Technical Information in the Application

LRA Section 2.3.4.3 describes the AFW system, which supplies adequate feedwater to the SGs to remove reactor decay heat under all circumstances, including loss of power and normal heat sink (e.g., condenser isolation or loss of CW flow), and identifies, as major components, the condensate storage tank (CST) and three AFW pumps—one steam turbine driven and two electric motor driven. Diverse AFW supplies come from two pumping systems using separate sources of motive power for their pumps. Each system supplies AFW to all four SGs. Two of the SGs can supply the steam turbine-driven pump. The AFW system operates during plant startup at low power levels before the main FW pump is available.

The CST is the safety-grade water source for the system, with a minimum water level maintained for an adequate inventory. The AFW pumps can draw an alternative supply from the city water storage tank for long-term cooling.

The AFW system contains safety-related components relied on to remain functional during and following DBEs. In addition, the AFW system performs functions that support fire protection, ATWS, and SBO.

Instrument air components included in the AFW system are reviewed with the compressed air systems (LRA Section 2.3.3.4). A small number of components are reviewed with the city water system (LRA Section 2.3.3.17).

LRA Table 2.3.4-3-IP2 identifies AFW system component types within the scope of license renewal and subject to an AMR, as well as their intended functions.

2.3A.4.3.2 Staff Evaluation

The staff reviewed LRA Section 2.3.4.3, UFSAR Section 10.2.6.3, and license renewal drawings using the evaluation methodology described in SER Section 2.3 and the guidance in SRP-LR Section 2.3.

During its review, the staff evaluated the system functions described in the LRA and UFSAR to verify that the applicant had not omitted from the scope of license renewal any components with intended functions, as required by 10 CFR 54.4(a). The staff then reviewed those components that the applicant identified as within the scope of license renewal to verify that it had not

omitted any passive and long-lived components subject to an AMR, in accordance with 10 CFR 54.21(a)(1).

During its review of LRA Section 2.3.4.3, the staff identified an area in which additional information was necessary to complete the review of the applicant's scoping and screening results. The applicant responded to the staff's RAI as discussed below.

In RAI 2.3A.4.2-2, dated February 13, 2008, the staff noted that LRA Section 2.3.4.3 states that the AFW system has no intended function under 10 CFR 54.4(a)(2). The staff indicated that the applicant had not highlighted components adjacent to safety-related systems on license renewal drawings; these components adjacent to safety-related systems may need to be considered under 10 CFR 54.4(a)(2) because of the potential for adverse spatial interaction. For IP2, these components include piping to the AFW pump bearing cooling line and the chemical FW line to AFW. The staff requested that the applicant confirm that it had evaluated the aforementioned components for inclusion within the scope of license renewal, in accordance with 10 CFR 54.4(a)(2).

In its response, dated March 12, 2008, the applicant stated that it assigned the bearing cooling lines to the AFW pumps identified by the staff to the city water system, and these lines are subject to an AMR based on the requirements of 10 CFR 54.4(a)(2). The applicant explained that several valves and components shown in dashed lines on one drawing indicate that they appear on the main drawing associated with that system. The applicant identified these components as part of the AFW system and as being within the scope of license renewal and subject to an AMR, in accordance with 10 CFR 54.4(a)(1). The applicant scoped the piping and components on the chemical feed line identified by the staff as part of the chemical feed system. The applicant included the chemical feed system components within the scope of license renewal under 10 CFR 54.4(a)(2), and these components are subject to an AMR.

During the review of the applicant's response to RAI 2.3A.4.2-2, the staff identified other piping lines on license renewal drawing LRA-9321-20183-001 that the applicant had not highlighted, but that were directly connected to highlighted lines. In a telephone conference held on May 30, 2008 (ADAMS Accession No. ML081720557), the staff asked the applicant to indicate whether these lines were within the scope of license renewal under 10 CFR 54.4(a)(2). The applicant explained that it had made a drawing error. The non-highlighted piping line for the AFW system, which includes valve CT-711, is within the scope of license renewal, in accordance with 10 CFR 54.4(a)(1), and should be highlighted. The applicant also explained that the non-highlighted short segments of piping coming off the highlighted valves, CT-709 and CT-710, are valve sealing water under the condensate system and are within the scope of license renewal under 10 CFR 54.4(a)(2).

Based on its review, the staff finds the applicant's response to RAI 2.3A.4.2-2 acceptable because it adequately explained that the components in question are within the scope of license renewal under 10 CFR 54.4(a)(2) because of their potential to adversely interact spatially with safety-related equipment; furthermore, these components are subject to an AMR, in accordance with 10 CFR 54.21(a)(1). The staff's concern described in RAI 2.3A.4.2-2 is resolved.

2.3A.4.3.3 Conclusion

The staff reviewed the LRA, UFSAR, RAI responses, and drawings to determine whether the applicant failed to identify any SSCs within the scope of license renewal. The staff found no such omissions. In addition, the staff sought to determine whether the applicant failed to identify any components subject to an AMR. The staff found no such omissions. On the basis of its review, the staff concludes that the applicant has appropriately identified the AFW system components within the scope of license renewal, as required by 10 CFR 54.4(a), and those subject to an AMR, as required by 10 CFR 54.21(a)(1).

2.3A.4.4 IP2 Steam Generator Blowdown System

2.3A.4.4.1 Summary of Technical Information in the Application

LRA Section 2.3.4.4 describes the SGBD system, which can control the concentration of solids in the shell side of the SGs. The system, which operates normally with a continuous blowdown and sample flow, has a drain connection and two blowdown connections (nozzles) at the bottom of each SG. Pipes from the connections (nozzles) join to form a stainless steel blowdown header. Four individual blowdown headers extend from each SG to the PAB through containment isolation valves. Blowdown flows normally to the flash tank, flashed vapor discharges to the atmosphere, and the condensate drains by gravity through an SW discharge line into the CW discharge canal. The system combines, cools, and monitors the sample flows for radiation.

The SGBD system contains safety-related components relied upon to remain functional during and following DBEs. It also contains nonsafety-related components whose failure potentially could prevent the satisfactory accomplishment of a safety-related function. In addition, the SGBD system performs functions that support fire protection, ATWS, and SBO.

The applicant reviewed a small number of SGBD components with the SW system in LRA Section 2.3.3.2.

LRA Tables 2.3.4-4-IP2 and 2.3.3-19-36-IP2 identify SGBD system component types within the scope of license renewal and subject to an AMR, as well as their intended functions.

2.3A.4.4.2 Staff Evaluation

The staff reviewed LRA Section 2.3.4.4, UFSAR Section 10.2.1.5, and a license renewal drawing using the evaluation methodology described in SER Section 2.3 and the guidance in SRP-LR Section 2.3.

During its review, the staff evaluated the system functions described in the LRA and UFSAR to verify that the applicant had not omitted from the scope of license renewal any components with intended functions, as required by 10 CFR 54.4(a). The staff then reviewed those components that the applicant identified as within the scope of license renewal to verify that it had not omitted any passive and long-lived components subject to an AMR, in accordance with 10 CFR 54.21(a)(1).

2.3A.4.4.3 Conclusion

The staff reviewed the LRA, UFSAR, and a drawing to determine whether the applicant failed to identify any SSCs within the scope of license renewal. The staff found no such omissions. In addition, the staff sought to determine whether the applicant failed to identify any components subject to an AMR. The staff found no such omissions. On the basis of its review, the staff concludes that the applicant has adequately identified the SGBD system components within the scope of license renewal, as required by 10 CFR 54.4(a), and those subject to an AMR, as required by 10 CFR 54.21(a)(1).

2.3A.4.5 IP2 Auxiliary Feedwater Pump Room Fire Event

2.3A.4.5.1 Summary of Technical Information in the Application

LRA Section 2.3.4.5 describes the IP2 AFW pump room fire event, which supplies and supports main FW flow to the SGs during a shutdown (IP2 only). The applicant credits this combination of systems for supplying makeup water to the SGs during a fire in the auxiliary boiler feed pump room for an assumed duration of at least 1 hour. This method was necessary because the current design and CLB assume that plant personnel cannot reenter the area for at least 1 hour following onset of a fire. A combination of secondary systems and components supplies the SGs if a fire in the AFW pump room makes it unavailable for operator actions. These plant systems and components supply FW flow through the main FW isolation valves to the SGs from the IP1 CSTs. Feedwater flows from the IP1 tanks through the hotwell dump, condensate transfer pump, condensate pumps, boiler feed pumps, and main FW isolation valves to the SGs. The following systems support this flowpath (the LRA section reference is included for those systems described elsewhere):

- auxiliary steam
- conventional closed cooling
- condensate (LRA Section 2.3.4.6)
- CW
- city water (LRA Section 2.3.3.17)
- FW (LRA Section 2.3.4.2)
- fresh water cooling (IP1 system)
- instrument air (LRA Section 2.3.3.4)
- instrument air closed cooling
- lube oil
- MS (LRA Section 2.3.4.1)
- river water service (IP1 system)
- SW (LRA Section 2.3.3.2)
- station air (IP1 system) (LRA Section 2.3.3.4)
- water treatment plant (IP1 system)
- wash water

These systems are normally in service and available prior to a fire in the auxiliary feed pump room. For those systems not described elsewhere in the LRA, a brief description is provided below.

2-118

Auxiliary Steam. The auxiliary steam system supplies steam for room and area heating, including the containment and the control room, and for various plant components, such as the RWST heating coil. The system includes IP1 and IP2 components. The heating function is not safety related. However, the system has several containment penetrations with safety-related components, and the RWST heating coil has a pressure boundary safety function. In the event of an AFW pump room fire, auxiliary steam supports the condenser water box priming steam jet air ejectors and preheats oil in the lube oil system.

Conventional Closed Cooling. The conventional closed cooling system supplies cooling water to various components, including condensate and heater drain pumps, main boiler feed pump pedestals, and station air compressors. This system has circulating pumps, heat exchangers (cooled by service water), a head tank, distribution piping valves, instruments, and controls. Cooling water from the conventional closed cooling system is not required to support any system safety function.

Circulating Water. The CW system supplies cooling water to the condenser to condense the steam exiting the low-pressure and main boiler feed pump turbines. The Hudson River supplies the condenser circulating water. The six condenser CW pumps are in the intake structure. The system pipes circulating water to the condensers and discharges it back into the river via the discharge canal. The system includes the CW pumps, condenser inlet and outlet water boxes, piping, valves, instruments, and controls.

Fresh Water Cooling. The fresh water cooling system cools miscellaneous, nonsafety-related heat loads, including IP1 air compressors and house service boiler components. The system includes the fresh water cooling recirculating tank, fresh water circulating pumps, heat exchangers cooled by river water, distribution piping, and valves. This system does not include any safety-related components.

Instrument Air Closed Cooling. The instrument air closed cooling system removes heat from the instrument air compressors and after-coolers. The system consists of a separate closed loop cooling water system of two small pumps, valves, piping, and heat exchangers that supply cooling water to the instrument air compressors and after-coolers and reject that heat to the SWS.

Lube Oil. The lube oil system, which supplies oil for lubrication and control of the main turbine and the main boiler FW pumps and turbines, includes the main lubricating/control oil reservoirs, pumps, coolers, piping, valves, indicators, and components of the main turbine controls. The applicant credits two turbine control components for turbine trip for Appendix R safe shutdown. The auto-stop trip solenoid has only an active function for turbine trip. The auto-stop oil turbine trip solenoid releases oil pressure to trip and need not maintain a pressure boundary. Neither of these components has a passive mechanical intended function.

River Water Service. The river water service system supplies cooling water from the Hudson River to the fresh water cooling system heat exchangers. This system consists primarily of IP1 equipment used to support IP2. The system provides backup to the SW system by providing nonessential loads. It includes four Class A pipe segments that support the SW system. The pipe segments are part of the SW supply and return from the instrument air cooling water heat exchanger.

Water Treatment Plant. The water treatment plant system supplies water for various uses throughout the plant. The water treatment plant consists primarily of IP1 equipment in the superheater building. The system, which takes city water through demineralization for all three units, includes demineralization and deaeration equipment, distribution piping, valves, instruments, controls, and the IP1 CSTs. In the event of an AFW pump room fire, the IP1 CSTs provide make-up water to the SGs. The make-up water flows from the IP1 CSTs to the IP2 hotwell dump and condensate transfer pump.

Wash Water. The wash water system washes fish and debris from the traveling screens for return to the river. The system includes the pumps, piping, strainers, valves, instruments, and controls for the screen wash function. Wash water components are not required to support SW system operation.

The IP2 AFW pump room fire event systems contain safety-related components relied on to remain functional during and following DBEs. They also contain nonsafety-related components whose failure could prevent the satisfactory accomplishment of a safety-related function. In addition, the IP2 AFW pump room fire event systems perform functions that support fire protection.

The IP2 AFW pump room fire event systems contain components that are evaluated with other systems. Auxiliary steam system components supporting the RWST pressure boundary are evaluated with the safety injection systems (LRA Section 2.3.2.4). River water system components supporting the SW system pressure boundary are evaluated with the SW system (LRA Section 2.3.3.2). Containment penetrations are evaluated with other containment penetrations (LRA Section 2.3.2.5).

Nonsafety-related components not evaluated with other systems whose failure could prevent satisfactory accomplishment of safety-related functions are evaluated with miscellaneous systems that are in scope under 10 CFR 54.4(a)(2) (LRA Section 2.3.3.19). For these systems, the following LRA tables identify IP2 AFW pump room fire event component types within the scope of license renewal under 10 CFR 54.4(a)(2), as well as their intended functions:

- LRA Table 2.3.3-19-1-IP2
- LRA Table 2.3.3-19-2-IP2
- LRA Table 2.3.3-19-6-IP2
- LRA Table 2.3.3-19-13-IP2
- LRA Table 2.3.3-19-19-IP2
- LRA Table 2.3.3-19-22-IP2
- LRA Table 2.3.3-19-32-IP2
- LRA Table 2.3.3-19-43-IP2

The staff notes that the LRA does not identify 10 CFR 54.4(a)(2) components of the wash water system. The applicant stated that it performed a review of the liquid-filled components that were not included in other AMRs and determined that the wash water system components are located where they cannot affect equipment with safety-related functions.

The intended function of the IP2 AFW pump room fire event component types within the scope of license renewal is primarily to provide pressure boundary integrity for adequate flow and pressure delivery. For license renewal, the primary intended function of AFW pump room fire event components is to maintain system pressure boundary integrity. Some components retain other functions (e.g., the heat exchangers have the function of heat transfer, and the filters provide filtration).

Aging management of the systems required to supply feedwater to the SGs during an AFW pump room fire is not based on the analysis of materials, environments, and aging effects. System components required to supply feedwater to the SGs during the short duration of such a fire are in service or available when the event occurs. Required components are separated from the AFW pump room; therefore, normal plant operation continuously confirms the integrity of the systems and components required for post-fire intended functions for at least 1 hour.

During the event, these systems and components must continue to perform their intended functions by supplying feedwater to the SGs for the 1-hour minimum duration assumed by the applicant. Significant degradation that could threaten the performance of intended functions will be apparent in the period immediately preceding the event, and corrective action will be required to sustain continued operation. For the minimal 1-hour period that these systems are required to supply make-up water to the SGs, further aging degradation apparent before the event is negligible; therefore, the applicant's evaluation considered no aging effects.

The IP1 CSTs are subject only to intermittent service; therefore, a daily check of tank level and intermittent usage of piping and valves from the IP1 CSTs to the IP2 condenser confirm availability. Significant degradation that could threaten the performance of the intended functions will be apparent in the period immediately preceding the event, and corrective action will be required to sustain continued operation.

Normal plant operation ensures adequate pressure boundary integrity, as well as the post-fire intended function to supply feedwater to the SGs; therefore, no specific AMP is required.

The intended function of the IP2 AFW pump room fire event component types within the scope of license renewal is to provide pressure boundary integrity for adequate flow and pressure delivery.

2.3A.4.5.2 Staff Evaluation

The staff reviewed LRA Section 2.3.4.5 and the UFSAR using the evaluation methodology described in SER Section 2.3 and the guidance in SRP-LR Section 2.3.

During its review, the staff evaluated the system functions described in the LRA and UFSAR to verify that the applicant had not omitted from the scope of license renewal any components with intended functions, as required by 10 CFR 54.4(a). The staff then reviewed those components that the applicant identified as within the scope of license renewal to verify that it had not omitted any passive and long-lived components subject to an AMR, in accordance with 10 CFR 54.21(a)(1).

During its review of LRA Section 2.3.4.5, the staff identified an area in which additional information was necessary to complete the review of the applicant's scoping and screening results. The applicant responded to the staff's RAI as discussed below.

In RAI 2.3A.4.5-1, dated December 7, 2007, the staff noted that, in LRA Section 2.3.4.5, the applicant stated that water treatment plant components are credited for the AFW pump fire event to support safe shutdown in the event of a fire in the IP2 AFW pump room. The applicant indicated that water from the IP1 CSTs is used as makeup water for the IP2 SGs. The applicant further described a combination of IP1 and IP2 systems that are used to complete this flowpath. The applicant stated that the current design and licensing bases requires this flowpath to be available for at least 1 hour following onset of the fire because the applicant assumes that personnel are unable to re-enter the area for at least 1 hour. The staff noted that, although the LRA states that the IP1 components comprising the required flowpath have an intended function under 10 CFR 54.4(a)(3) to support safe shutdown in a fire event, license renewal drawings do not identify the flowpath or its components.

The staff asked to applicant to identify those long-lived components comprising the required flowpath and to indicate whether they are subject to an AMR, in accordance with 10 CFR 54.21(a).

In its response, dated January 4, 2008, the applicant stated that it verifies the levels in the IP1 CSTs on a daily basis. The applicant also indicated that the majority of the components in this flowpath, as part of the water treatment plant system, are included within the scope of license renewal, in accordance with 10 CFR 54.4(a)(2), and are subject to an AMR. Finally, the applicant agreed that a few outdoor components (e.g., tanks, piping and valves) are not included in LRA Section 2.3.4.5. The applicant amended the LRA to include the components to provide further assurance that their intended functions can be performed. The applicant revised LRA Table 3.3.2-19-43-IP2 to add the line items that were not previously included (i.e., carbon steel for the IP1 CSTs).

Based on its review, the staff finds the response to RAI 2.3A.4.5-1 acceptable because the applicant adequately explained that the majority of the components in this flowpath are included within the scope of license renewal as part of the water treatment plant system. The applicant added the few outdoor components that had not been included in this LRA section as within the scope of license renewal and subject to an AMR. For these components, the staff's concern described in RAI 2.3A.4.5-1 is resolved. SER Sections 3.3.2.1 and 3.3A.2.3.33 document the staff's evaluation of new AMR results for the carbon steel CSTs.

In LRA Section 2.3.4.5, the applicant described systems not discussed elsewhere in the LRA that are credited for mitigating the consequences of an IP2 fire event in the AFW pump room. The intended function of each system listed is to support safe shutdown in the event of a fire in the auxiliary feed pump room (10 CFR 50.48), in accordance with 10 CFR 54.4(a)(3). The applicant stated that "no License renewal drawings are provided based on the intended function of supporting safe shutdown in the event of a fire in the auxiliary feed pump room." However, the applicant stated in LRA Section 2.2 that "[c]omponents subject to aging management review are highlighted on license renewal drawings, with the exception of components in scope for 10 CFR 54.4(a)(2)." Since the SCs that support mitigating the consequences of a fire event are in scope under 10 CFR 54.4(a)(3) and are subject to an AMR, in accordance with 10 CFR 54.21(a)(1), the applicant should have highlighted the components on the license renewal drawings. However, the applicant did not highlight the components or flowpaths needed to support this event. In addition, the applicant did not, in accordance with 10 CFR 54.21(a)(1), identify and list the SCs that are subject to an AMR. Therefore, based on the information provided in the LRA, the staff was unable to verify those components that are included within the scope of license renewal to perform the stated function and are subject to an AMR.

In RAI 2.3A.4.5-2, dated December 30, 2008, the staff asked the applicant to (a) identify the system support function for the AFW pump room fire event for each system that supports the flowpath, (b) clearly identify the portion of the systems' flowpaths that support these functions and are subject to an AMR, and (c) identify the portion of these flowpaths that are not already in scope under 10 CFR 54.4(a)(1) or 10 CFR 54.4(a)(2). This issue was identified as Open Item 2.3.4.5-1.

By letter dated January 27, 2009, the applicant explained that its mitigating strategy in the event of a fire in the AFW pump room is to use equipment that the plant typically uses during normal operation. The applicant assumed that if the equipment is available for normal operations, then it would be available in the event of a fire in the AFW pump room. In its response, the applicant identified those systems and their functions that it credits for use in an AFW pump room fire event.

The applicant amended LRA Section 2.2 to explain that it did not highlight those components required for the AFW pump room fire event, as described in Section 2.3.4.5, on license renewal drawings. The applicant described, for each system required to mitigate the AFW pump room fire event, the system's safety functions and the component types, along with their respective intended function. In addition, the applicant identified any components of these systems that it had not previously identified as within the scope of license renewal under 10 CFR 54.4(a)(1) or 10 CFR 54.4(a)(2).

Among those systems required for the AFW pump room fire event, the applicant identified four IP1 systems that it credited as continuously in service during normal plant operation: river water, station air, water treatment plant, and fresh water cooling. The normal condensate flowpath from the IP2 CST may be lost during a fire in the AFW pump room; therefore, the applicant credited the use of the IP1 CSTs, which are not typically in service. As described in its response to RAI 2.3A.4.5-1, the applicant added those components in the flowpath from the IP1 CSTs to the scope of license renewal, in accordance with 10 CFR 54.4(a)(3), that were not already included in the scope for license renewal under 10 CFR 54.4(a)(2). The other three IP1 systems supplement the respective IP2 systems and typically operate to support the normal operations of IP2.

Based on its review, the staff finds the applicant's response to RAI 2.3A.4.5-2 acceptable because it included the components required to support the safety function in the event of a fire in the AFW pump room within scope, in accordance with 10 CFR 54.4(a)(3), and identified passive long-lived components requiring an AMR, in accordance with 10 CFR 54.21. Therefore, the staff's concern described in RAI 2.3A.4.5-2 is resolved. (The staff evaluated the adequacy of the AMR performed for these components in its review of the applicant's response to RAI 3.4.2-1. SER Section 3.4.2 includes the results of this evaluation.)

2.3A.4.5.3 Conclusion

The staff reviewed the LRA, UFSAR, and RAI responses to determine whether the applicant failed to identify any SSCs within the scope of license renewal or failed to identify any components subject to an AMR. As described above, the applicant satisfactorily resolved the omission of components from an AMR. The staff found no further omissions. On the basis of its review, the staff concludes that the applicant has appropriately identified the IP2 AFW pump room fire event system components within the scope of license renewal, as required by 10 CFR 54.4(a), and those subject to an AMR, as required by 10 CFR 54.21(a)(1).

2.3A.4.6 IP2 Condensate System

2.3A.4.6.1 Summary of Technical Information in the Application

LRA Section 2.3.4.6 describes the condensate system, which transfers condensate and low-pressure heater drainage from the condenser hotwell through five stages of FW heating to the main FW pumps. Three condensate pumps, arranged in parallel, take suction from the bottom of the condenser hotwells and discharge into a common header that carries a portion of the condensate through three steam jet air ejector condensers arranged in parallel and one gland steam condenser. The condensate passes through the tube sides of three parallel strings of two low-pressure FW heaters. The flows from these heaters combine in a common line which divides to go to the remaining three strings of three low-pressure heaters. After the No. 5 FW heater, the three condensate lines join into a common header. The heater drain pump discharge enters this header and continues on to the suction of the main FW pumps.

The condensate system includes most components from the condenser to the outlet of the main boiler FW pumps, the main condensers, the condensate and main boiler FW pumps, low-pressure FW heaters, piping, valves, instruments, and controls. Most of the system is not safety related; however, the air ejector discharge to containment penetration is in this system code.

Some system components support the pressure boundary of the AFW system flowpath from the CST to the AFW pumps.

The condensate system contains safety-related components relied on to remain functional during and following DBEs. It also contains nonsafety-related components whose failure potentially could prevent the satisfactory accomplishment of a safety-related function.

Condensate system components that support safe shutdown in the event of an auxiliary feed pump room fire are evaluated in LRA Section 2.3.4.5. Components that support the AFW system flowpath pressure boundary are reviewed with the AFW systems (LRA Section 2.3.4.3). Containment penetration components are reviewed with containment penetrations (LRA Section 2.3.2.5).

2.3A.4.6.2 Staff Evaluation

The staff reviewed LRA Section 2.3.4.6 and UFSAR Section 10.2.6 using the evaluation methodology described in SER Section 2.3 and the guidance in SRP-LR Section 2.3.

During its review, the staff evaluated the system functions described in the LRA and UFSAR to verify that the applicant had not omitted from the scope of license renewal any components with intended functions, as required by 10 CFR 54.4(a). The staff then reviewed those components that the applicant has identified as within the scope of license renewal to verify that it had not

omitted any passive and long-lived components subject to an AMR, in accordance with 10 CFR 54.21(a)(1).

2.3A.4.6.3 Conclusion

The staff reviewed the LRA, UFSAR, and drawings to determine whether the applicant failed to identify any SSCs within the scope of license renewal. The staff found no such omissions. In addition, the staff sought to determine whether the applicant failed to identify any components subject to an AMR. The staff found no such omissions. On the basis of its review, the staff concludes that the applicant had adequately identified the condensate system components within the scope of license renewal, as required by 10 CFR 54.4(a), and those subject to an AMR, as required by 10 CFR 54.21(a)(1).

2.3B Scoping and Screening Results: IP3 Mechanical Systems

2.3B.1 Reactor Coolant System

LRA Section 2.3.1 identifies the RCS SCs subject to an AMR for license renewal.

The RCS includes mechanical components in the following subsystems:

- reactor vessel
- reactor vessel internals
- SGs
- RCPs
- pressurizer
- control rod drives
- in-core instrumentation
- reactor vessel level instrumentation
- SG secondary-side instrumentation
- SG level control

The applicant described the supporting SCs of the RCS in the following LRA sections:

- 2.3.1.1, "Reactor Vessel"
- 2.3.1.2, "Reactor Vessel Internals"
- 2.3.1.3, "Reactor Coolant Pressure Boundary"
- 2.3.1.4, "Steam Generators"

LRA Section 2.3.1 describes the following RCS subsystems:

Reactor Vessel. The cylindrical reactor vessel has a hemispherical bottom and a flanged and gasketed removable upper head. The upper reactor closure head and the reactor vessel flange are joined by studs. Two metallic O-rings seal the reactor vessel when the reactor closure head is bolted in place. A leak-off connection between the two O-rings monitors leakage across the inner O-ring. Vessel design was in accordance with ASME Code, Section III. Coolant enters the reactor vessel through inlet nozzles in a plane just below the vessel flange and above the core, flows downward through the annular space between the vessel wall and the core barrel into a plenum at the bottom of the vessel, reverses direction, and flows up through the core. After mixing in the upper plenum, the mixed coolant stream then flows out of the vessel through exit nozzles on the same plane as the inlet nozzles. The core instrumentation nozzles are on the lower head, and the control rod nozzle penetrations are on the upper head.

Reactor Vessel Internals. The reactor vessel internals direct the coolant flow, support the reactor core, and guide the control rods. The reactor vessel contains the core support assembly, upper plenum assembly, fuel assemblies, control cluster assemblies, surveillance specimens, and in-core instrumentation. The reactor vessel internals consist of three major parts: the lower core support structure, the upper core support structure, and the in-core instrumentation support structure. A one-piece thermal shield, concentric with the reactor core, is between the core barrel and the reactor vessel. The shield, cooled by the coolant on its downward pass, protects the vessel by attenuating much of the gamma radiation and some of the fast neutrons that escape from the core.

Steam Generators. Each loop has a vertical shell and a U-tube SG. Reactor coolant enters the inlet side of the channel head at the bottom of the SG through the inlet nozzle, flows through the U-tubes to an outlet channel, and exits the generator through another bottom nozzle. The inlet and outlet channels are separated by a partition. Feedwater to the SG enters just above the top of the U-tubes through an FW ring, flows downward through an annulus between the tube wrapper and the shell, and then flows upward through the tube bundle where it converts to a steam-water mixture that passes through a primary separator assembly that reduces the water content in the mixture. The separated water combines with the feedwater for another pass through the tube bundle. The remaining higher steam content mixture rises through additional secondary separators to further reduce its water content.

Reactor Coolant Pumps. Each reactor coolant loop has a vertical, single-stage centrifugal pump with a controlled leakage seal assembly. Reactor coolant pumped by the impeller attached to the bottom of the rotor shaft and drawn up through the impeller discharges through passages in the diffuser and out through a discharge nozzle in the side of the casing. A flywheel at the top of the rotor shaft extends the pump coastdown flow during any loss of power to the pump motor. A portion of the flow from the CVCS charging pumps is injected into the RCP between the impeller and the controlled leakage seal. CCW flows to the motor bearing oil coolers and the thermal barrier cooling coil.

The RCS contains safety-related components relied on to remain functional during and following DBEs. The failure of nonsafety-related SSCs in the RCS could prevent the satisfactory accomplishment of a safety-related function. In addition, the RCS performs functions that support fire protection, PTS, ATWS, and SBO.

Pressurizer. The pressurizer system maintains the required reactor coolant pressure during steady-state operation, limits the pressure changes of coolant thermal expansion and contraction during normal load transients, and prevents RCS pressure from exceeding design pressure. The pressurizer maintains pressure by electrical heaters and sprays. Steam can be formed by the heaters or condensed by a pressurizer spray to minimize pressure variations due to coolant contraction and expansion. The pressurizer design accommodates inflow and outflow surges caused by load transients. The surge line attached to the bottom of the pressurizer connects it to the hot leg of a reactor coolant loop. The pressurizer protects the RCS from overpressure by code relief valves connected to its top head. Two PORVs and three code safety valves protect against pressure surges beyond the pressure-limiting capacity of the pressurizer spray. The PORV also operates from the overpressure protection system to prevent RCS pressure from exceeding the limits found in ASME Code, Section III, Appendix G, during low-temperature operation. Steam and water discharge from the power relief and safety valves passes to the pressurizer relief tank partially filled with water at or near ambient containment conditions. The tank normally contains water in a predominantly nitrogen atmosphere. Steam discharged under the water level condenses and cools by mixing with the water. Rupture discs that discharge into the reactor containment protect the tank against a discharge exceeding the design value. The system includes the pressurizer, pressurizer relief valves, PORVs, spray line components, pressurizer relief tank, piping, valves, instruments, controls, and several containment penetrations supporting the pressurizer relief tank.

The pressurizer system contains safety-related components relied on to remain functional during and following DBEs. The failure of nonsafety-related SSCs in the pressurizer system could prevent the satisfactory accomplishment of a safety-related function. In addition, the pressurizer system performs functions that support fire protection and SBO.

Control Rod Drives. The control rod drive system positions the control rods within the core. The reactor uses the Westinghouse magnetic-type control rod drive assemblies on the upper reactor vessel head to insert or withdraw the rods to control generation of nuclear power. Control rod motion is accomplished through the sequential operation of three different magnetic coils. Upon a loss of power to the coils, the released rod cluster control assemblies with full-length absorber rods fall by gravity into the core. Each control rod drive assembly is designed as a hermetically-sealed unit to prevent leakage of reactor coolant. The design of all pressure-containing components meets the requirements of ASME Code, Section III, Division 1, for Class A vessels.

The control rod drive system contains safety-related components relied on to remain functional during and following DBEs.

In-Core Instrumentation. The in-core instrumentation system provides information on the neutron flux distribution and fuel assembly outlet temperatures at selected core locations to confirm the reactor core design parameters and calculated hot channel factors. The system acquires data and performs no operational plant control. The system consists of thermocouples positioned to measure fuel assembly coolant outlet temperature at preselected locations, flux thimbles running the length of selected fuel assemblies to measure the neutron flux distribution within the reactor core using moveable in-core detectors, and in-core drives, drive motors, positioning equipment, and instruments. The flux thimbles, seal table, and guide tube form part of the RCPB.

The in-core instrumentation system contains safety-related components relied on to remain functional during and following DBEs.

Reactor Vessel Level Instrumentation. The reactor vessel level instrumentation monitors the water level in the reactor vessel or relative voids in the RCS during accident conditions. The instrumentation indicates levels from the bottom of the reactor vessel to the top of the reactor head during natural circulation conditions and indicates reactor vessel liquid level for any combination of running RCPs. The instrumentation utilizes RCS penetrations leading to manual isolation valves at which sealed capillary impulse lines transmit pressure measurements to transmitters outside the containment building. Sensor bellows serving as hydraulic couplers seal the capillary impulse lines at the RCS end and at the penetrations. The impulse lines extend through the containment wall to hydraulic isolators which seal and isolate the lines and hydraulically couple them to capillary tubes going to the transmitters.

The reactor vessel level instrumentation system (RVLIS) contains safety-related components relied on to remain functional during and following DBEs. The failure of nonsafety-related SSCs in the RVLIS could prevent the satisfactory accomplishment of a safety-related function.

Steam Generator (Secondary-Side Instrumentation). The SG system has secondary-side instrumentation. The SG system code includes the passive mechanical instrument piping and valves for the SG secondary-side-level instrumentation. These components are safety related because they form part of the SG pressure boundary.

The SG system contains safety-related components relied on to remain functional during and following DBEs. In addition, the SG system performs functions that support fire protection and SBO.

Steam Generator Level Control. The SG level control system supports the control of FW flow to maintain SG secondary-side level. Primarily an electrical system, it includes several level instrument vent valves. These components are safety related because they form part of the SG pressure boundary.

The SG level control system contains safety-related components relied on to remain functional during and following DBEs. In addition, the SG level control system performs functions that support fire protection and SBO.

The RCS Class I piping evaluation boundary extends into portions of systems attached to the RCS. For both units, the RCS AMR includes the Class I components of the systems listed below. The applicant evaluated the non-Class 1 system portions in the LRA section indicated:

- CVCS (LRA Section 2.3.3.6)
- isolation valve seal water (LRA Section 2.3.2.3)
- primary sampling (LRA Section 2.3.3.19)
- RHR system (LRA Section 2.3.2.1)
- safety injection system (LRA Section 2.3.2.4)

IP3 RCS RCP lube oil collection system components are reviewed with the fire protection—CO_2, Halon, and RCP oil collection systems (LRA Section 2.3.3.12).

Components in the IP3 nitrogen supply to the PORVs are reviewed with the nitrogen systems (LRA Section 2.3.3.5). A small number of IP3 pressurizer components are reviewed with the primary water makeup systems (LRA Section 2.3.3.7).

The following components are evaluated with containment penetrations (LRA Section 2.3.2.5):

- IP3 pressurizer system containment penetration components
- certain mechanical IP3 RVLIS components

Fuel assemblies replaced after a limited number of fuel cycles are not subject to an AMR. The control rods are active components and, therefore, are not subject to an AMR.

The intended function of the RCS component types within the scope of license renewal is to provide pressure boundary integrity for adequate flow and pressure delivery.

Because the IP2 and IP3 RCS and supporting SCs are very similar, SER Sections 2.3A.1.1–2.3A.1.4, respectively, document the staff's review findings for LRA Sections 2.3.1.1–2.3.1.4 for IP3.

2.3B.2 Engineered Safety Features

LRA Section 2.3.2 identifies the engineered safety features SCs subject to an AMR for license renewal.

The applicant described the supporting SCs of the engineered safety features in the following LRA sections:

- 2.3.2.1, "Residual Heat Removal"
- 2.3.2.2, "Containment Spray System"
- 2.3.2.3, "Containment Isolation Support Systems"
- 2.3.2.4, "Safety Injection Systems"
- 2.3.2.5, "Containment Penetrations"

The staff summarized the findings of its review of LRA Sections 2.3.2.1–2.3.2.5 in SER Sections 2.3B.2.1–2.3B.2.5, respectively.

2.3B.2.1 IP3 Residual Heat Removal

2.3B.2.1.1 Summary of Technical Information in the Application

LRA Section 2.3.2.1 describes the RHR system, which provides emergency core cooling as part of the safety injection system and removes residual heat during later stages of plant cooldown. The RHR system is one of three (RHR, CCW, SFPC) auxiliary coolant systems. The RHR system consists of two RHR heat exchangers, two seal coolers, two RHR (low-head) pumps, and required piping, valves, and I&C components. The RHR system provides emergency core cooling during the injection phase of a LOCA. The RHR heat exchangers, in conjunction with the safety injection recirculation pumps, provide post-accident heat removal during the LOCA recirculation phase. Outlet flow from the RHR heat exchangers may be directed to the CS headers, to the RCS cold legs, or to the RCS hot legs via the high-head safety injection pumps. The RHR pumps also back up the safety injection system recirculation pumps during the LOCA recirculation phase. In this capacity, the RHR pumps may draw water from the containment sump and deliver it to the RCS cold leg injection lines, to the suction of the high-head safety injection pumps, or to the CS headers. The RHR system removes residual heat during later stages of plant cooldown, as well as during cold shutdown and refueling operations. After the RCS temperature and pressure have been reduced to 350 degrees F and less than 450 psig, alignment of the RHR pumps initiates decay heat cooling by taking suction from one reactor hot leg and discharging it through the RHR heat exchangers into the reactor cold legs.

The RHR system contains safety-related components relied on to remain functional during and following DBEs. In addition, the RHR system performs functions that support fire protection and SBO.

In the LRA, ASME Code Class 1 components with the intended function of RCPB maintenance are reviewed with the RCS (LRA Section 2.3.1). A small number of components are reviewed with the CCW system (LRA Section 2.3.3.3).

LRA Table 2.3.2-1-IP3 identifies RHR system component types within the scope of license renewal and subject to an AMR as well as their intended functions.

2.3B.2.1.2 Staff Evaluation

The staff reviewed LRA Section 2.3.2.1, the UFSAR, and license renewal drawings using the evaluation methodology described in SER Section 2.3 and the guidance in SRP-LR Section 2.3.

During its review, the staff evaluated the system functions described in the LRA and UFSAR to verify that the applicant had not omitted from the scope of license renewal any components with intended functions, as required by 10 CFR 54.4(a). The staff then reviewed those components that the applicant identified as within the scope of license renewal to verify that it had not omitted any passive and long-lived components subject to an AMR, in accordance with 10 CFR 54.21(a)(1).

2.3B.2.1.3 Conclusion

The staff reviewed the LRA, UFSAR, and drawings to determine whether the applicant failed to identify any SSCs within the scope of license renewal. The staff found no such omissions. In addition, the staff sought to determine whether the applicant failed to identify any components subject to an AMR. The staff found no such omissions. On the basis of its review, the staff concludes that the applicant has adequately identified the RHR system components within the scope of license renewal, as required by 10 CFR 54.4(a), and those subject to an AMR, as required by 10 CFR 54.21(a)(1).

2.3B.2.2 IP3 Containment Spray System

2.3B.2.2.1 Summary of Technical Information in the Application

LRA Section 2.3.2.2 describes the CS system, which cools the containment and removes iodine following an accident. The system consists of two trains of pumps, valves, and spray headers that spray borated water into the containment automatically when the system senses high containment pressure following a LOCA or MS line break accident. The CS system sprays a portion of the RWST contents into the containment atmosphere through nozzles connected to four ring headers in the containment dome. Each spray pump supplies two ring headers. The CS pumps take their suction from the RWST. After injection from the RWST has been terminated, the spray headers can be supplied with recirculated water from the recirculation sump or the containment sump by a diversion of a portion of the injection flow from the safety injection system. By letter dated June 30, 2009, the applicant submitted Amendment 8, Revision 1 to the LRA to reflect a modification to the containment spray system. The applicant stated that the buffer chemical in the containment spray system was changed from sodium hydroxide (liquid injection) to sump baskets containing sodium tetraborate. Retention of iodine during long-term post-accident conditions is assured by the sodium tetraborate baskets located in the containment that will be flooded under accident conditions, allowing the sodium tetraborate to dissolve into the fluid for pH control. The containment spray system also includes a dousing system for the carbon filter bank of each fan cooler unit of the containment air recirculation cooling and filtration system. Each dousing system can be started manually if high-temperature conditions occur.

The CS system contains safety-related components relied on to remain functional during and following DBEs. It also contains nonsafety-related components whose failure potentially could prevent the satisfactory accomplishment of a safety-related function. In addition, the CS system performs functions that support fire protection.

Containment spray system components that support the RHR system pressure boundary are reviewed with the RHR system (LRA Section 2.3.2.1). A small number of components are reviewed with the safety injection system (LRA Section 2.3.2.4).

LRA Tables 2.3.2-2-IP3 and 2.3.3-19-10-IP3 identify CS system component types within the scope of license renewal and subject to an AMR, as well as their intended functions.

2.3B.2.2.2 Staff Evaluation

The staff reviewed LRA Section 2.3.2.2, UFSAR Section 6.3, and license renewal drawings using the evaluation methodology described in SER Section 2.3 and the guidance in SRP-LR Section 2.3. In addition, the staff reviewed the applicant's letter dated June 30, 2009, which provided a modification to LRA Section 2.3.2.2 to reflect a change in the buffer chemical and the method of adding it to the containment spray system for pH control.

During its review, the staff evaluated the system functions described in the LRA and UFSAR to verify that the applicant had not omitted from the scope of license renewal any components with intended functions, as required by 10 CFR 54.4(a). The staff then reviewed those components that the applicant identified as within the scope of license renewal to verify that it had not omitted any passive and long-lived components subject to an AMR, in accordance with 10 CFR 54.21(a)(1).

2.3B.2.2.3 Conclusion

The staff reviewed the LRA, UFSAR and drawings to determine whether the applicant failed to identify any SSCs within the scope of license renewal. The staff found no such omissions. In addition, the staff sought to determine whether the applicant failed to identify any components subject to an AMR. The staff found no such omissions. On the basis of its review, the staff concludes that the applicant has adequately identified the CS system components within the scope of license renewal, as required by 10 CFR 54.4(a), and those subject to an AMR, as required by 10 CFR 54.21(a)(1).

2.3B.2.3 IP3 Containment Isolation Support Systems

2.3B.2.3.1 Summary of Technical Information in the Application

LRA Section 2.3.2.3 describes the containment isolation support systems, which include the isolation valve seal water systems and the weld channel and containment penetration pressurization systems. For IP3, this evaluation also includes the PAB system, which has a containment penetration. The containment isolation support systems consist of piping and valves routed to the various system piping that penetrates the containment. The isolation valve seal water, weld channel, and containment penetration pressurization systems isolate the containment from the outside environment for various systems with piping penetrating containment. The containment isolation support systems inject fluid or either air or gas into system lines between the containment isolation valves penetrating the containment to ensure pressure boundary integrity against leakage of radioactive fluids to the environment in the event of a LOCA. These barriers of piping and isolation valves systems are defined by individual lines. Besides satisfying containment isolation criteria, the valving facilitates normal operation and maintenance of the systems for reliable operation of other engineered safeguard systems.

The isolation valve seal water system provides sealing water or gas between the isolation and double-disk isolation valves of containment penetrations located in lines connected to the RCS or exposed to the containment atmosphere during any condition which requires containment isolation. This system limits fission product release from the containment. Although not credited in post-accident dose analyses, the system ensures a containment leak rate in an accident that is lower than that assumed in the accident analysis and the offsite dose calculations. System components form parts of the containment penetration isolation boundary.

The weld channel and containment penetration pressurization systems provide pressurized gas to all containment penetrations and most liner inner weld seams so that, in a LOCA, no leakage occurs through these potential paths from the containment to the atmosphere. The system also serves spaces between selected isolation valves. Although not credited in the post-accident dose analyses, weld channel and penetration pressurization systems maintained at some pressure level above the peak accident pressure will keep any postulated leakage in, rather than out of, the containment. The plant's compressed air systems supply regulated clean, dry compressed air outside the containment to all containment penetrations and most inner liner weld channels. The primary source of air for this system is the instrument air system backed up by the station air system and by a bank of nitrogen cylinders as a standby source of gas pressure.

The PAB houses and protects emergency safeguards equipment and other systems supporting safe reactor operation. This system code is primarily structural but, because it also includes the guard pipe and enclosure containment leakage boundary for a containment sump penetration, the system has a mechanical intended function which is discussed in this section. This enclosure (tank) is a second leakage boundary for the primary containment penetration from the containment sump.

The containment isolation support systems contain safety-related components relied on to remain functional during and following DBEs. They also contain nonsafety-related components whose failure potentially could prevent the satisfactory accomplishment of a safety-related function.

The isolation valve seal water system components with the intended function of maintaining the RCPB are reviewed with the RCS (LRA Section 2.3.1.3).

LRA Tables 2.3.2-3-IP3 and 2.3.3-19-62-IP3 identify containment isolation support systems component types within the scope of license renewal and subject to an AMR, as well as their intended functions.

2.3B.2.3.2 Staff Evaluation

The staff reviewed LRA Section 2.3.2.3; UFSAR Sections 6.2.2, 6.5, and 6.6; and license renewal drawings using the evaluation methodology described in SER Section 2.3 and the guidance in SRP-LR Section 2.3.

During its review, the staff evaluated the system functions described in the LRA and UFSAR to verify that the applicant had not omitted from the scope of license renewal any components with intended functions, as required by 10 CFR 54.4(a). The staff then reviewed those components that the applicant identified as within the scope of license renewal to verify that it had not omitted any passive and long-lived components subject to an AMR, in accordance with 10 CFR 54.21(a)(1).

During its review of LRA Section 2.3.2.3, the staff identified an area in which additional information was necessary to complete the review of the applicant's scoping and screening results. The applicant responded to the staff's RAI as discussed below.

In RAI 2.3B.2.3-1, dated November 9, 2007, the staff identified line-mounted components (valves PCV 1076 and PCV 1090) located on sensing lines that have a pressure boundary function. However, the applicant did not identify the sensing lines (i.e., those connecting these components to the main line) as being subject to an AMR. Therefore, the staff requested that the applicant clarify whether these sensing lines are subject to an AMR.

In its response, dated December 6, 2007, the applicant stated that the sensing lines are internal to the valve bodies and provide a control function for operation of the valves. The valves (with internal sensing lines) are subject to an AMR and are identified in LRA Table 2.3.2-3-1P3 as component type "valve body," with AMR results summarized in LRA Table 3.2.2-3-1P3.

Based on its review, the staff found the applicant's response to RAI 2.3B.2.3-1 acceptable because the applicant clarified that the subject sensing lines are within the scope of license renewal and subject to an AMR. The staff's concern described in RAI 2.3B.2.3-1 is resolved.

In RAI 2.3B.2.3-2, dated November 9, 2007, the staff identified several line-mounted components (valves PCV 1193 through PCV 1199) located in lines (i.e., 3/8-inch stainless steel tubing) with a pressure boundary function. However, the applicant did not identify the components themselves as being subject to an AMR. Therefore, the staff requested that the applicant clarify whether these components are subject to an AMR or justify their exclusion.

In its response, dated December 6, 2007, the applicant stated that the line-mounted components are aluminum pressure-regulating valves. These components are within the scope of license renewal and subject to an AMR. The applicant amended the application to add the line item "valve body" to LRA Table 3.2.2-3-IP3 to reflect the aluminum material.

Based on its review, the staff finds the applicant's response to RAI 2.3B.2.3-2 acceptable because the applicant clarified that the subject valves are within the scope of license renewal and subject to an AMR. In addition, the applicant added aluminum valve bodies to the AMR. The staff's concern described in RAI 2.3B.2.3-2 is resolved. SER Section 3.2.2.1 discusses the staff's evaluation of the added AMR for aluminum valve bodies.

2.3B.2.3.3 Conclusion

The staff reviewed the LRA, UFSAR, RAI responses, and drawings to determine whether the applicant failed to identify any SSCs within the scope of license renewal. The staff found no such omissions. In addition, the staff sought to determine whether the applicant failed to identify any components subject to an AMR. The staff found an instance in which the applicant omitted components that should have been subject to an AMR. The applicant has satisfactorily resolved this issue as discussed in the preceding staff evaluation. On the basis of its review, the staff concludes that the applicant has adequately identified the containment isolation support systems components within the scope of license renewal, as required by 10 CFR 54.4(a), and those subject to an AMR, as required by 10 CFR 54.21(a)(1).

2.3B.2.4 IP3 Safety Injection System

2.3B.2.4.1 Summary of Technical Information in the Application

LRA Section 2.3.2.4 describes the safety injection system, which automatically delivers cooling water to the reactor core in a LOCA to limit the fuel clad temperature so that the core remains intact and in place with its essential heat transfer geometry preserved. Components comprising the safety injection system code include the RWST, the three safety injection (high-head) pumps, the accumulators (one for each reactor loop), recirculation pumps and piping, valves, and other components of these subsystems. The three safety injection (high-head) pumps inject RWST borated water into the RCS for core cooling. The safety injection signal automatically opens the required safety injection system isolation valves and starts the safety injection pumps. The accumulators containing borated water pressurized with nitrogen are connected to the RCS by injection piping and valves. Two check valves isolate these tanks from the RCS during normal operation. When RCS pressure falls below accumulator pressure the check valves open, discharging the contents of the tanks into the RCS through the same injection piping used by the safety injection pumps.

After the injection, the recirculation system cools and returns the coolant spilled from the break and water collected from the CS to the RCS. The system recirculation pumps take suction from the recirculation sump in the containment floor and deliver spilled reactor coolant and borated refueling water back to the core through the RHR heat exchangers. For smaller RCS breaks in which recirculated water must be injected against higher pressures for long-term cooling, the system delivers the water from an RHR heat exchanger to the high-head safety injection pump suction and, by this external recirculation route, to the reactor coolant loops. The system also allows either of the RHR pumps to take over the recirculation function.

For IP3, the engineered safeguards initiation logic system was evaluated with the safety injection system. The system actuates (depending on the severity of the condition) the safety injection, containment isolation, containment air recirculation, and CS systems. The engineered safeguards initiation logic system is primarily electrical, but does include some mechanical components, specifically the piping and valves from the containment to the containment pressure transmitters, and has mechanical intended functions.

The safety injection system contains safety-related components relied on to remain functional during and following DBEs. It also contains nonsafety-related components whose failure potentially could prevent the satisfactory accomplishment of a safety-related function. In addition, the safety injection system performs functions that support fire protection.

ASME Code Class 1 components with the intended function of maintaining the RCPB are reviewed with the RCS (LRA Section 2.3.1.3). A small number of components are reviewed with the CS system (LRA Section 2.3.2.2), RHR systems (LRA Section 2.3.2.1), or nitrogen systems (LRA Section 2.3.3.5).

LRA Tables 2.3.2-4-IP3 and 2.3.3-19-53-IP3 identify safety injection system component types within the scope of license renewal and subject to an AMR, as well as their intended functions.

2.3B.2.4.2 Staff Evaluation

The staff reviewed LRA Section 2.3.2.4, the UFSAR, and license renewal drawings using the evaluation methodology described in SER Section 2.3 and the guidance in SRP-LR Section 2.3.

During its review, the staff evaluated the system functions described in the LRA and UFSAR to verify that the applicant had not omitted from the scope of license renewal any components with intended functions, as required by 10 CFR 54.4(a). The staff then reviewed those components that the applicant identified as within the scope of license renewal to verify that it had not omitted any passive and long-lived components subject to an AMR, in accordance with 10 CFR 54.21(a)(1).

2.3B.2.4.3 Conclusion

The staff reviewed the LRA, UFSAR, and drawings to determine whether the applicant failed to identify any SSCs within the scope of license renewal. The staff found no such omissions. In addition, the staff sought to determine whether the applicant failed to identify any components subject to an AMR. The staff found no such omissions. On the basis of its review, the staff concludes that the applicant has adequately identified the safety injection system components within the scope of license renewal, as required by 10 CFR 54.4(a), and those subject to an AMR, as required by 10 CFR 54.21(a)(1).

2.3B.2.5 IP3 Containment Penetrations

2.3B.2.5.1 Summary of Technical Information in the Application

LRA Section 2.3.2.5 describes the containment penetrations, which is not an independent system but a grouping of containment penetration components not evaluated with other systems. These penetrations include the following:

- electrical penetrations
- fuel core component handling system
- hydrogen recombiners
- BVS
- fuel handling
- integrated leak rate testing

The BVS system draws samples from the building ventilation to identify radioactive gases that may be present and verifies whether plant radioactive gaseous effluents are within technical specification limits. The system has several containment penetrations and a flowpath to two process radiation monitors.

The fuel-handling system defuels and refuels the reactor core and is designed to transport and handle fuel safely and effectively. The structural evaluations address most components shown in the database and the fuel storage racks and pools. The fuel transfer tube blind flange in this system code is a passive mechanical component for that containment penetration.

The integrated leak rate testing system, which tests containment integrated leak rates during shutdown conditions, has piping, valves, and equipment to pressurize containment, instrumentation to monitor containment parameters during the test, and containment penetrations isolated by blind flanges during normal operation.

The containment penetrations contain safety-related components relied on to remain functional during and following DBEs.

Components in the containment penetrations evaluated in this section are those that maintain the system pressure boundary inside containment from the first weld from the penetration to the class boundary change outside containment. Components in the Class 1 boundary are evaluated with the RCPB (LRA Section 2.3.1.3). Structural portions of the containment penetrations are evaluated with the containment building (LRA Section 2.4.1). Electrical portions of electrical penetration assemblies are evaluated with electrical components (LRA Section 2.5). Containment penetrations not included in other systems' AMRs are evaluated in LRA Section 2.3.2.5. This evaluation includes the BVS system process flowpath to the radiation monitors.

LRA Table 2.3.2-5-IP3 and newly created Table 2.3.3-19-63-IP3 (see evaluation below) identify containment penetration component types within the scope of license renewal and subject to an AMR, as well as their intended functions.

2.3B.2.5.2 Staff Evaluation

The staff reviewed LRA Section 2.3.2.5; UFSAR Sections 1.2.2, 5.1.4, 9.4.2, 9.5, and 11.2; and license renewal drawings using the evaluation methodology described in SER Section 2.3 and the guidance in SRP-LR Section 2.3.

During its review, the staff evaluated the system functions described in the LRA and UFSAR to verify that the applicant had not omitted from the scope of license renewal any components with

intended functions, as required by 10 CFR 54.4(a). The staff then reviewed those components that the applicant identified as within the scope of license renewal to verify that it had not omitted any passive and long-lived components subject to an AMR, in accordance with 10 CFR 54.21(a)(1).

During its review of LRA Section 2.3.2.5, the staff identified areas in which additional information was necessary to complete the review of the applicant's scoping and screening results. The applicant responded to the staff's RAIs as discussed below.

In RAI 2.3A.2.2-1, dated February 13, 2008, during review of license renewal drawings for the CS system, the staff identified portions of piping in the CS system that were not highlighted, indicating that a particular section of piping had no intended functions, in accordance with 10 CFR 54.4(a)(1) or 10 CFR 54.4(a)(3). LRA Section 2.3.2.2 states that the CS system has no intended function under 10 CFR 54.4(a)(2). This section of piping is directly connected to safety-related CS piping; therefore, the staff believed that it should be in scope, in accordance with 10 CFR 54.4(a)(2), as nonsafety-related piping that is structurally attached to safety-related piping. The staff asked the applicant to explain this apparent discrepancy. The staff also asked the applicant to indicate any portions of the CS system that it evaluated for inclusion in the scope of license renewal, in accordance with 10 CFR 54.4(a)(2), and to identify any other instances in which it identified a system as not having any 10 CFR 54.4(a)(2) components, but having nonsafety-related components that were not identified as within scope under 10 CFR 54.4(a)(2).

In its response, dated March 12, 2008, the applicant determined that the components identified by the staff do have an intended function to maintain integrity such that no physical interaction with safety-related components could prevent satisfactory accomplishment of a safety function. The applicant responded to the staff's request by performing a reevaluation of those safety-related systems that the LRA identified as only being in scope under 10 CFR 54.4(a)(1) and that have no 10 CFR 54.4(a)(2) components. The applicant explained that it should have included the IP3 BVS system within the scope of license renewal, in accordance with 10 CFR 54.4(a)(2).

For the BVS system, the applicant amended the LRA to reflect the changes described below:

a) LRA Table 2.3.3-19-A-IP3 would reflect the BVS system as a miscellaneous system within the scope of license renewal for 10 CFR 54.4(a)(2).

b) Removal of the BVS system from the LRA Section 2.3.3.19 table of areas excluded from an AMR based on their lack of potential for spatial interaction.

c) Revision of LRA Table 2.3.3-19-B-IP3 to reflect that the BVS system has components subject to an AMR for meeting 10 CFR 54.4(a)(2).

d) Creation of a new LRA Table 2.3.3-19-63-IP3 for the four added component types in the BVS system for nonsafety-related components potentially affecting safety function subject to AMR.

e) Creation of a new LRA Table 3.3.2-19-63-IP3 for the four added component types, their materials, environments, and AMPs.

Based on its review, the staff finds the applicant's response to RAI 2.3A.2.2-1 for the BVS system acceptable because it adequately explained that the applicant's reevaluation of safety-related systems identified components that should have been in scope for meeting the requirements of 10 CFR 54.4(a)(2). The staff reviewed the applicant's amended LRA to ensure that the new LRA tables include those components that have been brought into the scope of license renewal under 10 CFR 54.4(a)(2) because of their potential for spatial interaction with safety-related components. The staff finds the tables acceptable. Therefore, the staff's concern described in RAI 2.3A.2.2-1 for the BVS system is resolved. SER Sections 3.2.2.1 and 3.3B.2.3.41 document the staff's evaluation of new AMR results for the IP3 BVS system.

2.3B.2.5.3 Conclusion

The staff reviewed the LRA, UFSAR, RAI response, and drawings to determine whether the applicant failed to identify any SSCs within the scope of license renewal. The staff found no such omissions. In addition, the staff sought to determine whether the applicant failed to identify any components subject to an AMR. The staff found an instance in which the applicant omitted components that should have been subject to an AMR. The applicant has satisfactorily resolved this issue as discussed in the preceding staff evaluation. On the basis of its review, the staff concludes that that the applicant has adequately identified the containment penetrations' components within the scope of license renewal, as required by 10 CFR 54.4(a), and those subject to an AMR, as required by 10 CFR 54.21(a)(1).

2.3B.3 Scoping and Screening Results: IP3 Auxiliary Systems

LRA Section 2.3.3 identifies the auxiliary systems SCs subject to an AMR for license renewal.

The applicant described the supporting SCs of the auxiliary systems in the following LRA sections:

- 2.3.3.1, "Spent Fuel Pit Cooling"
- 2.3.3.2, "Service Water"
- 2.3.3.3, "Component Cooling Water"
- 2.3.3.4, "Compressed Air"
- 2.3.3.5, "Nitrogen Systems"
- 2.3.3.6, "Chemical and Volume Control"
- 2.3.3.7, "Primary Water Makeup"
- 2.3.3.8, "Heating, Ventilation and Air Conditioning"
- 2.3.3.9, "Containment Cooling and Filtration"
- 2.3.3.10, "Control Room Heating, Ventilation and Cooling"
- 2.3.3.11, "Fire Protection—Water"
- 2.3.3.12, "Fire Protection—CO_2, Halon, and RCP Oil Collection Systems"
- 2.3.3.13, "Fuel Oil"
- 2.3.3.14, "Emergency Diesel Generators"
- 2.3.3.15, "Security Generators"
- 2.3.3.16, "Appendix R Diesel Generators"
- 2.3.3.17, "City Water"
- 2.3.3.18, "Plant Drains"
- 2.3.3.19, "Miscellaneous Systems In-Scope for (a)(2)"

The applicant created LRA Section 2.3.3.19 to capture all systems or portions of systems that are within the scope of license renewal only under 10 CFR 54.4(a)(2). Among the subsections identified in LRA Section 2.3.3.19, the staff identified the following auxiliary systems for simplified Tier 1 reviews:

- ammonia morpholine addition
- CL
- CW
- extraction steam
- floor drains
- hydrazine addition
- heater drain/moisture separator drains/vents
- lube oil
- low pressure steam dump
- main turbine generator
- nuclear equipment drains
- river water service
- main generator seal oil
- secondary plant sampling
- turbine hall closed cooling water

The staff conducted a more detailed Tier 2 review for all of the remaining auxiliary systems.

Staff's RAIs

During its review, the staff noted that the applicant did not specifically identify components that were in scope under 10 CFR 54.4(a)(2) on the associated drawings. To determine that the applicant did not omit any components from scope under 10 CFR 54.4(a)(2), the staff asked the applicant to verify that it had included segments of the selected systems in scope under 10 CFR 54.4(a)(2). In the following RAIs, dated February 13, 2008, the staff asked that the applicant confirm its methodology for identifying nonsafety-related portions of systems with a potential for adversely affecting safety-related functions, in accordance with 10 CFR 54.4(a)(2), by describing the applicable portions of system piping that it included within the scope of license renewal under 10 CFR 54(a)(2):

- RAI 2.3B.3.1-2
- RAI 2.3B.3.2-1
- RAI 2.3B.3.3-1
- RAI 2.3B.3.13-1
- RAI 2.3B.3.14-2
- RAI 2.3B.3.18-1

In its response to the RAIs referenced above, dated March 12, 2008, the applicant stated that all of the component types identified by the staff on the license renewal drawings in question are within the scope of license renewal, in accordance with 10 CFR 54.4(a)(2), and are subject to an AMR.

Based on its review, the staff finds the applicant's response to these RAIs acceptable because the applicant adequately explained that all of the component types identified by the staff are within the scope of license renewal, in accordance with 10 CFR 54.4(a)(2), and subject to an AMR. The staff's concern described in these RAIs is resolved.

SER Sections 2.3B.3.1–2.3B.3.19, respectively, discuss the staff's review of the IP3 systems described in LRA Sections 2.3.3.1–2.3.3.19. The following sections discuss the staff's findings for these systems.

2.3B.3.1 IP3 Spent Fuel Pit Cooling System

2.3B.3.1.1 Summary of Technical Information in the Application

LRA Section 2.3.3.1 describes the SFPC system, which removes residual heat from the spent fuel pit. The SFPC loop consists of pumps (main and standby), a heat exchanger, filters, demineralizer, piping, valves, and instrumentation. The operating pump draws water from the pit for circulation through the heat exchanger and return. CCW cools the heat exchanger, which forms part of the CCW system pressure boundary. Loop piping is arranged so that any pipeline failure does not drain the spent fuel pit below the top of the stored fuel elements. The spent fuel pit pump suction line, which draws water from the pit, penetrates the spent fuel pit wall above the fuel assemblies. A purification loop circulates spent fuel pit water through the demineralizer and filter for purification. A portion of the system piping supporting the RWST purification loop with the spent fuel pit demineralizer forms part of the safety injection system pressure boundary. The system includes the spent fuel pit. Spent fuel storage racks at the bottom of the pit for spent fuel assemblies are a full-length, top-entry type made of stainless steel with Boral as a neutron absorber.

The SFPC system contains safety-related components relied on to remain functional during and following DBEs. It also contains nonsafety-related components whose failure potentially could prevent the satisfactory accomplishment of a safety-related function.

The spent fuel pit and the spent fuel racks are reviewed with the fuel storage buildings (LRA Section 2.4.3). Components supporting the CCW system pressure boundary are reviewed with the CCW systems (LRA Section 2.3.3.3). Components supporting the pressure boundary of the safety injection system are reviewed with the safety injection systems (LRA Section 2.3.2.4). A small number of components are reviewed with the primary water makeup systems (LRA Section 2.3.3.7).

LRA Tables 2.3.3-1-IP3 and 2.3.3-19-49-IP3 identify SFPC system component types within the scope of license renewal and subject to an AMR, as well as their intended functions.

2.3B.3.1.2 Staff Evaluation

The staff reviewed LRA Section 2.3.3.1, UFSAR Sections 9.3 and 9.5, and a license renewal drawing using the evaluation methodology described in SER Section 2.3 and the guidance in SRP-LR Section 2.3.

During its review, the staff evaluated the system functions described in the LRA and UFSAR to verify that the applicant had not omitted from the scope of license renewal any components with intended functions, as required by 10 CFR 54.4(a). The staff then reviewed those components that the applicant identified as within the scope of license renewal to verify that it had not

omitted any passive and long-lived components subject to an AMR, in accordance with 10 CFR 54.21(a)(1).

During its review of LRA Section 2.3.3.1, the staff identified areas in which additional information was necessary to complete the review of the applicant's scoping and screening results. The applicant responded to the staff's RAIs as discussed below.

In RAI 2.3B.3.1-1, dated December 7, 2007, the staff noted that the UFSAR for IP3 references a backup SFPC system that operates in parallel with the normal SFPC system during refueling activities. Further, the LRA stated that the normal SFPC system is within the scope of license renewal under 10 CFR 54.4(a)(1), with the intended function of providing a pressure boundary for the CCW system and the safety injection system, and under 10 CFR 54.4(a)(2) because of possible physical interaction. The staff noted that the scope of license renewal excludes the backup spent fuel cooling system and requested that the applicant explain the exclusion of these components from scope.

In its response, dated January 4, 2008, the applicant stated that the backup SFPC system is a nonsafety-related system that has no functions under 10 CFR 54.4(a)(1) and is not relied on to perform a function that demonstrates compliance with 10 CFR 54.4(a)(3). The applicant explained that the system is normally drained when the plant is in normal power operation, such that its failure cannot prevent satisfactory accomplishment of any 10 CFR 54.4(a)(1) functions through spatial interaction. Lastly, the applicant explained that no components in the backup SFPC system are directly connected to safety-related equipment, and none meet the scoping requirements of 10 CFR 54.4(a)(1), 10 CFR 54.4(a)(2), or 10 CFR 54.4(a)(3).

Based on its review, the staff finds the applicant's response to RAI 2.3B.3.1-1 acceptable because it adequately explained that the components in the backup SFPC system do not have intended functions under 10 CFR 54.4(a). The applicant adequately explained that the backup SFPC system is a nonsafety-related system, is normally drained when the plant is in normal power operation, and is not credited with performing functions identified in 10 CFR 54.4(a)(3). The staff's concern described in RAI 2.3B.3.1-1 is resolved.

The discussion of the staff's RAIs in SER Section 2.3B.3 details the disposition of RAI 2.3B.3.1-2, dated February 13, 2008.

2.3B.3.1.3 Conclusion

The staff reviewed the LRA, UFSAR, RAI responses, and a drawing to determine whether the applicant failed to identify any SSCs within the scope of license renewal. The staff found no such omissions. In addition, the staff sought to determine whether the applicant failed to identify any components subject to an AMR. The staff found no such omissions. On the basis of its review, the staff concludes that the applicant has appropriately identified the SFPC system components within the scope of license renewal, as required by 10 CFR 54.4(a), and those subject to an AMR, as required by 10 CFR 54.21(a)(1).

2.3B.3.2 IP3 Service Water System

2.3B.3.2.1 Summary of Technical Information in the Application

LRA Section 2.3.3.2 describes the SW system, which supplies cooling water from the Hudson River to various heat loads in both the primary and secondary portions of the plant in a continuous flow to systems and components necessary for plant safety during either normal operation or abnormal or accident conditions. Sufficient redundancy of active and passive components maintains short- and long-term cooling to vital loads in accordance with the single-failure criterion. Six identical, vertical, centrifugal sump-type pumps at the intake structure supply service water to two independent discharge headers, each supplied by three pumps. An automatic, self-cleaning, rotary-type strainer in the discharge of each pump removes solids. Each header connects to an independent supply line. Either of the two supply lines can supply the essential loads while the other line supplies the nonessential loads. Three nonseismic-class pumps independent of the intake structure can supply an SW system backup by drawing suction from the discharge canal. The applicant credits one of these pumps with supplying service water during a safe shutdown following a fire.

The SW system supplies cooling water to nonessential loads, including SGBD heat exchangers, CW pump seal coolers, the turbine building CCW system, hydrogen coolers, exciter air coolers, and the isolated phase bus heat exchangers, to support normal operation.

The SW system contains safety-related components relied on to remain functional during and following DBEs. It also contains nonsafety-related components whose failure potentially could prevent the satisfactory accomplishment of a safety-related function. In addition, the SW system performs functions that support fire protection.

LRA Tables 2.3.3-2-IP3 and 2.3.3-19-56-IP3 identify SW system component types within the scope of license renewal and subject to an AMR, as well as their intended functions.

2.3B.3.2.2 Staff Evaluation

The staff reviewed LRA Section 2.3.3.2, UFSAR Section 9.6.1, and license renewal drawings using the evaluation methodology described in SER Section 2.3 and the guidance in SRP-LR Section 2.3.

During its review, the staff evaluated the system functions described in the LRA and UFSAR to verify that the applicant had not omitted from the scope of license renewal any components with intended functions, as required by 10 CFR 54.4(a). The staff then reviewed those components that the applicant identified as within the scope of license renewal to verify that it had not omitted any passive and long-lived components subject to an AMR, in accordance with 10 CFR 54.21(a)(1).

The staff's review of LRA Section 2.3.3.2 identified an area in which additional information was necessary to complete the review of the applicant's scoping and screening results. The discussion of the staff's RAIs in SER Section 2.3B.3 details the disposition of RAI 2.3B.3.2-1, dated February 13, 2008.

2.3B.3.2.3 Conclusion

The staff reviewed the LRA, UFSAR, RAI response, and drawings to determine whether the applicant failed to identify any SSCs within the scope of license renewal. The staff found no such omissions. In addition, the staff sought to determine whether the applicant failed to identify any components subject to an AMR. The staff found no such omissions. On the basis of its review, the staff concludes that the applicant has appropriately identified the SW system components within the scope of license renewal, as required by 10 CFR 54.4(a), and those subject to an AMR, as required by 10 CFR 54.21(a)(1).

2.3B.3.3 IP3 Component Cooling Water System

2.3B.3.3.1 Summary of Technical Information in the Application

LRA Section 2.3.3.3 describes the CCW system, which removes RCS residual and sensible heat via the RHR loop during plant shutdown, cools the letdown flow to the CVCS during power operation, and dissipates waste heat from various primary plant components. It also cools engineered safeguards and safe-shutdown components. The system has pumps, heat exchangers, distribution and return piping and valves, instruments, and controls to cool the following:

- RHR heat exchangers
- RCPs
- non-regenerative heat exchanger
- excess letdown heat exchanger
- CVCS seal water heat exchanger
- sample heat exchangers
- waste gas compressors
- reactor vessel support pads
- RHR pumps
- safety injection pumps
- recirculation pumps
- spent fuel pit heat exchanger
- charging pumps, fluid drive coolers, and crankcase
- gross failed fuel detector

Some of the CCW-cooled heat exchangers in other systems have no safety function; however, these nonsafety-related heat exchangers form parts of the CCW system pressure boundary. These heat exchangers are within the scope of license renewal with an intended function to maintain the pressure boundary but not to transfer heat. The heat exchangers within the CCW system are safety-related components.

The CCW system contains safety-related components relied upon to remain functional during and following DBEs. In addition, the CCW system performs functions that support fire protection.

A few components in the CCW system support the RHR system pressure boundary and are reviewed with the RHR systems (LRA Section 2.3.2.1). Component cooling water system components that service the safety injection system are reviewed with the safety injection systems (LRA Section 2.3.2.4).

LRA Table 2.3.3-3-IP3 and newly created Table 2.3.3-19-64-IP3 (see evaluation below) identify CCW system component types within the scope of license renewal and subject to an AMR, as well as their intended functions.

2.3B.3.3.2 Staff Evaluation

The staff reviewed LRA Section 2.3.3.3, UFSAR Section 9.3, and license renewal drawings using the evaluation methodology described in SER Section 2.3 and the guidance in SRP-LR Section 2.3.

During its review, the staff evaluated the system functions described in the LRA and UFSAR to verify that the applicant had not omitted from the scope of license renewal any components with intended functions, as required by 10 CFR 54.4(a). The staff then reviewed those components that the applicant identified as within the scope of license renewal to verify that it had not omitted any passive and long-lived components subject to an AMR, in accordance with 10 CFR 54.21(a)(1).

During its review of LRA Section 2.3.3.3, the staff identified an area in which additional information was necessary to complete the review of the applicant's scoping and screening results. The applicant responded to the staff's RAI as discussed below.

During its review of license renewal drawings for the CS system, the staff identified portions of the system that were not highlighted, indicating that sections of piping had no intended functions under 10 CFR 54.4(a)(1) or 10 CFR 54.4(a)(3). In RAI 2.3A.2.2-1, dated February 13, 2008, the staff asked the applicant to identify any instances in which a system was identified as having no intended functions under 10 CFR 54.4(a)(1), but having nonsafety-related components not identified as within the scope of license renewal.

In its response, dated March 12, 2008, the applicant identified, in addition to the CS system, three other instances in which it had not identified nonsafety-related components as being within the scope of license renewal under 10 CFR 54.4(a)(2). SER Sections 2.3A.2.2, 2.3A.3.3, and 2.3B.2.5 discuss the staff's evaluation of the affected systems. The applicant further explained that it should have identified the CCW systems for IP2 and IP3 and the IP3 BVS system as meeting the requirements of 10 CFR 54.4(a)(2). In these instances, the applicant amended the LRA for IP3 CCW system to include the following:

a)   LRA Table 2.3.3-19-A-IP3 would reflect the CCW system as a miscellaneous system within the scope of license renewal pursuant to 10 CFR 54.4(a)(2).

b)   Removal of the CCW system from the list of IP3 systems not reviewed for spatial interaction, pursuant to 10 CFR 54.4(a)(2).

c)   Revision of LRA Table 2.3.3-19-B-IP3 to reflect that the CCW system now has components subject to an AMR pursuant to 10 CFR 54.4(a)(2).

d)   Creation of a new LRA Table 2.3.3-19-64-IP3 for the six added component types in the CCW system for nonsafety-related components potentially affecting safety function, subject to an AMR.

e)   Creation of a new LRA Table 3.3.2-19-64-IP3 for the six added component types, their materials, environments, and AMPs.

Based on its review, the staff finds the applicant's response to RAI 2.3A.2.2-1 for the IP3 CCW system acceptable because it adequately explained that the applicant's reevaluation of safety-related systems identified components that should have been within scope for meeting the requirements of 10 CFR 54.4(a)(2). Additionally, the applicant amended the LRA to include portions of the CCW system within the scope of license renewal under 10 CFR 54.4(a)(2). The staff reviewed the applicant's addition of new tables to the LRA to ensure that they include those components with the potential for spatial interaction with safety-related components as within the scope of license renewal, in accordance with 10 CFR 54.4(a)(2). The staff's concern described in RAI 2.3A.2.2-1 for the IP3 CCW system is resolved. SER Section 3.3.2.1 documents the staff's evaluation of new AMR results for the IP3 CCW system.

The discussion of the staff's RAIs in SER Section 2.3B.3 details the disposition of RAI 2.3B.3.3-1, dated February 13, 2008.

2.3B.3.3.3 Conclusion

The staff reviewed the LRA, UFSAR, RAI responses, and drawings to determine whether the applicant failed to identify any SSCs within the scope of license renewal. The staff found no such omissions. In addition, the staff sought to determine whether the applicant failed to identify any components subject to an AMR. The staff found an instance in which the applicant omitted components that should have been subject to an AMR. The applicant has satisfactorily resolved this issue as discussed in the preceding staff evaluation. On the basis of its review, the staff concludes that the applicant has appropriately identified the CCW system components within the scope of license renewal, as required by 10 CFR 54.4(a), and those subject to an AMR, as required by 10 CFR 54.21(a)(1).

2.3B.3.4 IP3 Compressed Air Systems

2.3B.3.4.1 Summary of Technical Information in the Application

LRA Section 2.3.3.4 describes the compressed air systems, including the instrument air and station air systems. The instrument air system continuously supplies dry, oil-free air from duplicate compressors with duplicate dryers and filters for pneumatic instruments and controls. Each compressor discharges into a common air receiver and takes a backup supply from the station air system. To meet current and future instrument air loads, a third compressor-dryer package is available to supply the conventional plant. This compressor also can supply the station air system with backup air, if necessary. The system has compressors, dryers, filters, receivers, distribution piping and valves, instruments, and controls. Items essential for safe operation and safe cooldown have air reserves or gas bottles that enable the equipment to function safely until its air supply resumes. The instrument air system includes piping, valves, and controls supporting this air reserve function, but does not include air or gas bottles, which are part of other systems.

The station air system, which supplies compressed air for pneumatic tools, CW pump priming, and miscellaneous cleaning and maintenance purposes throughout the primary and secondary plants, has diesel-driven and motor-driven air compressors, inter- and after-coolers, a receiver, piping, valves, instruments, and controls. Distribution piping to the containment includes containment isolation valves.

The compressed air system contains safety-related components relied on to remain functional during and following DBEs. It also contains nonsafety-related components whose failure potentially could prevent the satisfactory accomplishment of a safety-related function. In addition, the compressed air system performs functions that support fire protection and SBO.

LRA Tables 2.3.3-4-IP3, 2.3.3-19-29-IP3, and 2.3.3-19-48-IP3 identify compressed air system component types within the scope of license renewal and subject to an AMR, as well as their intended functions.

2.3B.3.4.2 Staff Evaluation

The staff reviewed LRA Section 2.3.3.4, UFSAR Section 9.6.3, and license renewal drawings using the evaluation methodology described in SER Section 2.3 and the guidance in SRP-LR Section 2.3.

During its review, the staff evaluated the system functions described in the LRA and UFSAR to verify that the applicant had not omitted from the scope of license renewal any components with intended functions, as required by 10 CFR 54.4(a). The staff then reviewed those components that the applicant identified as within the scope of license renewal to verify that it had not omitted any passive and long-lived components subject to an AMR, in accordance with 10 CFR 54.21(a)(1).

2.3B.3.4.3 Conclusion

The staff reviewed the LRA, UFSAR, and drawings to determine whether the applicant failed to identify any SSCs within the scope of license renewal. The staff found no such omissions. In addition, the staff sought to determine whether the applicant failed to identify any components subject to an AMR. The staff found no such omissions. On the basis of its review, the staff concludes that the applicant has adequately identified the compressed air system components within the scope of license renewal, as required by 10 CFR 54.4(a), and those subject to an AMR, as required by 10 CFR 54.21(a)(1).

2.3B.3.5 IP3 Nitrogen System

2.3B.3.5.1 Summary of Technical Information in the Application

LRA Section 2.3.3.5 describes the nitrogen system, which supplies motive gas as a backup to the instrument air supply and nitrogen to various components for process functions (including cover gas, calibration gas, purge gas, and gas for operation of level instrumentation). Nitrogen enters containment through several penetrations that must isolate for containment isolation capability under accident conditions. The containment penetration pressurization system also has nitrogen-filled components not included with this system code.

The nitrogen system contains safety-related components relied on to remain functional during and following DBEs. It also contains nonsafety-related components whose failure potentially could prevent the satisfactory accomplishment of a safety-related function. In addition, the nitrogen system performs functions that support fire protection.

A small number of nitrogen system components are reviewed with the AFW systems (LRA Section 2.3.4.3).

LRA Tables 2.3.3-5-IP3 and 2.3.3-19-37-IP3 identify nitrogen system component types within the scope of license renewal and subject to an AMR, as well as their intended functions.

2.3B.3.5.2 Staff Evaluation

The staff reviewed LRA Section 2.3.3.5; UFSAR Sections 7.3, 9.6.2.5, 9.9.2, and 10.2.6; and license renewal drawings using the evaluation methodology described in SER Section 2.3 and the guidance in SRP-LR Section 2.3.

During its review, the staff evaluated the system functions described in the LRA and UFSAR to verify that the applicant had not omitted from the scope of license renewal any components with intended functions, as required by 10 CFR 54.4(a). The staff then reviewed those components that the applicant identified as within the scope of license renewal to verify that it had not omitted any passive and long-lived components subject to an AMR, in accordance with 10 CFR 54.21(a)(1).

2.3B.3.5.3 Conclusion

The staff reviewed the LRA, UFSAR, and drawings to determine whether the applicant failed to identify any SSCs within the scope of license renewal. The staff found no such omissions. In addition, the staff sought to determine whether the applicant failed to identify any components subject to an AMR. The staff found no such omissions. On the basis of its review, the staff concludes that the applicant has adequately identified the nitrogen system components within the scope of license renewal, as required by 10 CFR 54.4(a), and those subject to an AMR, as required by 10 CFR 54.21(a)(1).

2.3B.3.6 IP3 Chemical and Volume Control System

2.3B.3.6.1 Summary of Technical Information in the Application

LRA Section 2.3.3.6 describes the CVCS, which controls RCS inventory (amounts of makeup and letdown) and chemistry (RCS boron concentration and other chemical additions). The system cleans up reactor coolant by degasification and purification, injects seal water to the RCPs, depressurizes the RCS via a pressurizer auxiliary spray flowpath, and injects control poison in the form of boric acid solution from the boric acid storage tanks.

During normal plant operation, reactor coolant letdown flows through the shell side of the regenerative heat exchanger, which reduces its temperature by transferring heat to the charging fluid. The coolant then flows through a letdown orifice, which regulates flow and reduces the coolant pressure. The cooled, low-pressure water leaves the reactor containment and enters the PAB. After passing through the nonregenerative heat exchanger and one of the mixed-bed demineralizers, the fluid flows through the reactor coolant filter and enters the VCT.

The coolant flows from the VCT to three positive-displacement, variable-speed charging pumps, which raise the pressure above that in the RCS. The high-pressure water flows from the PAB to the reactor containment along two parallel paths, one returning directly to the RCS through the tube side of the regenerative heat exchanger to the RCS cold leg, and the other injecting water into the RCP seals through seal injection filters. The RCP seal water returns to the CVCS through a seal water filter and heat exchanger back to the VCT.

The RWST and the boric acid storage tanks can provide borated water to the charging system. The RWST is available to the charging pumps for injection of borated water. The boric acid system has boric acid transfer pumps, a boric acid filter, and storage tanks to maintain a large inventory of concentrated boric acid solution.

The CVCS contains safety-related components relied on to remain functional during and following DBEs. It also contains nonsafety-related components whose failure potentially could prevent the satisfactory accomplishment of a safety-related function. In addition, the CVCS performs functions that support fire protection and ATWS.

CVCS components that maintain the RCS pressure boundary are reviewed with the RCS pressure boundary (LRA Section 2.3.1.3). A small number of system components are reviewed with the primary water makeup systems (LRA Section 2.3.3.7) and with the CCW systems (LRA Section 2.3.3.3).

LRA Tables 2.3.3-6-IP3 and 2.3.3-19-11-IP3 identify CVCS component types within the scope of license renewal and subject to an AMR, as well as their intended functions.

2.3B.3.6.2 Staff Evaluation

The staff reviewed LRA Section 2.3.3.6, UFSAR Section 9.2.2, and license renewal drawings using the evaluation methodology described in SER Section 2.3 and the guidance in SRP-LR Section 2.3.

During its review, the staff evaluated the system functions described in the LRA and UFSAR to verify that the applicant had not omitted from the scope of license renewal any components with intended functions, as required by 10 CFR 54.4(a). The staff then reviewed those components that the applicant identified as within the scope of license renewal to verify that the applicant had not omitted any passive and long-lived components subject to an AMR, in accordance with 10 CFR 54.21(a)(1).

2.3B.3.6.3 Conclusion

The staff reviewed the LRA, UFSAR, and drawings to determine whether the applicant failed to identify any SSCs within the scope of license renewal. The staff found no such omissions. In addition, the staff sought to determine whether the applicant failed to identify any components subject to an AMR. The staff found no such omissions. On the basis of its review, the staff concludes that the applicant has adequately identified the CVCS components within the scope of license renewal, as required by 10 CFR 54.4(a), and those subject to an AMR, as required by 10 CFR 54.21(a)(1).

2.3B.3.7 IP3 Primary Water System

2.3B.3.7.1 Summary of Technical Information in the Application

LRA Section 2.3.3.7 describes the primary water makeup system, which supplies makeup water to primary plant systems as required in support of normal plant operation. Among other components, this system includes tanks, piping, valves, and pumps. It is also a source of fire water to the containment. The system has a containment penetration and one safety-related component part of the RWST pressure boundary.

The demineralized water system is evaluated with the primary water system. The system supplies demineralized water for normal plant operation and refueling activities to the spent fuel pit, refueling cavity, and RWST; for decontamination, hydrostatic testing, and flushing during refueling outages; for condensate polisher regeneration through the sluice water pumps; and for fire protection in containment.

The system includes safety-related position indicators for the containment penetration isolation valves, which are in the primary water makeup system; therefore, this system has no safety-related mechanical function.

The primary water makeup system contains safety-related components relied on to remain functional during and following DBEs. It also contains nonsafety-related components whose failure potentially could prevent the satisfactory accomplishment of a safety-related function. In addition, the primary water makeup system performs functions that support fire protection.

Portions of the primary water makeup system that support the RWST pressure boundary are reviewed with the safety injection system (LRA Section 2.3.2.4).

LRA Tables 2.3.3-7-IP3, 2.3.3-19-15-IP3, and 2.3.3-19-42-IP3 identify primary water makeup system component types within the scope of license renewal and subject to an AMR, as well as their intended functions.

2.3B.3.7.2 Staff Evaluation

The staff reviewed LRA Section 2.3.3.7; UFSAR Sections 9.2.2, 9.6.2.3, and 9.11.1; and license renewal drawings using the evaluation methodology described in SER Section 2.3 and the guidance in SRP-LR Section 2.3.

During its review, the staff evaluated the system functions described in the LRA and UFSAR to verify that the applicant had not omitted from the scope of license renewal any components with intended functions, as required by 10 CFR 54.4(a). The staff then reviewed those components that the applicant identified as within the scope of license renewal to verify that it had not omitted any passive and long-lived components subject to an AMR, in accordance with 10 CFR 54.21(a)(1).

2.3B.3.7.3 Conclusion

The staff reviewed the LRA, UFSAR, and drawings to determine whether the applicant failed to identify any SSCs within the scope of license renewal. The staff found no such omissions. In addition, the staff sought to determine whether the applicant failed to identify any components subject to an AMR. The staff found no such omissions. On the basis of its review, the staff concludes that the applicant has adequately identified the primary water makeup system components within the scope of license renewal, as required by 10 CFR 54.4(a), and those subject to an AMR, as required by 10 CFR 54.21(a)(1).

2.3B.3.8 IP3 Heating, Ventilation and Air Conditioning Systems

2.3B.3.8.1 Summary of Technical Information in the Application

LRA Section 2.3.3.8 describes the HVAC systems, which maintain the area environment for personnel and equipment. HVAC systems for specific buildings or areas of buildings, including portions of ventilation systems serving various areas of the plant, generally have separate system codes. The HVAC system includes fans and dampers for the electrical tunnels, intake structure, and fire pump house and portable ventilation equipment for safe-shutdown requirements.

The IP3 HVAC systems evaluation includes the following HVAC systems:

- control building heating and ventilation
- fire barriers
- fuel storage building heating and ventilation
- HVAC
- PAB heating and ventilation
- plant vent
- security heating and ventilation
- vapor containment purge and supply
- vapor containment pressure relief
- Appendix R diesel generator heating and ventilation
- EDG building heating and ventilation

LRA Section 2.3.3.9 describes containment cooling and filtration, and LRA Section 2.3.3.10 discusses the control room HVAC.

The control building heating and ventilation system heats and ventilates the 15-foot and 33-foot elevations of the control building and ventilates battery rooms 31, 32, and 34 to maintain hydrogen concentrations below maximum acceptable limits during normal plant operation. The system includes dampers, ductwork, heaters, and fans.

The fire barriers system has structural barriers and components for penetrations to prevent or delay the spread of fire to adjoining areas. This system includes fire doors and fire dampers that also support the HVAC systems like that for the diesel generator building. The fire doors and fire dampers are evaluated with their respective structures for their fire barrier function. Fire damper housings that form part of an HVAC system pressure boundary within the scope of

2-150

license renewal are included in the HVAC evaluation to maintain the housing function of HVAC system support.

The fuel storage building heating and ventilation system heats and ventilates the fuel storage building, minimizes leakage of unfiltered air from the building during fuel-handling operations, and filters building exhaust. The system has two fresh-air-tempering units with supply fans and heaters; exhaust-roughing, HEPA, and carbon filters; an exhaust fan; motor-operated dampers; and ducts. During normal operation, the fresh-air-tempering units and exhaust fan ventilate and heat the fuel storage building with exhaust air, which passes through the roughing and HEPA filters. During fuel handling, the system maintains a slight negative pressure in the building and passes all ventilation exhaust through the roughing, HEPA, and charcoal filters before release through the plant vent. Originally credited in the fuel-handling accident, the system has no safety functions because the new analysis (described in UFSAR Section 14.2.1), which uses the alternate source term, no longer assumes operation of the ventilation system or any holdup of the radionuclides released from the spent fuel pit.

The HVAC system maintains the area environment for personnel and equipment. HVAC systems for specific buildings or areas of buildings generally have a separate system code. The HVAC system includes portions of various ventilation systems serving different areas of the plant. The HVAC system includes fans and dampers for various areas, such as the electrical tunnels, intake structure, and fire pump house. This system also includes portable ventilation equipment supporting safe-shutdown requirements.

The PAB heating and ventilation system heats and ventilates the waste hold-up tank pit and the PAB enclosed spaces. The waste hold-up tank pit contains the waste hold-up tanks which are central collection points for liquid radioactive waste. The PAB houses equipment and components required for normal plant operation as well as accident mitigation, including pumps for the CCW, safety injection, RHR, CS, and other systems. Also located in the PAB are tanks for the waste disposal system that collect radioactive liquids and gases. The PAB heating and ventilation system maintains an environment for personnel and equipment during normal operating and post-accident conditions. The PAB and tank pit are ventilated by a balanced flow between supply and exhaust, maintaining a slight negative pressure in the PAB. Air supplied to each building enters areas of low contamination. A set of fans exhausts air out the plant vent from areas of higher contamination after passing it through filters. No dose consequence analyses credit filtration.

The plant vent system with its plant vent duct and vent flow monitoring instrumentation provides a flowpath for plant ventilation systems to exhaust to the atmosphere. The offsite dose analyses do not credit the plant vent; however, this vent is the release point for control room dose calculations and its structural integrity must be maintained for this purpose.

The security heating and ventilation system heats and ventilates the security building and supports operation of the security propane generator. The system includes fans, heaters, and dampers.

The vapor containment purge and supply system filters, monitors, and purges containment air to the plant vent for exhaust to the environment and supplies makeup air to the containment. Operation of the purge system during reactor shutdown maintains radioactivity concentrations inside containment within acceptable limits. The purge system is isolated to maintain containment integrity whenever the plant is above the cold shutdown condition. The system has filters, heating coils, fans, penetration isolation valves, ductwork, instruments, and controls. Some system components share a common pressure boundary with PAB heating and ventilation system components.

The vapor containment pressure relief system relieves the normal pressure changes in containment during reactor power operation. This system consists of a pressure relief line equipped with three isolation valves, one inside and two outside the containment. The pressure relief line discharges through roughing, HEPA, and charcoal filters to the plant vent.

The IP3 Appendix R diesel generator has its own enclosure in the yard. Ventilation to the engine is by exhaust fans that draw outside air through covered intake dampers or louvers when required. Exhaust fans that draw outside air in through louvers provide ventilation to the electrical enclosure and the battery enclosure. This equipment is required to support operation of the IP3 Appendix R diesel generator credited for both 10 CFR Part 50, Appendix R, requirements and SBO response.

The IP3 EDG building houses and protects the EDGs. The rooms have outside-air fixed louvers, pneumatically-operated adjustable louvers, and exhaust fans with motor-operated discharge dampers. The pneumatically-operated dampers operate from control air supplied by the EDG starting air system. EDG building ventilation is relied on to support EDG operations during DBAs and regulated events.

The HVAC system contains safety-related components relied on to remain functional during and following DBEs. It also contains nonsafety-related components whose failure potentially could prevent the satisfactory accomplishment of a safety-related function. In addition, the HVAC system performs functions that support fire protection.

Instrument air volume tanks, tubing, and valves in the vapor containment pressure relief system needed for the containment penetration valves to close are reviewed with the compressed air systems (LRA Section 2.3.3.4).

LRA Tables 2.3.3-8-IP3, 2.3.3-19-21-IP3, 2.3.3-19-39-IP3, 2.3.3-19-60-IP3, and 2.3.3-19-61-IP3 identify HVAC system component types within the scope of license renewal and subject to an AMR, as well as their intended functions.

2.3B.3.8.2 Staff Evaluation

The staff reviewed LRA Section 2.3.3.8; UFSAR Sections 1.3.6, 5.3.2.3, 5.3.2.5, 9.5, 9.6.2.2, 9.8, and 14.2.1; and license renewal drawings using the evaluation methodology described in SER Section 2.3 and the guidance in SRP-LR Section 2.3.

During its review, the staff evaluated the system functions described in the LRA and UFSAR to verify that the applicant had not omitted from the scope of license renewal any components with intended functions, as required by 10 CFR 54.4(a). The staff then reviewed those components that the applicant identified as within the scope of license renewal to verify that it had not omitted any passive and long-lived components subject to an AMR, in accordance with 10 CFR 54.21(a)(1).

2.3B.3.8.3 Conclusion

The staff reviewed the LRA, UFSAR, and drawings to determine whether the applicant failed to identify any SSCs within the scope of license renewal. The staff found no such omissions. In addition, the staff sought to determine whether the applicant failed to identify any components subject to an AMR. The staff found no such omissions. On the basis of its review, the staff concludes that the applicant has adequately identified the HVAC system components within the scope of license renewal, as required by 10 CFR 54.4(a), and those subject to an AMR, as required by 10 CFR 54.21(a)(1).

2.3B.3.9 IP3 Vapor Containment Building Ventilation System

2.3B.3.9.1 Summary of Technical Information in the Application

LRA Section 2.3.3.9 describes the IP3 vapor containment building ventilation system, which by recirculation cooling and filtration removes normal heat losses from equipment and piping in containment during plant operation, ensures personnel access and safety during shutdown, and depressurizes the containment vessel following an accident. Air recirculation cooling and filtering during normal operation is achieved using all five air-handling units discharging to a common header ductwork distribution system. Each air-handling unit consists of cooling coils, a centrifugal fan with direct-drive motor, and a distribution header. In an accident, the system diverts the flowpath first through a compartment with moisture separators, HEPA filters, and charcoal filters. Dose analyses for some accidents credit the HEPA filters but not the charcoal filters for fission product removal.

The vapor containment building ventilation system contains safety-related components relied on to remain functional during and following DBEs. In addition, the vapor containment building ventilation system performs functions that support fire protection.

LRA Table 2.3.3-9-IP3 identifies vapor containment building ventilation system component types within the scope of license renewal and subject to an AMR, as well as their intended functions.

2.3B.3.9.2 Staff Evaluation

The staff reviewed LRA Section 2.3.3.9, UFSAR Sections 5.3.2.2 and 6.4.2, and license renewal drawings using the evaluation methodology described in SER Section 2.3 and the guidance in SRP-LR Section 2.3.

During its review, the staff evaluated the system functions described in the LRA and UFSAR to verify that the applicant had not omitted from the scope of license renewal any components with intended functions, as required by 10 CFR 54.4(a). The staff then reviewed those components that the applicant identified as within the scope of license renewal to verify that it had not omitted any passive and long-lived components subject to an AMR, in accordance with 10 CFR 54.21(a)(1).

2.3B.3.9.3 Conclusion

The staff reviewed the LRA, UFSAR, and drawings to determine whether the applicant failed to identify any SSCs within the scope of license renewal. The staff found no such omissions. In addition, the staff sought to determine whether the applicant failed to identify any components subject to an AMR. The staff found no such omissions. On the basis of its review, the staff concludes that the applicant has adequately identified the vapor containment building ventilation system components within the scope of license renewal, as required by 10 CFR 54.4(a), and those subject to an AMR, as required by 10 CFR 54.21(a)(1).

2.3B.3.10 IP3 Control Room Heating, Ventilation and Cooling System

2.3B.3.10.1 Summary of Technical Information in the Application

LRA Section 2.3.3.10 describes the control room HVAC, which maintains a safe, habitable environment in the central control room during both normal operation and accident conditions. The fresh air intake duct supplying make-up air to the control room has air-operated dampers to divert this air through carbon filters or close it off completely. The control room HVAC system consists of two air-conditioning units with fans and roughing filters, roughing, high-energy particulate air, and charcoal filters, charcoal filter booster fans, heaters, exhaust fans, and a duct system with dampers, controls, and instrumentation. The system also has five independent air-conditioning units that supplement the cooling capacity of the water-cooled air-conditioning units. Each supplemental unit has a wall-mounted evaporator and electric heater in the control room.

The control room HVAC system contains safety-related components relied on to remain functional during and following DBEs. In addition, the control room heating, ventilation and cooling system performs functions that support fire protection.

LRA Table 2.3.3-10-IP3 identifies control room HVAC component types within the scope of license renewal and subject to an AMR, as well as their intended functions.

2.3B.3.10.2 Staff Evaluation

The staff reviewed LRA Section 2.3.3.10, UFSAR Section 9.9, and license renewal drawings using the evaluation methodology described in SER Section 2.3 and the guidance in SRP-LR Section 2.3.

During its review, the staff evaluated the system functions described in the LRA and UFSAR to verify that the applicant had not omitted from the scope of license renewal any components with intended functions, as required by 10 CFR 54.4(a). The staff then reviewed those components that the applicant identified as within the scope of license renewal to verify that it had not omitted any passive and long-lived components subject to an AMR, in accordance with 10 CFR 54.21(a)(1).

2.3B.3.10.3 Conclusion

The staff reviewed the LRA, UFSAR, and drawings to determine whether the applicant failed to identify any SSCs within the scope of license renewal. The staff found no such omissions. In addition, the staff sought to determine whether the applicant failed to identify any components subject to an AMR. The staff found no such omissions. On the basis of its review, the staff concludes that the applicant has adequately identified the control room HVAC system components within the scope of license renewal, as required by 10 CFR 54.4(a), and those subject to an AMR, as required by 10 CFR 54.21(a)(1).

2.3B.3.11 IP3 Fire Protection – Water

2.3B.3.11.1 Summary of Technical Information in the Application

LRA Section 2.3.3.11 describes the fire protection system, which provides fire protection for the station through the use of water, foam, Halon 1301, detection and alarm systems, and rated fire barriers, doors, and dampers. The fire water system components include fire water and foam subsystem pumps, piping, hydrants, hose reels, valves, tanks, and drains. This system also includes the fuel oil supply to the fire pump house diesel. Fire protection systems include the fire detection and alarm system, as described below. LRA Section 2.3.3.12 describes the CO_2 and Halon 1301 systems. LRA Section 2.3.3.8 discusses the fire barrier system code.

The fire protection water distribution system has two ground-level storage tanks supplied by the city water distribution system. Heating provisions for the storage tanks consist of two sets of dual electric heaters and two sets of dual circulating pumps. The pumping facilities maintain system pressure and, using jockey pumps, supply makeup for system leakage. Two main fire pumps (one electric motor driven and the other diesel engine driven) provide an automatic water supply during a fire. The pumping facilities provide flow and pressure requirements for the water-based fire protection systems. The fire protection water distribution system consists of outdoor underground and aboveground piping and indoor distribution piping in all buildings except the containment building. Demineralized water piping is for fire protection inside containment. IP3 underground piping has two connections with the IP1 fire protection system, providing defense in depth for the IP3 fire protection systems in terms of both water supply and pumping capacity. The distribution system also has isolation valves, strainers, hose stations, and outdoor hydrants. The distribution piping delivers anticipated fire water requirements to individual suppression systems. The yard hydrants provide effective hose stream protection for exterior hazards and for supplementary use for fire conditions within the main buildings of the plant. The water-based fire suppression systems include the wet pipe sprinkler systems, preaction sprinkler systems, deluge water spray systems, foam water spray systems, hydrants, and hose stations. To prevent local flooding, areas with safety-related equipment, or equipment required for safe plant shutdown with automatically operated fire protection, have either gravity or pump drains to handle the maximum quantity of spray water. The fire water system includes plant drain components that protect safety-related equipment from the effects of Class III component failures. The fire water system can supply makeup to the spent fuel pit. While not a safety function, this feature of the fire water system is included as a license renewal intended function.

According to the LRA, the fire protection—water system has no intended function under 10 CFR 54.4(a)(1). The scoping and screening methodology identified the following fire water system intended functions, in accordance with 10 CFR 54.4(a)(2):

- Maintain integrity of nonsafety-related components such that no physical interaction with safety-related components can prevent satisfactory accomplishment of a safety function.

- Provide a backup source of makeup water to the spent fuel pit.

The scoping and screening methodology also identified the following fire protection—water system intended functions, in accordance with 10 CFR 54.4(a)(3):

- Provide fixed automatic and manual fire suppression (including hydrants, hose stations and portable extinguishers) to extinguish fires in vital areas of the plant (10 CFR 50.48).

- Ensure adequate protection of safety-related equipment from water damage in areas susceptible to flooding (10 CFR 50.48).

The fire detection and alarm system transmits fire alarm and supervisory signals to the control room audible and visual alarms. The system has signals for actuation of fire detectors, status indicators for most installed fire suppression systems, control and indicating lights for the fire pumps, level indicators for the fire water storage tanks, and door status indicator lights for operator notification of critical fire doors. The fire detection and alarm system is primarily electrical, but includes instrument air-operated valve and piping parts of an electrical tunnel fire alarm that actuates upon a loss of pressure within the piping.

The fire detection and alarm system has no intended function under 10 CFR 54.4(a)(2). The scoping and screening methodology identified the following fire detection and alarm system intended function, in accordance with 10 CFR 54.4(a)(3):

- Support a fire alarm in the electrical tunnel (10 CFR 50.48).

The mechanical portions of the fire detection and alarm system are within the scope of license renewal, but the pressure boundary for the instrument air piping is not required for the system to perform its intended function. Therefore, the components of the fire detection and alarm system are not subject to an AMR. The system drain portion is evaluated with plant drains (LRA Section 2.3.3.18). The fuel oil subsystem components are evaluated with fuel oil systems (LRA Section 2.3.3.13).

Nonsafety-related components not evaluated with other systems but whose failure could prevent satisfactory accomplishment of safety functions are evaluated with miscellaneous systems within the scope of license renewal, in accordance with 10 CFR 54.4(a)(2) (LRA Section 2.3.3.19). The remaining fire protection—water system and fire detection and alarm system components are evaluated in LRA Section 2.3.3.12.

LRA Tables 2.3.3-11-IP3 and 2.3.3-19-20-IP3 identify the fire protection—water system component types within the scope of license renewal and subject to an AMR, as well as their intended functions.

2.3B.3.11.2 Staff Evaluation

The staff reviewed LRA Section 2.3.3.11, UFSAR Sections 9.6.2.3 and 9.6.2.4, and license renewal drawings using the evaluation methodology described in SER Section 2.3 and the guidance in SRP-LR Section 2.3.

During its review, the staff evaluated the system functions described in the LRA and UFSAR to verify that the applicant had not omitted from the scope of license renewal any components with intended functions, as required by 10 CFR 54.4(a). The staff then reviewed those components that the applicant identified as within the scope of license renewal to verify that it had not omitted any passive and long-lived components subject to an AMR, in accordance with 10 CFR 54.21(a)(1).

The staff also reviewed NRC fire protection SERs for IP3, dated September 21, 1973; March 6, 1979; May 2, 1980; November 18, 1982; December 30, 1982; February 2, 1984; April 16, 1984; January 7, 1987; September 9, 1988; October 21, 1991; April 20, 1994; and January 5, 1995.

The staff also reviewed the IP3 commitments associated with 10 CFR 50.48 (i.e., an approved fire protection program) using its commitment responses to BTP APCSB 9.5-1 and Appendix A to BTP APCSB 9.5-1.

During its review of LRA Section 2.3.3.11, the staff identified areas in which additional information was necessary to complete the review of the applicant's scoping and screening results. The applicant responded to the staff's RAIs as discussed below.

In RAI 2.3B.3.11-1, dated October 24, 2007, the staff asked the applicant to explain why LRA license renewal drawings indicate that certain fire protection system components are not subject to an AMR. Specifically, license renewal drawing LRA-9321-40903-0 indicates that the following fire protection system component is not subject to an AMR (i.e., the component is not highlighted in green):

• FP-T-4 pneumatic tank and components

License renewal drawing LRA-9321-40913-001-0 indicates that the following fire protection system components are not subject to an AMR (i.e., these components are not highlighted in green):

• turbine generator building foam system
• turbine building wall spray system No. 3
• yard transformer separation spray system
• main transformer No. 31 deluge system
• main transformer No. 32 deluge system
• unit auxiliary transformer deluge system
• station auxiliary transformer deluge system
• north half sprinkler No. 6
• boiler room sprinkler system
• sprinkler system for AFW pump room
• lube oil storage tank

- lube oil reservoir
- manual/spray system No. 2 for the boiler feed pump
- hydrogen seal oil unit
- manual boiler feed pump oil accumulators Nos. 31 and 32
- boiler feed console pump

The staff requested that the applicant verify whether the above components are within the scope of license renewal, as required by 10 CFR 54.4(a), and subject to an AMR, in accordance with 10 CFR 54.21(a)(1). The staff requested that the applicant justify excluding any of these components from the scope of license renewal and an AMR.

In its response, dated November 16, 2007, the applicant provided scoping and screening results applicable to fire protection system components. With respect to the FP-T-4 pneumatic tank and components indicated on license renewal drawing LRA-9321-40903-0, the applicant stated the following:

> FP-T-4 pneumatic tank and components are not required for compliance with 10 CFR 50.48 and are not described in fire protection SERs, for the response to BTP APCSB 9.5-1, Appendix A. The pneumatic tank and components were used in the past to aid the two jockey pumps in maintaining fire loop pressure. They are no longer used and are isolated from the rest of the system by normally closed valve FP-84. Jockey pumps FP-P-5 and 6 maintain sufficient pressure on the fire protection piping system during non-fire conditions to prevent unnecessary starting of the main fire pumps.

Based on its review, the staff finds the applicant's response acceptable because the applicant clarified that it no longer relies on the FP-T-4 pneumatic tank and components to demonstrate compliance with 10 CFR 50.48. Jockey pumps FP-P-5 and FP-P-6 maintain sufficient pressure on the fire protection piping system during non-fire conditions to prevent the main fire pumps from unnecessarily starting.

With respect to license renewal drawing LRA-9321-40913-001-0, the applicant addressed each item in the staff's RAI. For the turbine generator building foam systems, the applicant stated the following:

> Fluid-containing portions of the turbine generator building foam systems are included with miscellaneous systems in-scope in compliance with 10 CFR 54.4(a)(2) and are subject to an AMR. The AMR for the fluid-containing portions of the systems are in LRA Table 3.3.2-19-20-IP3. Based on discussion in the SER for IP3 dated March 6, 1979, the foam suppression systems for various areas in the turbine building are considered to meet the scoping requirements of 10 CFR 54.4(a)(3), in addition to 10 CFR 54.4(a)(2). The AMR results in LRA Table 3.3.2-11 IP3 are applicable to the portions of the turbine generator building foam systems normally containing air.

Based on its review, the staff finds the applicant's response acceptable because it indicated that the generator building foam systems are included in the scope of license renewal with miscellaneous systems, in accordance with 10 CFR 54.4(a)(2), and are subject to an AMR.

For the turbine building wall spray system No. 3, the applicant stated the following:

> The turbine building wall spray system No. 3 is in-scope as shown on drawing LRA-9321- 40913-001-0, coordinates D5. The absence of boundary flags where highlighted piping enters a text box indicates that the portion of the system described in the text box is in-scope and subject to an AMR.

Based on its review, the staff finds the applicant's response acceptable because it indicated that the turbine building wall spray system No. 3 is within the scope of license renewal and subject to an AMR.

For the yard transformer separation spray system, the applicant stated the following:

> The yard transformer separation spray system is in-scope as shown on drawing LRA-9321- 40913-001-0, coordinates D5. The absence of boundary flags where highlighted piping enters a text box indicates that the portion of the system described in the text box is in-scope and subject to an AMR.

Based on its review, the staff finds the applicant's response acceptable because it indicated that the yard transformer separation spray system is within the scope of license renewal and subject to an AMR as shown on license renewal drawing LRA-9321- 40913-001-0.

For the main transformer No. 31 deluge system, the applicant stated the following:

> The deluge system and associated components for main transformer No. 31, adjacent to the control building, were initially determined to have no license renewal intended function. They were considered as required only to protect the transformer, to satisfy requirements of the plant insurance carrier. However, the spray systems provide for defense-in-depth, in addition to installed 3-hour rated fire barriers between the transformer and the turbine building, and are now considered in-scope and subject to an AMR. Applicable component types that are subject to an AMR are included in LRA Table 2.3.3-11-IP3, with the AMR results provided in LRA Table 3.3.2-11-IP3.

Based on its review, the staff finds the applicant's response acceptable because it clarified that (1) the main transformer No. 31 deluge system and its associated components have no license renewal intended function and (2) the water spray systems provide for defense in depth, in addition to the installed 3-hour-rated fire barriers, and are considered within the scope of license renewal and subject to an AMR.

For the main transformer No. 32 deluge system, the applicant stated the following:

> The deluge system and associated components for main transformer No. 32, adjacent to the control building, were initially determined to have no license renewal intended function. They were considered as required only to protect the transformer to satisfy requirements of the plant insurance carrier. However, the spray systems provide for defense-in-depth, in addition to installed 3-hour rated fire barriers between the transformer and the turbine building, and are now considered in-scope and subject to an AMR. Applicable component types that

2-159

are subject to an AMR are included in LRA Table 2.3.3-11-IP3, with the AMR results provided in LRA Table 3.3.2-11-IP3.

Based on its review, the staff finds the applicant's response acceptable because it clarified that (1) the main transformer No. 32 deluge system and its associated components have no license renewal intended function and (2) the water spray systems provide for defense in depth, in addition to the installed 3-hour-rated fire barriers, and are considered within the scope of license renewal and subject to an AMR.

For the unit auxiliary transformer deluge system, the applicant stated the following:

> The deluge system and associated components for the unit auxiliary transformer, adjacent to the control building, were initially determined to have no license renewal intended function. They were considered as required only to protect the transformer to satisfy requirements of the plant insurance carrier. However, the spray systems provide for defense-in-depth, in addition to installed 3-hour rated fire barriers between the transformer and the turbine building, and are now considered in-scope and subject to an AMR. Applicable component types that are subject to an AMR are included in LRA Table 2.3.3-11-IP3, with the AMR results provided in LRA Table 3.3.2-11-IP3.

Based on its review, the staff finds the applicant's response acceptable because it clarified that (1) the unit auxiliary transformer deluge systems and their associated components have no license renewal intended function and (2) the water spray systems provide for defense in depth, in addition to the installed 3-hour-rated fire barriers, and are considered within the scope of license renewal and subject to an AMR.

For the station auxiliary transformer deluge system, the applicant stated the following:

> The deluge system and associated components for the station auxiliary transformer, adjacent to the control building, were initially determined to have no license renewal intended function. They were considered as required only to protect the transformer to satisfy requirements of the plant insurance carrier. However, the spray systems provide for defense-in-depth, in addition to installed 3-hour rated fire barriers between the transformer and the turbine building, and are now considered in-scope and subject to an AMR. Applicable component types that are subject to an AMR are included in LRA Table 2.3.3-11-IP3, with the AMR results provided in LRA Table 3.3.2-11-IP3.

Based on its review, the staff finds the applicant's response acceptable because it clarified that (1) the station auxiliary transformer deluge systems and their associated components have no license renewal intended function and (2) the water spray systems provide for defense in depth, in addition to installed 3-hour-rated fire barriers, and are considered within the scope of license renewal and subject to an AMR.

For the north half sprinkler No. 6, the applicant stated the following:

> Turbine building north half sprinkler No. 6 system was initially determined to have no license renewal intended function, since a fire in the area protected by the system cannot disable the credited safe-shutdown equipment, which is located outside the area. However, based on discussion in the SER for IP3 dated March 6, 1979, the turbine building north half sprinkler No. 6 system provides defense-in-depth, in addition to hose stations throughout the turbine building, and fire barriers between the turbine building and control building. The system is therefore considered within the scope of license renewal and subject to an AMR, in accordance with 10 CFR 54.4(a)(3). The AMR results in LRA Table 3.3.2-11-IP3 are applicable to the turbine building north half sprinkler system No. 6.

Based on its review, the staff finds the applicant's response acceptable because it indicated that the turbine building north half sprinkler No. 6 system is within the scope of license renewal and subject to an AMR.

For the boiler room sprinkler system, the applicant stated the following:

> The boiler room sprinkler system is not required to satisfy the provisions of BTP APCSB 9.5-1, Appendix A and is not credited in fire protection SERs. The boiler room sprinkler system is maintained to satisfy requirements of the plant insurance carrier. The boiler room sprinkler system does not protect safety-related equipment and is not located near any building housing safety-related equipment. Fire in the area of the boiler room will be contained within that area and not affect safe-shutdown equipment, due to its location and limited amount of combustibles.

Based on its review, the staff finds the applicant's response acceptable because it clarified that the provisions of 10 CFR 50.48 do not require the boiler room sprinkler system and components because this system does not protect safety-related equipment and is not located near any building housing safety-related equipment.

For the sprinkler system for AFW pump room, the applicant stated the following:

> The sprinkler system for the auxiliary feedwater pump room is in-scope as shown on drawing LRA-9321-40913-001-0, coordinate (E8). The absence of boundary flags where the highlighted piping enters the text box indicates that the portion of the system described in the text box is in-scope and subject to an AMR.

Based on its review, the staff finds the applicant's response acceptable because it indicated that the sprinkler system for the AFW pump room is within the scope of license renewal and subject to an AMR.

For the lube oil storage tank foam system, the applicant stated the following:

> Fluid-containing portions of the LO storage tank foam suppression systems are included with miscellaneous systems in-scope pursuant to 10 CFR 54.4(a)(2)

and are subject to an AMR. The AMR results for the fluid-containing portions of the system are included in LRA Table 3.3.2-19-20-IP3. Based on discussion in the SER for IP3 dated March 6, 1979, the LO storage tank foam suppression system is considered as meeting the scoping requirements of 10 CFR 54.4(a)(3), in addition to 10 CFR 54.4(a)(2). The AMR results in LRA Table 3.3.2-11-IP3 are applicable to the portions of the LO storage tank foam suppression system normally containing air.

Based on its review, the staff finds the applicant's response acceptable because it clarified that lube oil storage tank foam suppression systems are included in the scope of license renewal with miscellaneous systems in accordance with 10 CFR 54.4(a)(2) and are subject to an AMR.

For the lube oil reservoir foam system, the applicant stated the following:

Fluid-containing portions of the LO reservoir foam suppression systems are included with miscellaneous systems in-scope pursuant to 10 CFR 54.4(a)(2) and are subject to an AMR. The AMR results for the fluid-containing portions of the system are provided in LRA Table 3.3.2-19-20-IP3. Based on discussion in the SER for IP3 dated March 6, 1979, the LO reservoir foam suppression system is considered as meeting the scoping requirements of 10 CFR 54.4(a)(3), in addition to 10 CFR 54.4(a)(2). The AMR results in LRA Table 3.3.2-11-IP3 are applicable to the portions of the LO reservoir foam suppression system normally containing air.

Based on its review, the staff finds the applicant's response acceptable because it clarified that the lube oil reservoir foam suppression systems are included in the scope of license renewal with miscellaneous systems, in accordance with 10 CFR 54.4(a)(2), and are subject to an AMR.

For the manual/spray system No. 2 for the boiler feed pump, the applicant stated the following:

The manual/spray system No. 2 for the boiler feed pump is not required to satisfy the provisions of BTP APCSB 9.5-1, Appendix A, and is not credited in fire protection SERs. The manual spray system No. 2 satisfies requirements of the plant insurance carrier. SER Section 5.9.1 states there is no safety-related equipment or electrical cables located within the turbine building. SER Section 5.9.6 discusses modifications to provide three-hour fire-rated doors and dampers in the barriers between the turbine building and the control building, as well as upgrading penetrations to a three-hour fire-rating.

Based on its review, the staff finds the applicant's response acceptable because it explained that the manual/spray system No. 2 for the boiler feed pump does not have a license renewal intended function. The manual/spray system No. 2 for the boiler feed pump does not provide a fire protection function as part of the applicant's approach to complying with 10 CFR 50.48; thus, the associated fire protection components are not within the scope of license renewal.

For the hydrogen seal oil unit foam system, the applicant stated the following:

Fluid-containing portions of the H_2 seal oil unit foam suppression systems are included with miscellaneous systems in-scope pursuant to 10 CFR 54.4(a)(2)

2-162

and are subject to an AMR. The AMR results for the fluid-containing portions of the system are included in LRA Table 3.3.2-19-20-IP3. Based on discussion in the SER for IP3 dated March 6, 1979, the H_2 seal oil unit foam suppression system is considered as meeting the scoping requirements of 10 CFR 54.4(a)(3), in addition to 10 CFR 54.4(a)(2). The AMR results in Table 3.3.2-11-IP3 are applicable to the portions of the H_2 seal oil unit foam suppression system normally containing air.

Based on its review, the staff finds the applicant's response acceptable because it clarified that hydrogen seal oil unit foam suppression systems are included in the scope of license renewal with miscellaneous systems, in accordance with 10 CFR 54.4(a)(2), and are subject to an AMR.

For the manual boiler feed pump oil accumulators Nos. 31 and 32 foam system, the applicant stated the following:

> Fluid-containing portions of the manual boiler feed pump oil accumulators No. 31 and 32 foam suppression systems are included with miscellaneous systems in-scope pursuant to 10 CFR 54.4(a)(2) and are subject to an AMR. The AMR results for the fluid-containing portions of the system are shown in LRA Table 3.3.2-19-20-IP3. Based on discussion in the SER for IP3 dated March 6, 1979, the manual boiler feed pump oil accumulators No. 31 and 32 foam suppression system is considered as meeting the scoping requirements of 10 CFR 54.4(a)(3), in addition to 10 CFR 54.4(a)(2). The AMR results in LRA Table 3.3.2-11-IP3 are applicable to the portions of the manual boiler feed pump oil accumulators No. 31 and 32 foam suppression system normally containing air.

Based on its review, the staff finds the applicant's response acceptable because it clarified that the manual boiler feed pump oil accumulators No. 31 and 32 foam suppression systems are included in the scope of license renewal with miscellaneous systems, in accordance with 10 CFR 54.4(a)(2), and are subject to an AMR.

For the boiler feed console pump foam system, the applicant stated the following:

> Fluid-containing portions of the boiler feed console pump foam suppression systems are included with miscellaneous systems in-scope pursuant to 10 CFR 54.4(a)(2) and are subject to an AMR. The AMR results for the fluid-containing portions of the system are shown in LRA Table 3.3.2-19-20-IP3. Based on discussion in the SER for IP3 dated March 6, 1979, the boiler feed console pump foam suppression system is considered as meeting the scoping requirements of 10 CFR 54.4(a)(3), in addition to 10 CFR 54.4(a)(2). The AMR results in LRA Table 3.3.2-11-IP3 are applicable to the portions of the boiler feed console pump foam suppression system normally containing air.

Based on its review, the staff finds the applicant's response acceptable because it clarified that the boiler feed console pump foam suppression systems are included in the scope of license renewal with miscellaneous systems in accordance with 10 CFR 54.4(a)(2) and are subject to an AMR.

Based on its review, the staff finds the applicant's response to RAI 2.3B.3.11-1 acceptable. The staff's concern described in RAI 2.3B.3.11-1 is resolved.

In RAI 2.3B.3.11-2, dated October 24, 2007, the staff stated that Section 3.1.8 of the fire protection SER for IP3, dated March 6, 1979, discusses dry-pipe, pre-action sprinkler systems for all cable trays in the electrical tunnels, electrical penetration areas, and cable trays in the motor control center areas. LRA Section 2.3.3.11 does not indicate that the dry-pipe pre-action sprinkler systems are within the scope of license renewal and subject to an AMR. The staff requested that the applicant verify whether the dry-pipe pre-action sprinkler systems for all cable trays in the electrical tunnels, electrical penetration areas, and cable trays in the motor control center areas are within the scope of license renewal, as required by 10 CFR 54.4(a), and subject to an AMR, in accordance with 10 CFR 54.21(a)(1). If they are excluded from the scope of license renewal and are not subject to an AMR, the staff asked the applicant to justify their exclusion.

In its response, dated November 16, 2007, the applicant stated that the dry-pipe, pre-action sprinkler systems for all cable trays in the electrical tunnels, electrical penetration areas, and cable trays in the motor control center areas are within the scope of license renewal and subject to an AMR. License renewal drawing LRA-9321-40913-001-0 shows the electrical tunnel dry pipe pre-action sprinkler systems 8, 8A, 9, and 9A at coordinates G6. The electrical tunnel sprinkler systems cover areas in the electrical penetration area and cable trays in the motor control center areas, in addition to the cable trays in the electrical tunnels. The absence of boundary flags where the highlighted piping enters the text box indicates that the portion of the system described in the text box is within scope and subject to an AMR.

Based on its review, the staff finds the response to RAI 2.3B.3.11-2 acceptable because the applicant identified the dry-pipe, pre-action sprinkler systems for all cable trays in the electrical tunnels, electrical penetration areas, and cable trays in the motor control center areas as within the scope of license renewal and subject to an AMR. Therefore, the staff concludes that the applicant correctly identified these dry-pipe, pre-action sprinkler systems and the associated components as within the scope of license renewal and subject to an AMR. The staff's concern described in RAI 2.3B.3.11-2 is resolved.

In RAI 2.3B.3.11-3, dated October 24, 2007, the staff stated that Section 5.9.1 of the March 6, 1979, fire protection SER for IP3 discusses automatic deluge foam suppression systems for various areas in the turbine building. LRA Section 2.3.3.11 does not indicate that the foam suppression systems are within the scope of the license renewal and subject to an AMR. The staff requested that the applicant verify whether the foam suppression systems for various areas in the turbine building are within the scope of license renewal, as required by 10 CFR 54.4(a), and subject to an AMR, in accordance with 10 CFR 54.21(a)(1). If the systems are excluded from the scope of license renewal and are not subject to an AMR, the staff asked the applicant to justify their exclusion.

In its response, dated November 16, 2007, the applicant stated that the fluid-containing portions of the foam suppression systems for various areas in the turbine building are included with miscellaneous systems, in accordance with 10 CFR 54.4(a)(2), and are subject to an AMR. LRA Table 3.3.2-19-20-IP3 summarizes the AMR results for the fluid-containing portions of the systems. Based on the discussion in the March 6, 1979, fire protection SER for IP3, the foam suppression systems for various areas in the turbine building meet the scoping requirements of

10 CFR 54.4(a)(3), in addition to 10 CFR 54.4(a)(2). The applicant further identified the system components that are subject to an AMR, in accordance with 10 CFR 54.21(a)(1). The applicant indicated that LRA Table 3.3.2-11-IP3 summarizes the AMR results.

Based on its review, the staff finds the applicant's response to RAI 2.3B.3.11-3 acceptable because fluid-containing portions of the foam systems for various areas in the turbine building were identified as being within the scope of license renewal and subject to an AMR. The AMR results are summarized in LRA Table 3.3.2-20-IP3.

In RAI 2.3B.3.11-4, dated October 24, 2007, the staff stated that Section 5.11.1 of the March 6, 1979, fire protection SER for IP3 discusses wet pipe automatic sprinklers in the diesel generator building sump area beneath each diesel engine and on the diesel day tank. On license renewal drawing LRA-9321-40913-0, at coordinate E3, the wet pipe automatic sprinkler system does not appear to be within the scope of the license renewal and subject to an AMR (i.e., the box surrounding the sprinklers in question is not highlighted). The staff requested that the applicant verify whether the wet pipe sprinkler system designed to protect the diesel generator building sump area and diesel day tank is within the scope of license renewal, as required by 10 CFR 54.4(a), and subject to an AMR, in accordance with 10 CFR 54.21(a)(1). If the system is excluded from the scope of license renewal and is not subject to an AMR, the staff asked the applicant to justify its exclusion.

In its response, dated November 16, 2007, the applicant stated that the IP3 wet pipe automatic sprinklers in the diesel generator building sump area beneath each diesel engine and on the diesel day tanks are in scope and subject to an AMR, as shown on license renewal drawing LRA-9321-40913-001-0, coordinate E3. The absence of boundary flags where the highlighted piping enters the text box indicates that the portion of the system described in the text box is within the scope of license renewal and subject to an AMR, along with the highlighted components on the drawing.

Based on its review, the staff finds the response to RAI 2.3B.3.11-4 acceptable because the applicant identified wet pipe automatic sprinklers in the diesel generator building sump area beneath each diesel engine and on the diesel day tank as within the scope of license renewal and subject to an AMR. Further, the applicant clarified that the absence of boundary flags where the highlighted piping enters the text box indicates that the portion of the system described in the text box is within the scope of license renewal and subject to an AMR, along with the highlighted components on license renewal drawing LRA-9321-40913-001-0. Therefore, the staff concludes that the applicant correctly identified the wet pipe automatic sprinklers in question as within the scope of license renewal and subject to an AMR. The staff's concern described in RAI 2.3B.3.11-4 is resolved.

In RAI 2.3B.3.11-5, dated October 24, 2007, the staff stated that Section 5.13.1 of the March 6, 1979, fire protection SER for IP3 discusses the charcoal filter manual water spray system. LRA Section 2.3.3.11 does not indicate that the manual water spray system and its associated components are within the scope of the license renewal and subject to an AMR. The staff requested that the applicant verify whether the charcoal filter manual water spray system and its associated components are within the scope of license renewal, as required by 10 CFR 54.4(a), and subject to an AMR, in accordance with 10 CFR 54.21(a)(1). If the system is excluded from the scope of license renewal and is not subject to an AMR, the staff asked the applicant to justify its exclusion.

In its response, dated November 16, 2007, the applicant stated that the IP3 charcoal filter manual water spray system is in scope, as shown on license renewal drawing LRA-9321-40913-001-0 at coordinates H8. The absence of boundary flags where the highlighted piping enters the text box indicates that the portion of the system described in the text box is in scope and subject to an AMR, along with the highlighted components on the drawing. License renewal drawing LRA-9321-40913-001-0 continues to an equipment arrangement drawing which is not available as a license renewal drawing.

Based on its review, the staff finds the response to RAI 2.3B.3.11-5 acceptable because the applicant identified the charcoal filter manual water spray system in question as within the scope of license renewal and subject to an AMR. Further, the applicant clarified that the absence of boundary flags where the highlighted piping enters the text box indicates that the portion of the system described in the text box is within the scope of license renewal and subject to an AMR, along with the highlighted components on license renewal drawing LRA-9321-40913-001-0.

In RAI 2.3B.3.11-6, dated October 24, 2007, the staff stated that Section 5.15.1 of the March 6, 1979, fire protection SER for IP3 discusses automatic water spray systems for oil-filled transformers located adjacent to the control building. LRA Section 2.3.3.11 does not indicate that the automatic water spray systems and their associated components are within the scope of license renewal and subject to an AMR. The staff requested that the applicant verify whether the automatic water spray systems for oil-filled transformers are within the scope of license renewal, as required by 10 CFR 54.4(a), and subject to an AMR, in accordance with 10 CFR 54.21(a)(1). If the systems are excluded from the scope of license renewal and are not subject to an AMR, the staff asked the applicant to justify their exclusion.

In its response, dated November 16, 2007, the applicant stated that it initially determined that the automatic water spray systems and their associated components for the oil-filled transformers located adjacent to the control building did not have a license renewal intended function. The applicant believed that they were only required to protect the transformers, satisfying requirements of the plant insurance carrier. However, the spray systems provide for defense in depth, in addition to the installed 3-hour-rated fire barriers between the control building and the transformer yard, and are considered in scope and subject to an AMR. LRA Table 2.3.3-11-IP3 includes the applicable component types subject to an AMR, and LRA Table 3.3.2-11-IP3 provides the AMR results.

Based on its review, the staff finds the response to RAI 2.3B.3.11-6 acceptable because the applicant concluded that the automatic spray system for the oil-filled transformer performs a defense-in-depth function and, therefore, is within the scope of license renewal and subject to an AMR. The staff confirmed that LRA Table 3.3.2-11-IP3 provides the AMR results. Therefore, the staff finds that the applicant correctly identified the automatic water spray systems and their associated components for the oil-filled transformers as within the scope of license renewal and subject to an AMR. The staff's concern described in RAI 2.3B.3.11-6 is resolved.

2.3B.3.11.3 Conclusion

The staff reviewed the LRA, UFSAR, RAI responses, and drawings to determine whether the applicant failed to identify any SSCs within the scope of license renewal. The staff found no such omissions. In addition, the staff sought to determine whether the applicant failed to identify any components subject to an AMR. The staff found no such omissions. On the basis of its review, the staff concludes that the applicant has adequately identified the fire protection - water system components within the scope of license renewal, as required by 10 CFR 54.4(a), and those subject to an AMR, as required by 10 CFR 54.21(a)(1).

2.3B.3.12 IP3 Fire Protection—Carbon Dioxide, Halon, and RCP Oil Collection Systems

2.3B.3.12.1 Summary of Technical Information in the Application

LRA Section 2.3.3.12 describes the fire protection—CO_2, Halon, and RCP oil collection system, which is listed under the following system codes:

- CO_2 system: system code CO2
- Halon: system code HAL
- RCP oil collection components: system code RCS

The CO_2 system provides fire protection and supplies CO_2 gas to purge the main generator. The CO_2 fire protection system has two 10-ton-capacity, low-pressure tanks, a distribution header, piping, and valves. An automatic total-flooding CO_2 fire suppression system protects the 480-V switchgear room, cable spreading room, diesel generator rooms, and the turbine generator exciter enclosure. A local application CO_2 fire suppression system protects the turbine building, including the main boiler FW pumps, turbine governor, MS and reheat valves, and generator bearings. Before maintenance work on the main generator, the hydrogen gas must be evacuated from the system. Inert CO_2 gas from a CO_2 gas-vaporizing system purges the generator. The IP2 CO_2 gas-vaporizing system also may operate through a supply line from the IP1 intake structure area.

The Halon 1301 system suppresses fires in the administration/service building technical support center/computer room, in the Appendix R diesel enclosure, and in the meteorological building. The Halon system does not protect any safety-related plant equipment. Protection of the Appendix R diesel enclosure from fire is not a required function under Appendix R. For IP3, the Halon 1301 system has no intended functions under 10 CFR 54.4(a)(1), 10 CFR 54.4(a)(2), or 10 CFR 54.4(a)(3).

The RCP oil collection system is designed, engineered, and installed so an RCP lube oil system failure will not lead to fire during normal or DBA conditions or impact any safety-related system capability during a safe-shutdown earthquake. The collection system can collect lube oil from all pressurized and unpressurized potential leakage sites in the RCP lube oil systems and drain it to a vented closed tank that can hold the required lube oil system inventory. A flame arrester in each tank vent prevents fire flashback. The collection system consists of leakproof enclosures or pans under oil-bearing components to contain leaks.

The fire protection—CO_2 and RCP oil collection systems have no intended functions under 10 CFR 54.4(a)(1).

The scoping and screening methodology identified the following RCP oil collection system intended function, in accordance with 10 CFR 54.4(a)(2):

- Maintain integrity of nonsafety-related components such that no physical interaction with safety-related components could prevent satisfactory accomplishment of a safety function.

The scoping and screening methodology also identified the following CO_2 and RCP oil collection systems intended functions, in accordance with 10 CFR 54.4(a)(3):

- Provide automatic and manual CO_2 flooding for areas of the plant that (1) contain safety-related equipment or (2) pose significant hazards to plant areas containing safety-related equipment (10 CFR 50.48) or both.

- Provide each RCP with an oil collection system that is designed to contain and direct the oil to remote storage containers in the event of an oil leak.

LRA Table 2.3.3-12-IP3 identifies fire protection—CO_2 and RCP oil collection systems component types within the scope of license renewal and subject to an AMR, as well as their intended functions.

2.3B.3.12.2 Staff Evaluation

The staff reviewed LRA Section 2.3.3.11, UFSAR Sections 9.6.2.3 and 9.6.2.4, and license renewal drawings using the evaluation methodology described in SER Section 2.3 and the guidance in SRP-LR Section 2.3.

During its review, the staff evaluated the system functions described in the LRA and UFSAR to verify that the applicant had not omitted from the scope of license renewal any components with intended functions, as required by 10 CFR 54.4(a). The staff then reviewed those components that the applicant identified as within the scope of license renewal to verify that the applicant had not omitted any passive and long-lived components subject to an AMR, in accordance with 10 CFR 54.21(a)(1).

The staff also reviewed the following IP3 fire protection CLB documents listed in the IP3 Operating License Condition 2.H: NRC fire protection SERs for IP3 dated September 21, 1973; March 6, 1979; May 2, 1980; November 18, 1982; December 30, 1982; February 2, 1984; April 16, 1984; January 7, 1987; September 9, 1988; October 21, 1991; April 20, 1994; and January 5, 1995.

The staff also reviewed IP3 commitments associated with 10 CFR 50.48 (i.e., an approved fire protection program), using its commitment responses to BTP APCSB 9.5-1 and BTP APCSB 9.5-1, Appendix A.

During its review of LRA Section 2.3.3.12, the staff identified areas in which additional information was necessary to complete the review of the applicant's scoping and screening results. The applicant responded to the staff's RAIs as discussed below.

In RAI 2.3B.3.12-1, dated October 24, 2007, the staff asked the applicant to explain why license renewal drawing LRA-9321-24403-0 indicated that the following fire protection system components were not subject to an AMR (i.e., they are not highlighted in brown):

- Appendix R diesel generator Halon 1301 system
- technical support center/plant computer Halon system
- IP3 record room vault Halon 1301 system

The staff requested that the applicant verify whether the above components are within the scope of license renewal, as required by 10 CFR 54.4(a), and subject to an AMR, in accordance with 10 CFR 54.21(a)(1). If these components are excluded from the scope of license renewal and are not subject to an AMR, the staff asked the applicant to justify their exclusion.

In its response, dated November 16, 2007, the applicant addressed each system individually. For the Appendix R diesel generator Halon 1301 system, the applicant stated that the Appendix R diesel generator is located in a standalone structure separated from other plant structures and equipment. The applicant further explained that the technical support center/plant computer and the record room vault are located in an administration building attached to the turbine building. The applicant added that a sprinkler system had replaced the IP3 record room vault Halon 1301 system.

The applicant stated that the areas referenced in the RAI response do not contain systems or components required for safe shutdown of the plant, do not provide an exposure hazard to any building or area required for safe shutdown, and are not located in safety-related areas. The applicable IP3 fire protection SER, dated March 6, 1979, credits no fire suppression systems for these areas. The Halon systems are not required for compliance with 10 CFR 50.48. The fire protection SER does not stipulate the addition of suppression systems for the Appendix R diesel generator, technical support center/plant computer, or the IP3 record room vault.

Based on its review, the staff finds the applicant's response to RAI 2.3B.3.12-1 acceptable. The applicant does not credit the Halon 1301 systems for the Appendix R diesel generator room, technical support center/plant computer room, and record room vault toward meeting the requirements of Appendix R to 10 CFR Part 50 for achieving safe shutdown in the event of a fire. Although the IP3 March 6, 1979, fire protection SER addresses the Halon 1301 systems for the Appendix R diesel generator room, technical support center/plant computer room, and record room vault, NRC fire protection regulations do not require these systems. The Appendix R diesel generator room, technical support center/plant computer room, and record room vault are not safety related and cannot affect safety-related equipment by spatial interaction. Furthermore, they are not required for safe shutdown. Therefore, they have no intended function under 10 CFR 54.4(a)(2). In addition, the staff reviewed commitments made by the applicant to satisfy BTP APCSB 9.5-1, Appendix A, which discusses Halon 1301 systems and found no intended function associated with 10 CFR 54.4(a)(2). Therefore, the staff finds that the applicant correctly excluded the Halon 1301 systems for the Appendix R diesel generator room, technical support center/plant computer room, and record room vault from the scope of license renewal and an AMR. The staff's concern described in RAI 2.3B.3.12-1 is resolved.

2.3B.3.12.3 Conclusion

The staff reviewed the LRA, UFSAR, RAI responses, and drawings to determine whether the applicant failed to identify any SSCs within the scope of license renewal. The staff found no such omissions. In addition, the staff sought to determine whether the applicant failed to identify any components subject to an AMR. The staff found no such omissions. On the basis of its review, the staff concludes that the applicant has adequately identified the fire protection CO_2, Halon, and RCP oil collection system components within the scope of license renewal, as required by 10 CFR 54.4(a), and those subject to an AMR, as required by 10 CFR 54.21(a)(1).

2.3B.3.13 IP3 Fuel Oil Subsystems

2.3B.3.13.1 Summary of Technical Information in the Application

LRA Section 2.3.3.13 describes the IP3 fuel oil subsystems, which include the IP3 EDGs, the IP3 fire protection diesel engines, and the IP3 Appendix R diesel generator.

Each diesel fuel oil storage and transfer system supplying fuel to the EDGs has its own fuel oil day tank and an underground storage tank. The day tanks are within the diesel generator buildings. An engine-driven fuel oil pump supplies the fuel from the day tank to the engine. The day tank fills automatically during engine operation from its dedicated underground storage tank, which is adjacent to the diesel generator building. Each underground storage tank has a motor-driven transfer pump to transfer fuel to the day tank.

Independent diesel fuel oil storage and transfer systems supply fuel to the IP2 and IP3 fire protection diesel engines. The IP3 fuel oil storage tank and components are located in the IP3 fire protection pump house.

An independent diesel fuel oil storage and transfer system supplies fuel to the IP3 Appendix R diesel generator, which has its own fuel oil day tank and underground storage tank. The day tank supplies fuel directly to the engine. A transfer pump fills the fuel oil day tank automatically from its storage tank during engine operation.

The fuel oil subsystems contain safety-related components relied on to remain functional during and following DBEs. They also contain nonsafety-related components whose failure potentially could prevent the satisfactory accomplishment of a safety-related function. In addition, the fuel oil subsystems perform functions that support fire protection and SBO.

LRA Table 2.3.3-13-IP3 identifies fuel oil subsystem component types within the scope of license renewal and subject to an AMR, as well as their intended functions.

2.3B.3.13.2 Staff Evaluation

The staff reviewed LRA Section 2.3.3.13; UFSAR Sections 1.3.1, 8.2, and 16.1.3; and license renewal drawings using the evaluation methodology described in SER Section 2.3 and the guidance in SRP-LR Section 2.3.

During its review, the staff evaluated the system functions described in the LRA and UFSAR to verify that the applicant had not omitted from the scope of license renewal any components with

intended functions, as required by 10 CFR 54.4(a). The staff then reviewed those components that the applicant identified as within the scope of license renewal to verify that it had not omitted any passive and long-lived components subject to an AMR, in accordance with 10 CFR 54.21(a)(1).

During its review of LRA Section 2.3.3.13, the staff identified an area in which additional information was necessary to complete the review of the applicant's scoping and screening results. The discussion of the staff's RAIs in SER Section 2.3B.3 details the disposition of RAI 2.3B.3.13-1, dated February 13, 2008.

2.3B.3.13.3 Conclusion

The staff reviewed the LRA, UFSAR, RAI response, and drawings to determine whether the applicant failed to identify any SSCs within the scope of license renewal. The staff found no such omissions. In addition, the staff sought to determine whether the applicant failed to identify any components subject to an AMR. The staff found no such omissions. On the basis of its review, the staff concludes that the applicant has appropriately identified the fuel oil system components within the scope of license renewal, as required by 10 CFR 54.4(a), and those subject to an AMR, in accordance with 10 CFR 54.21(a)(1).

2.3B.3.14 IP3 Emergency Diesel Generator System

2.3B.3.14.1 Summary of Technical Information in the Application

LRA Section 2.3.3.14 describes the EDG system, which supplies emergency shutdown power upon loss of all other alternating current auxiliary power. The system consists of three EDG sets, each with a diesel engine coupled to a 480-V generator. Each emergency diesel is started automatically by two redundant air motors and has an air storage tank and compressor system, its own starting air subsystem, fuel oil subsystem, intake air subsystem, exhaust subsystem, lube oil subsystem, and jacket water cooling subsystem. The EDG system also has ventilation equipment for the diesel generator building.

The EDG system contains safety-related components relied on to remain functional during and following DBEs. It also contains nonsafety-related components whose failure could prevent the satisfactory accomplishment of a safety-related function. In addition, the EDG system performs functions that support fire protection.

The HVAC component parts of this system code are reviewed with HVAC systems (LRA Section 2.3.3.8). Fuel oil subsystem components are evaluated with fuel oil (LRA Section 2.3.3.13). Nonsafety-related components not evaluated with other systems and whose failure could prevent satisfactory accomplishment of safety functions are evaluated with miscellaneous systems (LRA Section 2.3.3.19). Remaining components are evaluated in LRA Section 2.3.3.14.

LRA Tables 2.3.3-14-IP3, 2.3.3-19-16-IP3, and 2.3.3-19-17-IP3 identify EDG system component types within the scope of license renewal and subject to an AMR, as well as their intended functions.

2.3B.3.14.2 Staff Evaluation

The staff reviewed LRA Section 2.3.3.14, UFSAR Sections 8.2 and 16.1.3, and license renewal drawings using the evaluation methodology described in SER Section 2.3 and the guidance in SRP-LR Section 2.3.

During its review, the staff evaluated the system functions described in the LRA and UFSAR to verify that the applicant had not omitted from the scope of license renewal any components with intended functions, as required by 10 CFR 54.4(a). The staff then reviewed those components that the applicant identified as within the scope of license renewal to verify that it had not omitted any passive and long-lived components subject to an AMR, in accordance with 10 CFR 54.21(a)(1).

During its review of LRA Section 2.3.3.14, the staff identified an area in which additional information was necessary to complete the review of the applicant's scoping and screening results. The applicant responded to the staff's RAI as discussed below.

In RAI 2.3B.3.14-1, dated December 7, 2007, the staff noted that a license renewal drawing for the IP3 jacket water to EDGs identified that the jacket water pumps for diesel engine Nos. 31, 32, and 33 are not subject to an AMR, in accordance with 10 CFR 54.21(a), because they are not long-lived components. The staff noted that SRP-LR, Table 2.3-2, provides examples of passive, long-lived components, such as diesel engine jacket water skid-mounted equipment. To complete its review, the staff requested that the applicant confirm that the jacket water pumps are short-lived components and describe its method for periodic replacement of these components.

In its response, dated January 4, 2008, the applicant stated that IP3 EDG maintenance procedures specify that the jacket water pumps in question are scheduled for replacement every 16 years, in accordance with station maintenance procedures, and, therefore, they are not subject to an AMR.

Based on its review, the staff finds the response to RAI 2.3B.3.14-1 acceptable because the applicant adequately explained that the practice of replacing the jacket water pumps meets the intent of 10 CFR 54.21(a)(1)(ii) for short-lived components and that the maintenance procedures control the pumps' periodic replacement. Therefore, the staff agrees that the jacket water pumps are not subject to an AMR. The staff's concern described in RAI 2.3B.3.14-1 is resolved.

In RAI 2.3A.3.14-2, dated December 7, 2007, the staff noted that license renewal drawings for the EDG jacket water cooling systems and EDG fuel oil systems for IP2 and IP3 label multiple "flexible conn [connections]" as not long-lived components. By letter dated January 4, 2008, the applicant responded to the staff's RAI. SER Section 2.3A.3.14 documents the RAI, the applicant's response, and the staff's evaluation.

The discussion of the staff's RAIs in SER Section 2.3B.3 details the disposition of RAI 2.3B.3.14-2, dated February 13, 2008.

2.3B.3.14.3 Conclusion

The staff reviewed the LRA, UFSAR, RAI responses, and drawings to determine whether the applicant failed to identify any SSCs within the scope of license renewal. The staff found no such omissions. In addition, the staff sought to determine whether the applicant failed to identify any components subject to an AMR. The staff found no such omissions. On the basis of its review, the staff concludes that the applicant has appropriately identified the EDG system components within the scope of license renewal, as required by 10 CFR 54.4(a), and those subject to an AMR, as required by 10 CFR 54.21(a)(1).

2.3B.3.15 IP3 Security Generator System

2.3B.3.15.1 Summary of Technical Information in the Application

LRA Section 2.3.3.15 describes the security propane generator system, which supplies power for the security lighting system and other security functions. The applicant credits a portion of this security lighting under Appendix R, Section III.J (emergency lighting), to illuminate ingress and egress to the Appendix R diesel generator, main and backup SW pumps, CST, and RWST.

The security propane generator system performs functions that support fire protection.

LRA Table 2.3.3-15-IP3 identifies security propane generator system component types within the scope of license renewal and subject to an AMR as well as their intended functions.

2.3B.3.15.2 Staff Evaluation

The staff reviewed LRA Section 2.3.3.15 and UFSAR Section 9.6.2.6 using the evaluation methodology described in SER Section 2.3 and the guidance in SRP-LR Section 2.3.

During its review, the staff evaluated the system functions described in the LRA and UFSAR to verify that the applicant had not omitted from the scope of license renewal any components with intended functions, as required by 10 CFR 54.4(a). The staff then reviewed those components that the applicant identified as within the scope of license renewal to verify that it had not omitted any passive and long-lived components subject to an AMR, in accordance with 10 CFR 54.21(a)(1).

2.3B.3.15.3 Conclusion

The staff reviewed the LRA and UFSAR to determine whether the applicant failed to identify any SSCs within the scope of license renewal. The staff found no such omissions. In addition, the staff sought to determine whether the applicant failed to identify any components subject to an AMR. The staff found no such omissions. On the basis of its review, the staff concludes that the applicant has adequately identified the security propane generator system components within the scope of license renewal, as required by 10 CFR 54.4(a), and those subject to an AMR, as required by 10 CFR 54.21(a)(1).

2.3B.3.16 IP3 Appendix R Diesel Generator System

2.3B.3.16.1 Summary of Technical Information in the Application

LRA Section 2.3.3.16 describes the Appendix R diesel generator system, which supplies power to selected equipment and power supplies relied on in Appendix R and SBO events. The Appendix R diesel generator complies with SBO requirements and can supply sufficient power for safe-shutdown loads through the 6.9-kV distribution and the emergency 480-V buses and motor control centers or the turbine building switchgear and motor control centers. Located in a separate structure in the yard area, the Appendix R diesel generator installation is a self-contained package that operates upon a complete loss of power and includes a starting air compressor, batteries, battery charger, jacket water heater, lube oil heater, fuel oil pump and lube oil pumps, and necessary filters and strainers.

The Appendix R diesel generator system performs functions that support fire protection and SBO.

Fuel oil subsystem components are reviewed with fuel oil (LRA Section 2.3.3.13). Ventilation for the Appendix R diesel generator system is reviewed with HVAC systems (LRA Section 2.3.3.8). Remaining components are evaluated in LRA Section 2.3.3.16.

LRA Table 2.3.3-16-IP3 identifies Appendix R diesel generator system component types within the scope of license renewal and subject to an AMR, as well as their intended functions.

2.3B.3.16.2 Staff Evaluation

The staff reviewed LRA Section 2.3.3.16, UFSAR Sections 8.1.1 and 8.2.3, and license renewal drawings using the evaluation methodology described in SER Section 2.3 and the guidance in SRP-LR Section 2.3.

During its review, the staff evaluated the system functions described in the LRA and UFSAR to verify that the applicant had not omitted from the scope of license renewal any components with intended functions, as required by 10 CFR 54.4(a). The staff then reviewed those components that the applicant has identified as within the scope of license renewal to verify it had not omitted any passive and long-lived components subject to an AMR, in accordance with 10 CFR 54.21(a)(1).

2.3B.3.16.3 Conclusion

The staff reviewed the LRA, UFSAR, and drawings to determine whether the applicant failed to identify any SSCs within the scope of license renewal. The staff found no such omissions. In addition, the staff sought to determine whether the applicant failed to identify any components subject to an AMR. The staff found no such omissions. On the basis of its review, the staff concludes that the applicant has adequately identified the Appendix R diesel generator system components within the scope of license renewal, as required by 10 CFR 54.4(a), and those subject to an AMR, as required by 10 CFR 54.21(a)(1).

2.3B.3.17 IP3 City Water System

2.3B.3.17.1 Summary of Technical Information in the Application

LRA Section 2.3.3.17 describes the city water system, which supplies water to various components throughout the plant. The city water supply was installed originally for IP1, but now has functions for all three units. The IP2 city water description includes the city water tank and many of the shared site components. This system includes only the IP3 components. City water is used for a variety of purposes throughout IP3, such as supplying water to fire protection systems, to equipment for makeup or cooling, and to sanitary and potable facilities (e.g., emergency showers, eye wash stations, hose connections, sinks, water coolers, water heaters, and lavatories). The system also supplies a backup, but not a safety-grade, source of water to the AFW pumps and can supply makeup to the spent fuel pit.

The city water system contains nonsafety-related components whose failure could potentially prevent the satisfactory accomplishment of a safety-related function. In addition, the city water makeup performs functions that support fire protection.

Components of the city water system that provide water to the AFW system are reviewed with the AFW systems (LRA Section 2.3.4.3).

LRA Tables 2.3.3-17-IP3 and 2.3.3-19-13 identify city water system component types within the scope of license renewal and subject to an AMR, as well as their intended functions.

2.3B.3.17.2 Staff Evaluation

The staff reviewed LRA Section 2.3.3.17, UFSAR Sections 6.1.1 and 10.3.1, and license renewal drawings using the evaluation methodology described in SER Section 2.3 and the guidance in SRP-LR Section 2.3.

During its review, the staff evaluated the system functions described in the LRA and UFSAR to verify that the applicant had not omitted from the scope of license renewal any components with intended functions, as required by 10 CFR 54.4(a). The staff then reviewed those components that the applicant identified as within the scope of license renewal to verify that it had not omitted any passive and long-lived components subject to an AMR, in accordance with 10 CFR 54.21(a)(1).

During its review of LRA Section 2.3.3.17, the staff identified areas in which additional information was necessary to complete the review of the applicant's scoping and screening results. The applicant responded to the staff's RAIs as discussed below.

In RAI 2.3B.3.17-1, dated December 7, 2007, the staff noted that the LRA states that the IP3 city water system has the intended function under 10 CFR 54.4(a)(3) of providing water to the fire protection tanks. The staff further noted that the applicant did not highlight on a license renewal drawing for the city water system a portion of the city water system piping upstream of the eight isolation valves to fire water storage tanks 1 and 2 to indicate that it is within the scope of license renewal. This piping connects to the 16-inch water main from the Village of Buchanan and provides makeup water for the fire water supply function. The staff asked the applicant to explain why it considered all of the city water system piping from the 16-inch water main for the

Village of Buchanan to the fire water storage tanks to be outside the scope of license renewal under 10 CFR 54.4(a)(3) and not subject to an AMR.

In its response, dated January 4, 2008, the applicant stated that the 16-inch water main from the Village of Buchanan is a source of makeup water for the city water system. The applicant explained that city water is the normal source of makeup water to the two fire water storage tanks; however, the city water source is not required to support any fire scenarios or Appendix R events, since each of the storage tanks has a sufficient reserve for fire fighting, without makeup, available to handle all fire scenarios. Therefore, although the city water system can provide a water supply to the fire water tanks, it is not a license renewal intended function, since makeup is not required for compliance with 10 CFR 50.48 fire scenarios or Appendix R events. As a result, the applicant changed LRA Section 2.3.3.17, for IP3, to delete the intended function bullet item, "provide water supply to the fire protection tanks (10 CFR 50.48)," as a 10 CFR 54.4(a)(3) function.

Based on its review, the staff finds the applicant's response to RAI 2.3B.3.17-1 acceptable because it adequately explained that, although city water is the normal source of makeup water to the two fire water storage tanks, the source is not required to support any fire scenarios or Appendix R events. Each of the storage tanks has a sufficient reserve for firefighting that can handle all fire scenarios without the need for continued makeup. Since makeup is not required for 10 CFR 50.48 fire scenarios or Appendix R events, the applicant has changed LRA Section 2.3.3.17, for IP3, to delete the intended function bullet item, "provide water supply to the fire protection tanks (10 CFR 50.48)," as a 10 CFR 54.4(a)(3) function. The staff's concern described in RAI 2.3B.3.17-1 is resolved.

In RAI 2.3B.3.17-2, dated December 7, 2007, the staff noted that the LRA states that the IP3 city water system has no intended functions, in accordance with 10 CFR 54.4(a)(1). However, the staff noted that, on a license renewal drawing for the city water system under "General Notes," the applicant stated under the heading "Class I Piping," "(1) above ground city water make-up to closed cooling water system—expansion tank in control room and EDG jacket water expansion tank," and "(2) city water from Unit 1 tie into AFW pumps suction." The staff also noted that under the heading "Class III Piping," the LRA states, "(1) above ground city water make-up to closed cooling water system—head tank in turbine building," and "(2) above ground city water supply to nuclear services."

In addition, the staff found that a license renewal drawing for the condensate and boiler feed pump suction system shows a small portion of the city water system piping. This portion of city water system piping is highlighted in purple, indicating that it is within the scope of license renewal and subject to an AMR. The drawing identifies this portion of city water system piping as Class I. By definition, all Class I and Class III piping should have intended functions under 10 CFR 54.4(a)(1). The staff requested that the applicant address the following:

(a)   Explain why the Class I and Class III city water system piping on the two drawings do not have an intended function, in accordance with 10 CFR 54.4(a)(1).

(b)   Explain why the city water piping up to the closed cooling water system expansion tank, EDG jacket water expansion tank, closed cooling water system head tank, and nuclear services on the one city water system license renewal drawing is not highlighted in purple, indicating that it is within the scope of license renewal and subject to an AMR.

(c)   Explain why the city water system piping that continues from one city water license renewal drawing onto another drawing for supplying the 40-gallon EDG jacket water expansion tanks is also not highlighted in purple, indicating that it is within the scope of license renewal and subject to an AMR.

In its response, dated January 4, 2008, the applicant stated the following:

(a)   Class I and Class III refer to seismic classification; not to ASME safety class, and that Class I components include safety-related equipment. The applicant further stated that Class I SSCs also include components that do not perform a safety function. The applicant explained Class III is the designation for SSCs which are not directly related to reactor operation and containment, and which do not have to maintain structural integrity during or following an SSE. Further, when defining the city water system components required to support a 10 CFR 54.4(a)(1) system intended functions for license renewal, the seismic classification boundaries were not used, since they do not accurately reflect the portions of the system required to meet system intended functions. Finally, the applicant explained that all components needed to accomplish system intended functions were included within scope regardless of the class breaks on the drawings.

(b)   The license renewal drawings only highlight portions of systems within scope and subject to an aging management review for 10 CFR 54.4(a)(1) or (a)(3). The city water piping up to the closed cooling water system expansion tank, EDG jacket water expansion tank, closed cooling water system head tank, and nuclear services on the city water license renewal drawing is not required to meet any system intended functions described in 10 CFR 54.4(a)(1) or (a)(3); therefore, the piping is not highlighted. However, this piping and valves are within scope for 10 CFR 54.4(a)(2) due to the potential for spatial interaction and are included in LRA tables for components subject to an AMR.

(c)   The LRA drawings only reflect portions of systems in scope and subject to aging management review for 10 CFR 54.4(a)(1) or (a)(3). The city water piping up to the diesel generator jacket water expansion tank on drawings LRA-9321-20343-001 and 9321-H-20283 is not required to meet any system intended functions described in 10 CFR 54.4(a)(1) or (a)(3) and therefore is not highlighted. However, this piping and valves are in scope for 10 CFR 54.4(a)(2) due to the potential for spatial interaction. They are included in LRA Tables 2.3.3-19-13-1P3 and 3.3.2-19-13-1P3.

City water is the source of makeup water to the 40-gallon diesel generator jacket water expansion tanks. Makeup water is not required for the EDGs to perform their intended function.

Based on its review, the staff finds the response to RAI 2.3B.3.17-2(a) acceptable because the applicant adequately explained that Class I and Class III on the license renewal drawing refer to

2-177

seismic classification, rather than ASME safety class. Class I SSCs at IP2 and IP3 include components that do not perform a safety function. At IP2 and IP3, Class III is the designation for SSCs that are not directly related to reactor operation and containment and that do not have to maintain structural integrity during or following a safe-shutdown earthquake. The applicant did not use the seismic classification boundaries when defining the city water system components that are required to comply with 10 CFR 54.4(a)(1) system intended functions for license renewal, since they do not accurately reflect the portions of the system required to meet system intended functions. The applicant included all components needed to accomplish system intended functions within the scope of license renewal, regardless of the seismic class breaks on the drawings. The staff's concern described in RAI 2.3B.3.17-2(a) is resolved.

Based on its review, the staff finds the applicant's response to RAI 2.3B.3.17-2(b) acceptable because it adequately explained that the license renewal drawings reflect only the portions of systems within scope and subject to an AMR, in accordance with 10 CFR 54.4(a)(1) or 10 CFR 54.4(a)(3). The city water piping up to the closed cooling water system expansion tank, EDG jacket water expansion tank, closed cooling water system head tank, and nuclear services, as depicted on the city water license renewal drawing, is not required to meet any system intended functions under 10 CFR 54.4(a)(1) or 10 CFR 54.4(a)(3); therefore, it was not highlighted. Although not highlighted, the applicant has included the piping and valves within the scope of license renewal, in accordance with 10 CFR 54.4(a)(2), because of the potential for spatial interaction. The applicant also included the piping and valves in city water LRA tables for components subject to an AMR. The staff's concern described in RAI 2.3B.3.17-2(b) is resolved.

Based on its review, the staff finds the applicant's response to RAI 2.3B.3.17-2(c) acceptable because it adequately explained that the license renewal drawings reflect only the portions of systems within the scope of license renewal and subject to an AMR, under 10 CFR 54.4(a)(1) or 10 CFR 54.4(a)(3). The city water piping up to the EDG jacket water expansion tank is not required to meet any system intended functions, in accordance with 10 CFR 54.4(a)(1) or 10 CFR 54.4(a)(3); therefore, it was not highlighted. Although not highlighted, the applicant considered the piping and valves to be within scope, in accordance with 10 CFR 54.4(a)(2), because of the potential for spatial interaction. The applicant included the piping and valves in city water LRA tables for components subject to an AMR. City water, as a makeup water source to the EDG jacket water expansion tanks, is not required for the EDGs to perform their intended function. The staff's concern described in RAI 2.3B.3.17-2(c) is resolved.

2.3B.3.17.3 Conclusion

The staff reviewed the LRA, UFSAR, and drawings to determine whether the applicant failed to identify any SSCs within the scope of license renewal. The staff found no such omissions. In addition, the staff sought to determine whether the applicant failed to identify any components subject to an AMR. The staff found no such omissions. On the basis of its review, the staff concludes that the applicant has appropriately identified the city water system components within the scope of license renewal, as required by 10 CFR 54.4(a), and those subject to an AMR, as required by 10 CFR 54.21(a)(1).

2.3B.3.18 IP3 Plant Drains

2.3B.3.18.1 Summary of Technical Information in the Application

LRA Section 2.3.3.18 describes the plant drains, which are passive fire protection features required for adequate protection of safety-related equipment from water damage in areas with fixed suppression systems. Plant drain components also prevent drain systems in areas with combustible materials from spreading fires into other areas of the plant. Some plant drains protect safety-related equipment from flooding effects.

Plant drain components are included in various systems, but grouped for this evaluation. SRP-LR Section 2.1.3.1 indicates that it is appropriate to group similar components from various plant systems into one consolidated review.

To prevent local flooding, areas with automatically operated fire protection have either gravity or pump drains to handle the maximum quantity of spray water. Plant drains protect safety-related equipment in the diesel generator rooms, electrical tunnels, PAB, and auxiliary feed pump room from the effects of Class III component failure. Either floor drains remove fire suppression water adequately or the water flows through other passages to protect safety-related equipment. When safety-related equipment may be lost as a result of inadvertent actuation of a fire system, redundant systems are available for safe shutdown.

The floor drains, fire water, and liquid waste disposal systems include plant drain components. Other sections do not address the waste disposal and liquid waste disposal systems. The floor drains system is not required for regulated events. Other systems provide drainage for flooding protection.

The liquid waste disposal system collects and processes liquid wastes from throughout the plant, including wastes from equipment drains, radioactive chemical laboratory drains, decontamination drains, demineralizer regeneration, and floor drains. The system also collects and transfers liquid drained from the RCS directly to the CVCS for processing. The system includes piping, valves, pumps, collection tanks, instruments, and controls. The system includes several containment penetrations and accompanying isolation components.

SER Section 2.3B.3.19 describes the floor drains system. SER Section 2.3B.3.11 describes the fire water system.

The plant drains system contains safety-related components relied on to remain functional during and following DBEs. It also contains nonsafety-related components whose failure potentially could prevent the satisfactory accomplishment of a safety-related function. In addition, the plant drains system performs functions that support fire protection.

A small number of liquid waste disposal system components are reviewed with the safety injection systems (LRA Section 2.3.2.4) and the primary water makeup systems (LRA Section 2.3.3.7).

LRA Tables 2.3.3-18-IP3 and 2.3.3-19-33-IP3 identify plant drains system component types within the scope of license renewal and subject to an AMR, as well as their intended functions.

2.3B.3.18.2 Staff Evaluation

The staff reviewed LRA Section 2.3.3.18 and UFSAR Sections 9.6.2.3, 11.1, and 16.1.3 using the evaluation methodology described in SER Section 2.3 and the guidance in SRP-LR Section 2.3.

During its review, the staff evaluated the system functions described in the LRA and UFSAR to verify that the applicant had not omitted from the scope of license renewal any components with intended functions, as required by 10 CFR 54.4(a). The staff then reviewed those components that the applicant identified as within the scope of license renewal to verify that it had not omitted any passive and long-lived components subject to an AMR, in accordance with 10 CFR 54.21(a)(1).

During its review of LRA Section 2.3.3.18, the staff identified an area in which additional information was necessary to complete the review of the applicant's scoping and screening results. The discussion of the staff's RAIs in SER Section 2.3B.3 details the disposition of RAI 2.3B.3.18-1, dated February 13, 2008.

2.3B.3.18.3 Conclusion

The staff reviewed the LRA and UFSAR to determine whether the applicant failed to identify any SSCs within the scope of license renewal. The staff found no such omissions. In addition, the staff sought to determine whether the applicant failed to identify any components subject to an AMR. The staff found no such omissions. On the basis of its review, the staff concludes that the applicant has appropriately identified the plant drains system components within the scope of license renewal, as required by 10 CFR 54.4(a), and those subject to an AMR, as required by 10 CFR 54.21(a)(1).

2.3B.3.19 IP3 Miscellaneous Systems in Scope for 10 CFR 54.4(a)(2)

2.3B.3.19.1 Summary of Technical Information in the Application

In LRA Section 2.3.3.19, the applicant described those systems that it included within the scope of license renewal because of their potential for physical interactions with safety-related components, as required by 10 CFR 54.4(a)(2). In this section, the applicant also described the components in these systems that are subject to an AMR. LRA Table 2.3.3-19-A-IP3 lists all of these systems and the LRA section in which the applicant evaluated these systems. LRA Section 2.3.3.19 describes in detail those systems, which are listed below, that do not have correlating LRA sections:

- ammonia/morpholine addition
- boron and layup chemical addition
- CL
- CW
- extraction steam
- floor drains
- gaseous waste disposal
- hydrazine addition
- heater drain/moisture separator drain/vent

2-180

- instrument air closed cooling
- lube oil
- low-pressure steam dump
- main turbine generator
- nuclear equipment drains
- process radiation monitoring
- primary plant sampling
- river water service
- main generator seal oil
- secondary plant sampling
- turbine hall closed cooling
- vapor containment hydrogen analyzer
- hydrogen (added by applicant by letter dated March 12, 2008)

Also in LRA Section 2.3.3.19, the applicant identified the following IP3 systems that it did not review under 10 CFR 54.4(a)(2) for spatial interaction because the applicant included all of the system's passive mechanical components under either 10 CFR 54.4(a)(1), another function of 10 CFR 54.4(a)(2), or 10 CFR 54.4(a)(3):

- AFW
- control building HVAC
- CCW
- control rod drive
- control room HVAC
- engineered safeguards initiation logic
- isolation valve seal water
- RHR
- reactor protection and control
- SG
- SG level control
- security propane generator

The following are brief descriptions of IP3 systems that are included within the scope of license renewal and subject to an AMR, based only on the criterion of 10 CFR 54.4(a)(2).

Ammonia/Morpholine Addition System. The purpose of the ammonia/morpholine addition system is to provide ammonia or morpholine for pH control for the condensate system. LRA Table 2.3.3-19-1-IP3 identifies ammonia/morpholine addition system component types within the scope of license renewal and subject to an AMR as well as their intended functions.

Boron and Layup Chemical Addition System. The boron and layup chemical addition system supplies chemicals to the SGs for chemistry control, even during periods of wet layup. Components in the boron and layup chemical addition system that support the AFW system pressure boundary are evaluated with the AFW systems (LRA Section 2.3.4.3). LRA Table 2.3.3-19-3-IP3 identifies boron and layup chemical addition system component types within the scope of license renewal and subject to an AMR, as well as their intended functions.

Chlorination System. The chlorination system supplies sodium hypochlorite to limit microorganism fouling in the intake bays and river water systems. LRA Table 2.3.3-19-5-IP3 identifies chlorination system component types within the scope of license renewal and subject to an AMR, as well as their intended functions.

Circulating Water System. The CW system supplies the condenser with Hudson River water to cool the steam exiting the low-pressure turbines. LRA Table 2.3.3-19-12-IP3 identifies CW system component types within the scope of license renewal and subject to an AMR, as well as their intended functions.

Extraction Steam System. The extraction steam system utilizes steam to preheat feedwater. LRA Table 2.3.3-19-18-IP3 identifies extraction steam system component types within the scope of license renewal and subject to an AMR, as well as their intended functions.

Floor Drains System. The floor drains system removes any water collected in the nonradioactive floor drains in the turbine building, intake structure, and diesel generator building. LRA Table 2.3.3-19-19-IP3 identifies floor drains system component types within the scope of license renewal and subject to an AMR, as well as their intended functions.

Gaseous Waste Disposal System. The gaseous waste disposal system collects, compresses, stores, samples, and releases gaseous waste from the primary and auxiliary systems. LRA Table 2.3.3-19-25-IP3 identifies gaseous waste disposal system component types within the scope of license renewal and subject to an AMR, as well as their intended functions.

Hydrazine Addition System. The hydrazine addition system injects hydrazine into the secondary system for oxygen control. LRA Table 2.3.3-19-26-IP3 identifies hydrazine addition system component types within the scope of license renewal and subject to an AMR, as well as their intended functions.

Heater Drain/Moisture Separator Drain/Vent System. The heater drain/moisture separator drain/vent system collects and transfers FW heater and moisture separator-reheater drainage to the suction of the main boiler FW pumps. LRA Table 2.3.3-19-27-IP3 identifies heater drain/moisture separator drains/vents system component types within the scope of license renewal and subject to an AMR, as well as their intended functions.

Instrument Air Closed Cooling System. The instrument air closed-cooling system is a separate closed-loop cooling water system. This system supplies cooling water to the instrument air compressors and aftercoolers and rejects heat to the SW system. LRA Table 2.3.3-19-30-IP3 identifies instrument air closed-cooling system component types within the scope of license renewal and subject to an AMR, as well as their intended functions.

Lube Oil System. The lube oil system supplies oil for lubrication and control of the main turbine and the main boiler FW pumps and turbines. The lube oil system includes components that make up the main turbine controls. LRA Table 2.3.3-19-31-IP3 identifies lube oil system component types within the scope of license renewal and subject to an AMR, as well as their intended functions.

Low-Pressure Steam Dump System. The low-pressure steam dump system prevents turbine overspeed by discharging steam from the high-pressure turbine exhaust to the condenser upon turbine trip. LRA Table 2.3.3-19-32-IP3 identifies low-pressure steam dump system component types within the scope of license renewal and subject to an AMR, as well as their intended functions.

Main Turbine Generator System. The main turbine generator system, which receives steam from the SGs, converts a portion of the steam thermal energy to electricity, and supplies extraction steam for FW heating, consists of the turbine, generator, and instrumentation. This system does not include the control valves, moisture separator/reheaters, condensers, and generator cooling components. LRA Table 2.3.3-19-36-IP3 identifies main turbine generator system component types within the scope of license renewal and subject to an AMR, as well as their intended functions.

Nuclear Equipment Drains System. The nuclear equipment drains system collects leakage and drainage from the primary plant equipment (e.g., charging pumps, containment fan cooler units). LRA Table 2.3.3-19-38-IP3 identifies nuclear equipment drains system component types within the scope of license renewal and subject to an AMR, as well as their intended functions.

Process Radiation Monitoring System. The process radiation monitoring system monitors fluid streams for increasing radiation levels and generates an alarm or automatic action under abnormal conditions. LRA Table 2.3.3-19-40-IP3 identifies process radiation monitoring system component types within the scope of license renewal and subject to an AMR, as well as their intended functions.

Primary Plant Sampling System. The primary plant sampling system obtains samples for laboratory analysis of reactor coolant and other reactor auxiliary systems during normal operation. The system also includes the post-accident reactor coolant sampling system, which obtains pressurized coolant samples following accidents. LRA Table 2.3.3-19-41-IP3 identifies primary plant sampling system component types within the scope of license renewal and subject to an AMR, as well as their intended functions.

River Water Service System. The river water service system functionally supports the CW system to supply cooling water from the Hudson River to the main condensers. LRA Table 2.3.3-19-47-IP3 identifies river water system component types within the scope of license renewal and subject to an AMR, as well as their intended functions.

Main Generator Seal Oil System. The main generator seal oil system supplies oil to the main generator shaft seals to prevent hydrogen leakage from the generator into the turbine building. LRA Table 2.3.3-19-54-IP3 identifies seal oil system component types within the scope of license renewal and subject to an AMR, as well as their intended functions.

Secondary Plant Sampling System. The secondary plant sampling system collects and transports samples to the sample room for laboratory analysis of the condensate, FW, and MS systems during normal operation. LRA Table 2.3.3-19-55-IP3 identifies secondary plant sampling system component types within the scope of license renewal and subject to an AMR, as well as their intended functions.

Turbine Hall Closed Cooling System. The turbine hall closed cooling system supplies cooling water to condensate pumps; heater drain pumps; main boiler feed pumps; and station, instrument, and administration building air compressors. LRA Table 2.3.3-19-58-IP3 identifies turbine hall closed cooling system component types within the scope of license renewal and subject to an AMR, as well as their intended functions.

Vapor Containment Hydrogen Analyzer System. The vapor containment hydrogen analyzer system monitors hydrogen and oxygen concentrations and post-LOCA hydrogen concentration in the containment atmosphere. Since a recent license amendment (License Amendment No. 228), hydrogen monitoring is no longer required as a safety function; however, the system remains available. LRA Table 2.3.3-19-59-IP3 identifies vapor containment hydrogen analyzer system component types within the scope of license renewal and subject to an AMR, as well as their intended functions.

Hydrogen System (added by applicant by letter dated March 12, 2008). The hydrogen system provides hydrogen to the main generator for cooling and to the CVCS for the VCT cover gas. LRA Table 2.3.3-19-65-IP3 identifies hydrogen system component types within the scope of license renewal and subject to an AMR, as well as their intended functions.

2.3B.3.19.2 Staff Evaluation

The staff reviewed LRA Section 2.3.3.19 and the following UFSAR sections that were associated with these systems:

• ammonia/morpholine addition[3]	Section 10.2.6.
• auxiliary steam and condensate return[4]	Section 9.6.4
• circulating water[3]	Section 10.2.4
• extraction steam[3]	Section 10.2
• floor drains[3]	Sections 9.6.2.3 and 16.1.3
• gaseous waste disposal[4]	Sections 11.1 and 14.2.3
• hydrazine addition[3]	Section 10.2.6
• heater drain/moisture separator drain/vent[3]	Section 10.2.6
• instrument air closed cooling[4]	Section 9.6.3
• main turbine generator[3]	Section 10.2
• nuclear equipment drains[3]	Section 6.7.1.2
• process radiation monitoring[4]	Section 11.2.3.1
• primary plant sampling[4]	Section 9.4
• river water service[3]	Section 10.2.4
• main generator seal oil[3]	Section 10.2.2
• secondary plant sampling[3]	Section 9.4
• vapor containment hydrogen analyzer[4]	Section 6.8
• boron and layup chemical addition[3]	—
• chlorination[3]	—
• lube oil[3]	—
• low pressure steam dump[3]	—
• turbine hall closed cooling[3]	—

For those systems receiving a simplified Tier 1 evaluation, the staff reviewed the applicable LRA and UFSAR sections using the evaluation methodology described in SER Section 2.3 and

---

[3] The staff conducted a simplified Tier 1 system review for these systems as described in SER Section 2.3

[4] The staff conducted a detailed Tier 2 system review for these systems as described in SER Section 2.3.

the guidance in SRP-LR Section 2.3. For those systems receiving a detailed Tier 2 evaluation, the staff reviewed the applicable LRA sections, applicable UFSAR sections, and license renewal drawings (system components are shown on other associated system drawings). Based upon information provided in the UFSAR and the LRA, the staff evaluated the system functions described in LRA Section 2.3.3.19 to verify that the applicant had not omitted from the scope of license renewal any components with intended functions pursuant to 10 CFR 54.4(a). The staff then reviewed those components that the applicant identified as within the scope of license renewal to verify that the applicant had not omitted any passive and long-lived components subject to an AMR, in accordance with 10 CFR 54.21(a)(1).

The staff reviewed the list of IP3 systems the applicant identified in LRA Section 2.3.3.19 as not having any components in scope for 10 CFR 54.4(a)(2) for spatial interaction because they were already included in scope under 10 CFR 54.4(a)(1), functional (a)(2), or (a)(3). In RAI 2.3A.2.2-1, dated February 13, 2008, the staff asked the applicant to explain why it did not highlight on boundary drawings those piping segments directly attached to the IP2 CS system 10 CFR 54.4(a)(1) piping to indicate that they were included within the scope of license renewal. SER Section 2.3A.2.2.2 documents the staff's review of the applicant's response, dated March 12, 2008.

LRA Table 2.2-2-IP3 indicates that the hydrogen gas system is not within the scope of license renewal. This system, along with the nitrogen system, provides the VCT with gas for oxygen scavenging. Since the piping is directly connected to the VCT, the staff questioned whether the applicant should include the system within scope, in accordance with 10 CFR 54.4(a)(2), because of the potential for physical interaction between the nonsafety- and safety-related equipment. In its response, dated March 12, 2008, the applicant stated that the hydrogen system should be within scope, as required by 10 CFR 54.4(a)(2). The applicant amended the LRA to include the hydrogen system. SER Section 2.2B.3 documents the staff's review of the applicant's response, dated March 12, 2008.

During its review, the staff noted that the applicant did not specifically identify components on the license renewal drawings that are within the scope of license renewal under 10 CFR 54.4(a)(2). To determine that the applicant did not omit any components from scope under 10 CFR 54.4(a)(2), the staff used a sampling approach recommended in SRP-LR Section 2.3.3.1. In multiple RAIs, dated February 13, 2008, the staff asked the applicant to verify that it had included various segments of selected systems within the scope of license renewal, in accordance with 10 CFR 54.4(a)(2). This sampling approach allowed the staff to confirm that the applicant had properly implemented its methodology for identifying the nonsafety-related portions of systems with a potential to adversely affect safety-related functions, in accordance with 10 CFR 54.4(a)(2).

In its response, dated March 12, 2008, the applicant stated that all components identified by the staff on the license renewal drawings are within the scope of license renewal, in accordance with 10 CFR 54.4(a)(2), and subject to an AMR. Based on a review of its response, the staff finds that the applicant has adequately identified the components required to be within the scope of license renewal, in accordance with 10 CFR 54.4(a)(2), and subject to an AMR.

2.3B.3.19.3 Conclusion

For each system described above, the staff reviewed LRA Section 2.3.3.19, the applicable UFSAR section and license renewal drawings to determine whether the applicant failed to identify any SSCs within the scope of license renewal. In addition, the staff sought to determine whether the applicant failed to identify any components subject to an AMR. The staff found instances in which the applicant omitted systems and components that should have been included within the scope of license renewal. The applicant has satisfactorily resolved these issues as discussed in the preceding staff evaluation. On the basis of its review, the staff finds that, for all the systems identified in LRA Section 2.3.3.19 the applicant has appropriately identified the components within the scope of license renewal as required by 10 CFR 54.4(a), and those subject to an AMR, as required by 10 CFR 54.21(a)(1).

2.3B.4 Scoping and Screening Results: IP3 Steam and Power Conversion Systems

LRA Section 2.3.4 identifies the IP3 steam and power conversion systems SCs subject to an AMR for license renewal.

The applicant described the supporting SCs of the steam and power conversion systems in the following LRA sections:

- 2.3.4.1, "Main Steam"
- 2.3.4.2, "Main Feedwater"
- 2.3.4.3, "Auxiliary Feedwater"
- 2.3.4.4, "Steam Generator Blowdown"
- 2.3.4.5, "IP2 AFW Pump Room Fire Event"
- 2.3.4.6, "Condensate"

SER Sections 2.3B.4.1 through 2.3B.4.6, respectively, provide the staff's reviews of IP3 systems described in LRA Sections 2.3.4.1 through 2.3.4.6. The staff's findings for these systems are discussed below.

2.3B.4.1 IP3 Main Steam System

2.3B.4.1.1 Summary of Technical Information in the Application

LRA Section 2.3.4.1 describes the MS system, which includes the auxiliary steam and condensate return, condenser air removal, gland seal steam, high-pressure steam dump, reactor protection and control, reheat steam, and turbine generator hydraulic control systems.

The MS system conducts steam from the four SGs inside the containment structure to the turbine generator unit in the turbine generator building. The system has four MS pipes, one from each SG to the turbine stop and control valves, which are interconnected near the turbine. Each steam pipe has an MSIV and a non-return valve outside the containment. Five code safety valves and one PORV are located on each MS line outside the reactor containment and upstream of the isolation and non-return valves. A flow venturi upstream of the isolation valve measures steam flow. Steam pressure is also measured upstream of the isolation valve. The MS system supplies steam to the main boiler FW pump turbines and the AFW pump turbine. The MS system includes the main boiler FW pump turbines and the turbine steam bypass and low-pressure steam dump systems, which channel excess steam flow to the condenser. The SGBD flowpath includes MS system components.

The auxiliary steam and condensate return system supplies auxiliary steam to plant components for IP3 heating and for the recovery of condensate via the condensate return lines. The system supplies steam for heating throughout the plant to room and area heating units, refueling water and primary water storage tanks, boric acid batch mixing tank, and other areas. The system also supplies minor steam loads, such as the condenser waterbox air ejectors. System supply by the house service boiler or steam reboiler includes heaters, air ejectors, steam distribution piping and valves, condensate return piping, valves, pumps, tanks, instruments, and controls.

The condenser air removal system removes air and non-condensable gases from the condensers to prevent gas buildup that would interfere with steam condensation. Each condenser has a four-element, two-stage air ejector with a separate inter-condenser and common after-condensers. Normal air removal requires one air ejector unit per condenser. For initial condenser shell-side air removal, three non-condensing priming ejectors use steam from the MS system supplied through a pressure-reducing valve. The system monitors the air ejector exhaust for radioactivity. In an SG leak and the subsequent presence of radioactively contaminated steam in the secondary system, this radiation monitor detects the radioactive non-condensable gases that concentrate in the air ejector effluent. A high-activity-level signal automatically diverts the exhaust gases from the vent stack to the containment.

The gland seal steam system supplies steam to the main turbine and boiler FW pump turbine gland seals. The system includes pressure-regulating valves and distribution piping and valves.

The high-pressure steam dump system provides an MS flowpath, bypassing the turbine to the main condenser when the turbine generator cannot accept the steam flow. Two MS bypass lines, one on either side of the turbine, divert excess steam from the four MS lines directly to the condensers, when necessary, before they reach the turbine stop valves. From each of the MS bypass lines, six lines, each with a bypass control valve, discharge into the condenser. The system includes the bypass control valves and its piping, controls, and instruments.

The reactor protection and control system monitors primary and secondary plant parameters and trips the reactor to protect the reactor core and RCS. The reactor protection and control system is primarily electrical, but includes a small number of mechanical instrumentation components that form parts of the SG secondary-side pressure boundary.

The reheat steam system supplies reheated steam to the low-pressure turbines and steam from the MS system to the main boiler FW pump turbines. Steam from the high-pressure turbine exhaust passes through the moisture separator reheaters, which remove moisture and reheat the steam by main steam extracted before it reaches the turbine MS stop valves. Part of the extracted main steam goes to the main boiler FW pump turbines. The system includes the moisture separator reheaters, piping, valves, instruments, and controls.

The turbine generator hydraulic control system directly controls the main turbine. The system has electrical and mechanical components of the turbine hydraulic control system, including the main turbine stop valves, and parts of the MS system pressure boundary for Appendix R safe shutdown.

The MS system contains safety-related components relied on to remain functional during and following DBEs. It also contains nonsafety-related components whose failure could prevent the satisfactory accomplishment of a safety-related function. In addition, the MS system performs functions that support fire protection and SBO.

Main steam components supporting the AFW system are reviewed with the AFW systems (LRA Section 2.3.4.3). Components containing air are reviewed with the compressed air systems (LRA Section 2.3.3.4). Condenser air removal system components in the containment penetration are reviewed with containment penetrations (LRA Section 2.3.2.5). Reactor protection and control components supporting the mechanical intended function are reviewed with the SGs (LRA Section 2.3.1.4).

The following LRA tables identify IP3 MS system component types that are within the scope of license renewal and subject to an AMR, as well as their intended functions:

- LRA Table 2.3.4-1-IP3
- LRA Table 2.3.3-19-2-IP3
- LRA Table 2.3.3-19-4-IP3
- LRA Table 2.3.3-19-24-IP3
- LRA Table 2.3.3-19-28-IP3
- LRA Table 2.3.3-19-35-IP3
- LRA Table 2.3.3-19-45-IP3
- LRA Table 2.3.3-19-57-IP3

2.3B.4.1.2 Staff Evaluation

The staff reviewed LRA Section 2.3.4.1; UFSAR Sections 7.2, 9.6.4, 10.2, 10.2.1, 10.2.2, and 10.2.5; and license renewal drawings using the evaluation methodology described in SER Section 2.3 and the guidance in SRP-LR Section 2.3.

During its review, the staff evaluated the system functions described in the LRA and UFSAR to verify that the applicant had not omitted from the scope of license renewal any components with intended functions, pursuant to 10 CFR 54.4(a). The staff then reviewed those components that the applicant identified as within the scope of license renewal to verify that it had not omitted any passive and long-lived components subject to an AMR, in accordance with 10 CFR 54.21(a)(1).

During its review of LRA Section 2.3.4.1, the staff identified an area in which additional information was necessary to complete the review of the applicant's scoping and screening results. The applicant responded to the staff's RAI as discussed below.

In RAI 2.3B.4.1-1, dated December 7, 2007, the staff noted that license renewal drawings for the IP3 MS system show the following valves within the scope of license renewal and subject to an AMR: PCV-1134, PCV-1135, PCV-1136, PCV-1137, MS-1-31, MS-1-32, MS-1-33, MS-1-34, PCV-1120, PCV-1121, PCV-1122, PCV-1123, PCV-1124, PCV-1125, PCV-1126, PCV-1127, PCV-1128, PCV-1129, PCV-1130, PCV-1131. The staff also noted that these valves are air operated and have associated air cylinders and air tubing that were excluded from the scope of license renewal. Since some of these valves appear to rely on pressurized air (pneumatic

2-188

operation) to change position and fulfill their intended function, the staff asked the applicant to explain why it did not include the instrument air system, its tubing, and associated SOVs to the valves in question within the scope of license renewal, in accordance with 10 CFR 54.4(a).

In its response dated January 4, 2008, the applicant stated that the air operators are active components; therefore, they are not subject to an AMR, in accordance with 10 CFR 54.21(a)(1)(i) and NEI 95-10, Appendix B. The applicant further explained that the SOVs and air tubing associated with the air-operated valves in the MS system are within the scope of license renewal, but are not subject to an AMR because the majority of the air-operated valves shown on the MS license renewal drawings as within the scope of license renewal fail to their required position for accident mitigation. As such, these valves do not require pressurized air to fulfill their intended function, and pressure boundary of the air tubing is not necessary. The applicant stated that the atmospheric dump valves and MSIVs are an exception. These valves close upon loss of air, but are credited with being re-opened, as necessary, in an accident scenario, using standby nitrogen in bottles or compressed air stored in accumulators. The applicant explained that components used to re-open the MS system valves are subject to an AMR.

Based on its review, the staff finds the response to RAI 2.3B.4.1-1 acceptable because the applicant adequately explained that, for most of the air-operated valves, a failure of the air supply system will not result in a loss of the intended function because the MS valves fail to their safe positions. This explanation is consistent with NEI 95-10, Revision 6, Section 5.2.3.1, which governs fail-safe components. For those air-operated valves that rely on an air supply system (i.e., those MS system valves that do not fail to their safe position), the passive pneumatic components (accumulator tanks, tubing, and valves) of those air-operated valves are included within the scope of license renewal and are subject to an AMR, in accordance with 10 CFR 54.21(a)(1). The staff's concern described in RAI 2.3B.4.1-1 is resolved.

2.3B.4.1.3 Conclusion

The staff reviewed the LRA, UFSAR, RAI responses, and drawings to determine whether the applicant failed to identify any SSCs within the scope of license renewal. The staff found no such omissions. In addition, the staff sought to determine whether the applicant failed to identify any components subject to an AMR. The staff found no such omissions. On the basis of its review, the staff concludes that the applicant has appropriately identified the MS system components within the scope of license renewal, as required by 10 CFR 54.4(a), and those subject to an AMR, as required by 10 CFR 54.21(a)(1).

2.3B.4.2 IP3 Main Feedwater System

2.3B.4.2.1 Summary of Technical Information in the Application

LRA Section 2.3.4.2 describes the FW system, which transfers condensate and heater drain flow through the final stage of FW heating to the SGs. Two half-size, steam-driven main FW pumps increase the pressure of the condensate for delivery through the final stage of FW heating and the FW regulating valves to the SGs.

The main FW system includes the high-pressure FW heaters and piping and valves from the main feed pumps through the heaters to the SGs. The FW system also includes the main feed pump turbine drip tank drain pumps. The main FW pumps and services system supports the main FW system by increasing the pressure of the condensate for delivery through the final stage of FW heating and the FW regulating valves to the SGs.

The main FW system contains safety-related components relied on to remain functional during and following DBEs. It also contains nonsafety-related components whose failure potentially could prevent the satisfactory accomplishment of a safety-related function. In addition, the main FW system performs functions that support fire protection.

Feedwater system components supporting the AFW system are reviewed with such systems (LRA Section 2.3.4.3).

LRA Tables 2.3.4-2-IP3, 2.3.3-19-22-IP3, and 2.3.3-19-23-IP3 identify main FW system component types within the scope of license renewal and subject to an AMR, as well as their intended functions.

2.3B.4.2.2 Staff Evaluation

The staff reviewed LRA Section 2.3.4.2, UFSAR Section 10.2.6, and a license renewal drawing using the evaluation methodology described in SER Section 2.3 and the guidance in SRP-LR Section 2.3.

During its review, the staff evaluated the system functions described in the LRA and UFSAR to verify that the applicant had not omitted from the scope of license renewal any components with intended functions, as required by 10 CFR 54.4(a). The staff then reviewed those components that the applicant identified as within the scope of license renewal to verify that the applicant had not omitted any passive and long-lived components subject to an AMR, in accordance with 10 CFR 54.21(a)(1).

During its review of LRA Section 2.3.4.2, the staff identified an area in which additional information was necessary to complete the review of the applicant's scoping and screening results. The applicant responded to the staff's RAI as discussed below.

In RAI 2.3B.4.2-1, dated December 7, 2007, the staff noted that license renewal drawings identify valves FCV-417-L, FCV-417, FCV-427-L, FCV-427, FCV-437-L, FCV-437, FCV-447-L, FCV-447, BF2-31, and BF2-32 for the IP3 main FW system as within the system evaluation boundary. The staff noted that, although the aforementioned valves are passive and long lived, they are not highlighted, indicating that they are not subject to an AMR. The staff asked the applicant to explain the valves' exclusion from an AMR.

In its response, dated January 4, 2008, the applicant explained that, although these FW system valves are located upstream of the containment isolation check valves in nonsafety-related piping, they are classified as safety related because of their active function to provide FW isolation. The applicant also stated that these valves "have no passive intended function for 54.4(a)(1) or (a)(3) because their failure would accomplish the safety function of isolating feedwater flow to the SGs." The applicant further stated that these valves perform their function with moving parts; therefore, in accordance with 10 CFR 54.21(a)(1)(i), they are not subject to an AMR and are not highlighted on the license renewal drawing. However, the applicant did indicate that the valves in question are within the scope of license renewal under

10 CFR 54.4(a)(2) because of their potential for spatial interaction with safety-related equipment; therefore, they are subject to an AMR.

The staff disagreed with the applicant's rationale that the valves do not have a passive intended function in accordance with 10 CFR 54.4(a)(1). The staff discussed the applicant's view during a telephone call on March 7, 2008. The applicant subsequently amended its RAI response by letter dated March 24, 2008, and reiterated that the FW system valves are safety related. The applicant also stated that, although not highlighted, these valves and the remainder of the FW system components on the associated license renewal drawing are in scope and subject to an AMR under 10 CFR 54.4(a)(2) because of their potential for spatial interaction with safety-related equipment.

Based on its review, the staff finds the applicant's amended response to RAI 2.3B.4.2-1 acceptable because the applicant confirmed that the valves in question are within the scope of license renewal pursuant to 10 CFR 54.4(a), and subject to an AMR pursuant to 10 CFR 54.21(a)(1). Although the staff does not agree with the applicant's basis for determining how the valve bodies are subject to an AMR, the staff's concern is resolved because the AMR was performed, and the AMR results were provided in LRA Table 3.3.2-19-34-IP3. The staff's concern described in RAI 2.3B.4.2-1 is resolved.

In RAI 2.3B.4.2-2, dated December 30, 2007, the staff noted that UFSAR Section 14.2.5, Rupture of a Steam Pipe, states in the event of a main steam line break incident, the motor-operated valves (MOVs) associated with each of the feedwater regulating valves (FRVs) will close. UFSAR Section 14.2.5.1 states that redundant isolation of the main feedwater lines is necessary, because sustained high feedwater flow would cause additional cooldown; therefore, in addition to the normal control action which will close the main feedwater valves, any safety injection signal will rapidly close all feedwater control valves (including the motor-operated block valves and low-flow bypass valves), trip the main feedwater pumps, and close the feedwater pump discharge valves. In addition, license renewal drawing 9321-20193 shows a "HIGH STEAM FLOW SI LOGIC" signal going to these motor-operated isolation valves. The motor-operated block valves shown on license renewal drawings are BFD-5's and BFD-90's for the main FRVs, and the low flow bypass regulating valves, respectively.

The feedwater isolation valves, BFD-5's and BFD-90's, are not included within the "system intended function boundary," nor are they highlighted on the license renewal drawings as having an intended function in accordance with 10 CFR 54.4(a)(1). By letter dated December 30, 2008, the staff requested the applicant to justify the exclusion of the BFD-5 and BFD-90 isolation valves from scope of license renewal in accordance with 10 CFR 54.4(a)(1). This issue was also identified as Open Item 2.3.4.2-1.

By letter dated January 27, 2009, the applicant stated that based upon a review of the qualifications of the isolation valves, the BFD-5 and BFD-90 valves are classified as nonsafety-related in the site component database and are located outside the Class I boundary [as corrected by letter dated March 13, 2009] on license renewal drawing LRA-9321-2019-0. As indicated in the IP3 UFSAR, these valves provide a backup isolation function for feedwater in the event of such accidents as a feedwater or steamline break. Credit for nonsafety-related components as a backup to safety-related components in mitigating breaks in seismically-qualified steam line piping is consistent with regulatory guidance provided in Standard Review Plan (NUREG-0800), Section 15.1.5, "Steam System Piping Failures Inside and Outside of

Containment (PWR)," and is also consistent with the allowance for feedwater regulating and bypass valves to be nonsafety-related, as discussed in NUREG-0138, "Staff Discussion of Fifteen Technical Issues Listed in Attachment to November 3, 1976 Memorandum from Director, NRR to NRR Staff." The applicant concluded that, consistent with the CLB, regulatory guidance, and NUREG-0138, the BFD-5 and BRD-90 valves are classified as nonsafety-related, and as such, meet the criteria to be included in scope for license renewal under 10 CFR 54.4(a)(2).

Based on the information provided by the applicant, the staff finds applicant's response to RAI 2.3B.4.2-2 acceptable because the BFD-5 and BFD-90 isolation valves are nonsafety-related components, and the valves are included in the scope for license renewal under 10 CFR 54.4(a)(2). Therefore, the staff's concern described in RAI 2.3B.4.2-2 is resolved. As a result, Open Item 2.3.4.2-1 is closed.

2.3B.4.2.3 Conclusion

The staff reviewed the LRA, UFSAR, RAI response, and a drawing to determine whether the applicant failed to identify any SSCs within the scope of license renewal. The staff found no such omissions. In addition, the staff sought to determine whether the applicant failed to identify any components subject to an AMR. The staff found no such omissions. The staff concludes that the applicant has appropriately identified the main FW system components within the scope of license renewal, as required by 10 CFR 54.4(a), and those subject to an AMR, as required by 10 CFR 54.21(a)(1).

2.3B.4.3 IP3 Auxiliary Feedwater System

2.3B.4.3.1 Summary of Technical Information in the Application

LRA Section 2.3.4.3 describes the AFW system, which supplies a flow of water from the CST to the SGs when the main FW pumps are unavailable. One steam turbine-driven and two electric motor-driven AFW pumps supply adequate feedwater to the SGs to remove reactor decay heat under all circumstances, including loss of power and normal heat sink (e.g., condenser isolation or loss of CW flow). The system can supply all four SGs. The steam-turbine-driven pump can be supplied from two of the SGs. The AFW system operates during plant startup at low power levels before the main FW pump is available. The system includes the AFW pumps, the turbine for the turbine-driven pump, piping from both CST and city water supply (an alternate source) through the pumps to the FW line supplying the SGs, valves, instruments, and controls. However, the system does not include the CST, which is part of the condensate transfer system.

The AFW system contains safety-related components relied on to remain functional during and following DBEs. In addition, the AFW system performs functions that support fire protection, ATWS, and SBO. Instrument air components included in the AFW system are reviewed with the compressed air systems (LRA Section 2.3.3.4).

LRA Table 2.3.4-3-IP3 identifies AFW system component types within the scope of license renewal and subject to an AMR, as well as their intended functions.

2.3B.4.3.2 Staff Evaluation

The staff reviewed LRA Section 2.3.4.3, UFSAR Sections 7.2.2 and 10.2.6, and license renewal drawings using the evaluation methodology described in SER Section 2.3 and the guidance in SRP-LR Section 2.3.

During its review, the staff evaluated the system functions described in the LRA and UFSAR to verify that the applicant had not omitted from the scope of license renewal any components with intended functions, as required by 10 CFR 54.4(a). The staff then reviewed those components that the applicant identified as within the scope of license renewal to verify that it had not omitted any passive and long-lived components subject to an AMR, in accordance with 10 CFR 54.21(a)(1).

During its review, the staff identified an area in which additional information was necessary to complete the review of the applicant's scoping and screening results. The applicant responded to the staff's RAI as discussed below.

In RAI 2.3A.4.2-2, dated February 13, 2008, the staff noted that LRA Section 2.3.4.3 states that the AFW system has no intended function under 10 CFR 54.4(a)(2). However, the staff identified an instance in which components adjacent to safety-related components were not highlighted on license renewal drawings, but should have been considered for inclusion within the scope of license renewal because of their potential adverse spatial interaction, in accordance with 10 CFR 54.4(a)(2). For IP3, a license renewal drawing showed a section of piping extending from the AFW system piping (which includes valve SS-189) that was not highlighted. The staff asked the applicant to confirm that it evaluated the aforementioned components for inclusion within the scope of license renewal, in accordance with 10 CFR 54.4(a)(2).

In its response, dated March 12, 2008, the applicant stated that the section of piping extending from the AFW system piping, which includes valve SS-189 is included within the scope of license renewal, in accordance with 10 CFR 54.4(a)(2), and is subject to an AMR.

Based on its review, the staff finds the applicant's response to RAI 2.3A.4.2-2 acceptable because it adequately explained that the components in question are within the scope of license renewal, in accordance with 10 CFR 54.4(a)(2), and subject to an AMR because of their potential to adversely interact spatially with safety-related equipment. The staff's concern described in RAI 2.3A.4.2-2 is resolved.

2.3B.4.3.3 Conclusion

The staff reviewed the LRA, UFSAR, RAI responses, and drawings to determine whether the applicant failed to identify any SSCs within the scope of license renewal. The staff found no such omissions. In addition, the staff sought to determine whether the applicant failed to identify any components subject to an AMR. The staff found no such omissions. On the basis of its review, the staff concludes that the applicant has appropriately identified the AFW system components within the scope of license renewal, as required by 10 CFR 54.4(a), and those subject to an AMR, as required by 10 CFR 54.21(a)(1).

2.3B.4.4 IP3 Steam Generator Blowdown System

2.3B.4.4.1 Summary of Technical Information in the Application

LRA Section 2.3.4.4 describes the SGBD system, which includes the SGBD recovery and the SG sampling systems.

The SGBD system can control the concentration of solids in the shell side of the SGs. The system, which operates normally with a continuous blowdown and sample flow, has a drain connection and two blowdown connections (nozzles) at the bottom of each SG. Pipes from the connections (nozzles) join to form a stainless steel blowdown header. Four individual blowdown headers are routed from each SG to the PAB through containment isolation valves.

Downstream of the containment isolation valves, blowdown flow can be diverted to either the SGBD recovery system (during normal operation) or the blowdown flash tank. The SGBD recovery system consists of two heat exchangers, a filter and demineralizer package, piping, valves, and instrumentation.

The SG sampling system obtains representative secondary-side water samples for laboratory analysis of chemical and radiochemical conditions. The system has sample capability for each SG from its blowdown line inside containment. Each line to the sample room, where the liquid is cooled and the pressure reduced, has a containment penetration. Each sample is split into two routes—one to the sample sink for periodic chemical analysis and one to a conductivity cell, a radiation monitor, and then to the blowdown flash tank. The second line handles a continuous flow for constant conductivity reading and radiation monitoring.

The SGBD system contains safety-related components relied on to remain functional during and following DBEs. It also contains nonsafety-related components whose failure potentially could prevent the satisfactory accomplishment of a safety-related function. In addition, the SGBD system performs functions that support fire protection, ATWS, and SBO.

A small number of SGBD components are reviewed with the SW system (LRA Section 2.3.3.2). The SG sample heat exchangers (SG sampling system) are safety-related only for their cooling water pressure boundary function (heat transfer is not a required function). These heat exchangers are reviewed with the CCW system (LRA Section 2.3.3.3).

LRA Tables 2.3.4-4-IP3, 2.3.3-19-50-IP3, 2.3.3-19-51-IP3, and 2.3.3-19-52-IP3 identify SGBD system component types within the scope of license renewal and subject to an AMR, as well as their intended functions.

2.3B.4.4.2 Staff Evaluation

The staff reviewed LRA Section 2.3.4.4 and UFSAR Sections 9.4.1 and 10.2.1 using the evaluation methodology described in SER Section 2.3 and the guidance in SRP-LR Section 2.3.

During its review, the staff evaluated the system functions described in the LRA and UFSAR to verify that the applicant had not omitted from the scope of license renewal any components with intended functions, as required by 10 CFR 54.4(a). The staff then reviewed those components that the applicant identified as within the scope of license renewal to verify that it had not

omitted any passive and long-lived components subject to an AMR, in accordance with 10 CFR 54.21(a)(1).

2.3B.4.4.3 Conclusion

The staff reviewed the LRA and UFSAR to determine whether the applicant failed to identify any SSCs within the scope of license renewal. The staff found no such omissions. In addition, the staff sought to determine whether the applicant failed to identify any components subject to an AMR. The staff found no such omissions. On the basis of its review, the staff concludes that the applicant has adequately identified the SGBD system components within the scope of license renewal, as required by 10 CFR 54.4(a), and those subject to an AMR, as required by 10 CFR 54.21(a)(1).

2.3B.4.5 IP2 Auxiliary Feedwater Pump Room Fire Event (Not Applicable to IP3)

In the LRA, the applicant evaluated systems that in combination provide and support feedwater flow to the steam generators during a shutdown, and stated that the evaluation applies to IP2 only. During its review, the staff considered whether a similar evaluation was needed for IP3.

Similar to IP2, the IP3 AFW pump room contains redundant trains of safe shutdown systems and equipment separated by 20 feet with intervening combustibles. The NRC granted an exemption from the technical requirements of Section III.G.2 of 10 CFR Part 50, Appendix R on January 7, 1987. However, the AFW pump room fire event is not an issue at IP3 because the AFW pump room has area-wide coverage via automatic fire detection and a sprinkler system. This area is also equipped with manual hose stations and portable fire extinguishers. The NRC SER dated January 7, 1987, documents the staff's determination that fire protection features in the IP3 AFW Pump Room are adequate.

The staff finds that the applicant has demonstrated that the IP3 AFW pump room contains sufficient automatic fire suppression, the fire hazard within this area is low, and alternate shutdown capability exists. Therefore, an alternate feedwater flowpath is not required in the event of a fire in the IP3 AFW pump room.

2.3B.4.6 IP3 Condensate System

2.3B.4.6.1 Summary of Technical Information in the Application

LRA Section 2.3.4.6 describes the condensate system, which consists of components in the following systems: condensate, condensate polisher, condensate pump discharge, condensate pump suction, and condensate transfer.

The condensate system transfers condensate and low-pressure heater drainage from the condenser hotwell through the condensate polisher and five stages of FW heating to the main FW pump suctions. The condensate system is also the primary source of water to the AFW pumps. As part of the main condensate flowpath, three condensate pumps, arranged in parallel, take suction from the bottom of the condenser hotwells and discharge into a common header to the condensate polisher system. From the polisher system, a portion of the condensate passes through three steam jet air ejector condensers, arranged in parallel, and one gland steam condenser. The condensate passes through the tube sides of three parallel strings of two low-pressure FW heaters. The flows from these heaters combine in a common line, which then divides to flow into the remaining three strings of three low-pressure heaters. After the No. 5 FW heater, the three condensate lines join into a common header. The heater drain pump discharge enters this header and continues on to the suction of the main FW pumps.

The condensate system contains mostly valves, including a large number of small valves supplying condensate as gland seal water to various secondary plant valves. Within the condensate system, one valve has a safety function as part of the pressure boundary for the flowpath from the CST to the AFW pumps.

The condensate polishing system removes dissolved and suspended solids from the condensate to maintain FW quality required for the SGs. The polishers are within the existing condensate system between the condensate pumps and the first stage of FW heaters. The condensate polishing system consists of six service vessels, six condensate post-filters, three condensate booster pumps, piping, valves, instrumentation, and controls.

The condensate pump discharge system supports sampling of the condensate pump discharge. Components in this system code include the small sampling piping and valves at the discharge of the condensate pumps.

The condensate pump suction system supplies water to the condensate pumps from the main condenser. Components in this system code include the expansion joints, piping, and valves between the condenser and the condensate pumps.

The condensate transfer system transfers condensate from the condenser to the suction of the main boiler FW pumps and from the CST to the AFW pumps. This system includes condensate system components from the condensate pumps to the suction of the main boiler FW pumps (except the condensate polishers and their support equipment), the CST and piping and components to the AFW pump suction header, the main condensers, the condensate and low-pressure FW heaters, piping, valves, instruments, controls, and other condensate system components.

The condensate system contains safety-related components relied on to remain functional during and following DBEs. The failure of nonsafety-related SSCs in the condensate system potentially could prevent the satisfactory accomplishment of a safety-related function. In addition, the condensate system performs functions that support fire protection and SBO.

Components that support the pressure boundary of the AFW system flowpath are evaluated with the AFW systems (LRA Section 2.3.4.3).

LRA Tables 2.3.3-19-6-IP3, 2.3.3-19-7-IP3, 2.3.3-19-8-IP3, 2.3.3-19-9-IP3, and 2.3.3-19-14-IP3 identify condensate system component types within the scope of license renewal and subject to an AMR, as well as their intended functions.

2.3B.4.6.2 Staff Evaluation

The staff reviewed LRA Section 2.3.4.6, UFSAR Section 10.2.6, and license renewal drawings (condensate system components are shown on drawings of other system) using the evaluation methodology described in SER Section 2.3 and the guidance in SRP-LR Section 2.3.

During its review, the staff evaluated the system functions described in the LRA and UFSAR to verify that the applicant had not omitted from the scope of license renewal any components with intended functions, as required by 10 CFR 54.4(a). The staff then reviewed those components that the applicant identified as within the scope of license renewal to verify that it had not omitted any passive and long-lived components subject to an AMR, in accordance with 10 CFR 54.21(a)(1).

2.3B.4.6.3 Conclusion

The staff reviewed the LRA and UFSAR to determine whether the applicant failed to identify any SSCs within the scope of license renewal. The staff found no such omissions. In addition, the staff sought to determine whether the applicant failed to identify any components subject to an AMR. The staff found no such omissions. On the basis of its review, the staff concludes that the applicant has adequately identified the condensate system components within the scope of license renewal, as required by 10 CFR 54.4(a), and those subject to an AMR, as required by 10 CFR 54.21(a)(1).

2.4 Scoping and Screening Results: Structures

This section documents the staff's review of the applicant's scoping and screening results for structures. Specifically, this section discusses the following:

- containment buildings
- water control structures
- turbine buildings, auxiliary buildings, and other structures
- bulk commodities

In accordance with the requirements of 10 CFR 54.21(a)(1), the applicant must list passive, long-lived SCs within the scope of license renewal and subject to an AMR. To verify that the applicant properly implemented its methodology, the staff's review focused on the implementation results. This focus allowed the staff to confirm that there were no omissions of SCs that meet the scoping criteria and are subject to an AMR.

The staff's evaluation of the information in the LRA for all structures sought to determine whether the applicant had identified, in accordance with 10 CFR 54.4, the components and supporting structures, for structures that appear to meet the license renewal scoping criteria. Similarly, the staff evaluated the applicant's screening results to verify that all passive, long-lived SCs were subject to an AMR in accordance with 10 CFR 54.21(a)(1).

In its scoping evaluation, the staff reviewed the applicable LRA sections, focusing on components that had not been identified as within the scope of license renewal. The staff reviewed relevant licensing basis documents, including the UFSAR, for each structure to determine whether the applicant had omitted from the scope of license renewal SCs with license renewal intended functions in accordance with 10 CFR 54.4(a). The staff also reviewed the licensing basis documents to determine whether the LRA specified all license renewal intended functions, in accordance with 10 CFR 54.4(a). The staff requested additional information to resolve any omissions or discrepancies identified.

After its review of the scoping results, the staff evaluated the applicant's screening results. For those SCs with intended functions, the staff sought to determine whether (1) the functions are performed with moving parts or a change in configuration or properties or (2) the SCs are subject to replacement after a qualified life or specified time period, in accordance with 10 CFR 54.21(a)(1). For those meeting neither of these criteria, the staff sought to confirm that these SCs were subject to an AMR, as required by 10 CFR 54.21(a)(1). The staff requested additional information to resolve any omissions or discrepancies identified.

During its review of LRA Section 2.4, the staff identified areas in which additional information was necessary to complete the evaluation of the applicant's scoping and screening results for structures. Therefore, the staff issued issue-specific RAIs by letter dated January 28, 2008, to determine or confirm whether the applicant properly applied the scoping criteria of 10 CFR 54.4(a) and the screening criteria of 10 CFR 54.21(a) to structures and structural components. The applicant provided its responses to the staff's RAIs by letter dated February 27, 2008, and supplemented it by Amendment 3 to the LRA, dated March 24, 2008. The applicant further provided responses to the staff's followup RAIs by letter dated June 11, 2008, and submitted Amendment 5 to the LRA, dated June 11, 2008.

The following discussion describes the staff's RAI related to the scoping of structures in LRA Section 2.4, and the applicant's responses. Relative to the applicant's scoping and screening results for structures documented in LRA Section 2.4, the staff also reviewed LRA Table 2.2-3, which lists the plant-level structures that are within the scope of license renewal, and LRA Table 2.2-4, which lists the plant-level structures that are not within the scope of license renewal. The staff performed these reviews to determine if there were any omissions in the structures scoped at the plant-level and to verify that all the scoped structures were addressed in LRA Section 2.4.

Based on its review of the UFSAR, the staff identified certain structural components that do not appear to be included in LRA Tables 2.2-3 and 2.2-4 or in LRA Section 2.4. In the first part of RAI 2.4-1, the staff requested that the applicant explain whether or not the structures listed below are within the scope of license renewal and subject to an AMR:

(i)   pipe penetration tunnel (Reference: IP2 final safety analysis report (FSAR), Section 1.11.4.10)

(ii)   liquid waste storage building (Reference: IP3 FSAR, Sections 16.1.2 and 9.6.4)

(iii)   condenser tube withdrawal/removal pit (Reference: IP3 FSAR, Chapter 1; Site Plan Drawing 64513; and IP2 FSAR, Figure 10.2-3)

(iv)   fuel oil storage tank and its foundation at Buchanan Substation (this tank provides backup fuel oil for emergency diesels and gas turbines)

In its response to RAI 2.4-1, Item (i), dated February 27, 2008, the applicant stated that the pipe penetration tunnel is located in the IP2 fan house and is included within the scope of license renewal as part of the fan house structure, identified in LRA Table 2.2-3 as "fan house (IP2)." The staff verified that LRA Table 2.2-3, as well as LRA Section 2.4.3, identified the "fan house (IP2)" as a structure. Therefore, the staff finds the applicant's response acceptable. The staff's concern described in RAI 2.4-1, Item (i), is resolved.

In its response to RAI 2.4-1, Item (ii), dated February 27, 2008, the applicant stated that the liquid waste storage building is located within the administration building. The applicant stated that the liquid waste storage building is not within the scope of license renewal because it does not perform a license renewal intended function, as required by 10 CFR 54.4(a). Therefore, LRA Table 2.2-4 lists the liquid waste storage building as part of the line item "administration building (IP3) (service admin complex)." The staff verified that LRA Table 2.2-4 lists "administration building (IP3) (service admin complex)" as a structure that is not within the scope of license renewal. The staff further confirmed from UFSAR Section 16.1.2 for IP3 that the liquid waste storage building is a seismic Class III component of the waste disposal system. Its failure will not result in offsite doses in excess of the limits required by 10 CFR Part 20, "Standards for Protection against Radiation." Based on the above, the staff finds that the liquid waste storage building does not perform a license renewal intended function, as detailed in 10 CFR 54.4(a). Therefore, the staff finds the applicant's response acceptable. The staff's concern described in RAI 2.4-1, Item (ii), is resolved.

In its response to RAI 2.4-1, Item (iii), dated February 27, 2008, the applicant stated that the condenser tube withdrawal/removal pits are located in the lower level of the turbine buildings. The applicant included these components in the scope of license renewal as part of the structures identified in LRA Table 2.2-3 as "turbine building and heater bay (IP2)" and "turbine building and heater bay (IP3)." The staff verified that LRA Table 2.2-3, as well as LRA Section 2.4.3, identifies the "turbine building and heater bay (IP2)" and "turbine building and heater bay (IP3)" as structures. Therefore, the staff finds the applicant's response acceptable. The staff's concern described in RAI 2.4-1, Item (iii), is resolved.

In its response to RAI 2.4-1, Item (iv), dated February 27, 2008, the applicant stated that the "fuel oil storage tank foundation" at Buchanan Substation is within the scope of license renewal and included within the line item "gas turbine generator No. 2 and 3, enclosure and fuel tanks foundation" in LRA Table 2.2-3. The staff verified that LRA Table 2.2-3, as well as LRA Section 2.4.3, identifies the line item "gas turbine generator No. 2 and 3, enclosure and fuel tanks foundation." Further, the staff verified that the fuel oil storage tanks are scoped and screened as a mechanical fuel oil system component in LRA Section 2.3.3.13 and not in LRA Section 2.4.3. The staff finds that the applicant's response addressed the staff's question and, therefore, is acceptable. The staff's concern described in RAI 2.4-1, Item (iv), is resolved.

In its response, dated February 27, 2008, the applicant concluded that, as a result of this RAI, the applicant did not have to revise LRA Tables 2.2-3 or 2.2-4. The staff finds that the applicant appropriately confirmed and justified the license renewal scoping of the specific structures and structural components that were in question in the first part of RAI 2.4-1; therefore, the applicant's response to the first part of RAI 2.4-1 is acceptable. The staff's concerns described in the first part of RAI 2.4-1 are resolved. SER Section 2.4.3.2 discusses the second part of RAI 2.4-1.

Based on the information provided in the LRA, the RAI response discussed above, and the UFSAR, the staff concludes that, in LRA Section 2.4, the applicant identified, without omissions, the structures that are within the scope of license renewal for IP2 and IP3, in accordance with 10 CFR 54.4(a).

2.4.1 Containment Buildings

2.4.1.1 Summary of Technical Information in the Application

LRA Section 2.4.1 describes the containment buildings, which completely enclose the entire reactor and the RCS and ensures that essentially no leakage of radioactive materials to the environment would result even if a design basis LOCA occurs. The reactor containment structure is a seismic Class I, reinforced concrete vertical right cylinder with a flat base and hemispherical dome. A welded steel liner attached to the inside face of the concrete shell ensures a high degree of leak-tightness. The liner has accommodations for penetrations and personnel access. For IP2, the steel liner plate is covered by polyvinyl chloride insulation in a stainless steel jacket. For IP3, the steel liner plate is covered by urethane foam insulating material covered with gypsum board and a stainless steel jacket and backed with a fire-retardant paper on the unexposed side. The containment liner is anchored to the concrete shell by stud anchors. The bottom liner plate on top of the reinforced concrete base mat is covered with additional concrete, the top of which forms the floor of the containment. Internal structures consist of equipment supports, shielding, reactor cavity and canal for fuel transfer, manipulator crane, containment crane, and miscellaneous concrete and steel for floors and stairs. All internal structures are supported on the mat except equipment supports which are secured to the intermediate floors.

The containment buildings contain safety-related components relied on to remain functional during and following DBEs. The failure of nonsafety-related SSCs in the containment buildings potentially could prevent the satisfactory accomplishment of a safety-related function. In addition, the containment buildings perform functions that support fire protection.

LRA Table 2.4-1 identifies containment buildings component types, grouped by material (steel/other metals, concrete, other materials) within the scope of license renewal and subject to an AMR, as well as their intended functions.

2.4.1.2 Staff Evaluation

The staff reviewed LRA Section 2.4.1; IP2 UFSAR Sections 1.2.2, 1.11.2, and 5.1.2; and IP3 UFSAR Sections 1.3.5, 5.1.2, and 16.1.2 using the evaluation methodology described in SER Section 2.4 and the guidance in SRP-LR Section 2.4, "Scoping and Screening Results: Structures."

During its review, the staff evaluated the structural component functions described in the LRA and UFSAR to verify that the applicant had not omitted from the scope of license renewal any SCs with intended functions, as required by 10 CFR 54.4(a). The staff then reviewed those SCs that the applicant identified as within the scope of license renewal to verify that it had not omitted any passive and long-lived SCs subject to an AMR, in accordance with the requirements of 10 CFR 54.21(a)(1).

During its review of LRA Section 2.4.1, the staff identified areas in which additional information was necessary to complete the review of the applicant's scoping and screening results. The applicant responded to the staff's RAIs as discussed below.

In RAI 2.4.1-1, dated January 28, 2008, the staff noted that UFSAR Section 5.1.2.1 (IP2 and IP3) states that the containment structure serves as both a biological shield and a pressure boundary component. Since the biological shield function was not explicitly listed among the intended functions for containment buildings in LRA Section 2.4.1 and LRA Table 2.4-1, in RAI 2.4.1-1, the staff requested that the applicant clarify and include biological shield function as an intended function for containment buildings in the LRA.

In its response, dated February 27, 2008, the applicant stated that the biological shield function is an intended function for the IP2 and IP3 containment buildings. The applicant further stated that this intended function is implicit in the definition of the shelter or protection function *EN* in LRA Table 2.0-1, which includes "radiation shielding." The staff verified that the definition of the *EN* function in LRA Table 2.0-1 does include "radiation shielding" within parenthesis. The staff finds the response to be acceptable since it refers to an intended function. Therefore, the applicant's response to RAI 2.4.1-1 has adequately addressed the issue raised by the staff and is acceptable. The staff's concern described in RAI 2.4.1-1 is resolved.

A lack of clarity in LRA Table 2.4-1 prompted the staff to seek clarification. In RAI 2.4.1-2, the staff requested that the applicant confirm and/or clarify whether the following components associated with the containment buildings are included as components subject to an AMR in LRA Table 2.4-1 or justify their exclusion. For the components that are subject to an AMR, the applicant was requested to provide the appropriate AMR results in LRA Section 3.5.

(i)   primary shield wall around the reactor

(ii)   control rod drive missile shield

(iii)   retaining wall at the equipment hatch entrance and its missile shield (fixed and removable)

(iv)   blowout shield plug

(v)   insulation for the containment building liner (limits temperature rise in liner under accident conditions)

(vi)   protective coating for liner

(vii)   water proofing around fuel transfer tube

(viii) waterproof membrane for containment wall against backfill

(ix)   reactor cavity seal ring (see UFSAR Figures 5.1-6 and 5.1-7)

(x)   Seismic Class I debris screens at containment purge (Ref. UFSAR Section 5.1.4.2.4)

(xi)   stud anchors that anchor the containment liner plate to the concrete shell

In its response, dated February 27, 2008, the applicant addressed each of the components identified in the RAI with respect to whether they are subject to an AMR, as indicated below:

(i)   The primary shield wall around the reactor is included as part of "beams,

2-201

columns, interior walls, slabs" listed in LRA Table 2.4-1. AMR results are provided in Table 3.5.2-1.

(ii)   The control rod drive missile shield is included with the line item "missile shields" listed in Table 2.4-4. AMR results are provided in Table 3.5.2-4.

(iii)   The retaining wall at the equipment hatch entrance is included as part of "beams, columns, interior walls, slabs" listed in LRA Table 2.4-1. AMR results are provided in Table 3.5.2-1. The equipment hatch missile shield (fixed and removable) is included with the line item "missile shields" listed in Table 2.4-4. AMR results are provided in Table 3.5.2-4.

(iv)   Components/commodities identified in scope that provide missile protection are addressed in LRA Section 2.4-4 and Table 2.4-4. The "blowout shield plug" is included with the line item "missile shields" listed in LRA Table 2.4-4. AMR results are provided in Table 3.5.2-4.

(v)   The insulation for the containment building liner is included in Table 2.4-1 with line item "liner insulation jacket." AMR results are provided in Table 3.5.2-1.

(vi)   Protective coatings are not in the scope of license renewal because they do not perform an intended function. Their failure will not prevent satisfactory accomplishment of a safety function.

(vii)   The waterproofing material around the fuel transfer tube is not in scope. Waterproofing membranes have no license renewal intended function.

(viii) The waterproof membrane for containment wall against backfill is not in scope. Waterproofing membranes have no license renewal intended function.

(ix)   The reactor cavity seal ring identified in UFSAR Figures 5.1-6 and 5.1-7 has no license renewal intended function. This component is not safety-related and is not required to demonstrate compliance with regulations identified in 10 CFR 54.4(a)(3). Failure of the seal ring will not prevent satisfactory accomplishment of a safety function. The seal is provided to prevent leakage during refueling operations. This component is not listed in LRA Table 2.2-4 since it does not meet the threshold of a major structural component.

(x)   The seismic Class I debris screens at containment purge identified in UFSAR Section 5.1.4.2.4 do not perform a license renewal intended function. The primary containment isolation valves in the containment purge and pressure relief exhaust ducts are closed during normal plant operation. Failure of the screens will not prevent the ventilation systems from performing their intended function. These components are not required during design basis accidents or for any regulated event. The structural support of this component is included in scope and is included with line item "Structural steel: beams, columns, plates, trusses" listed in LRA Table 2.4-1.

(xi)   The stud anchors that anchor the containment liner plate to the concrete
         shell are included in the line item "anchorages/embedments" listed in
         LRA Table 2.4-4. AMR results are provided in Table 3.5.2-4.

In its response, dated February 27, 2008, the applicant has confirmed/clarified the screening of
each of the components in question and provided justification of the components that are not
subjected to an AMR. The staff finds the applicant's response to Items (ii), (iv), and (xi)
acceptable because the applicant explicitly clarified that the components in question are within
the scope of license renewal and are subject to an AMR. The staff finds the applicant's
response to Items (vii), (viii), and (x) acceptable because the applicant explicitly clarified that the
components in question do not have an intended function that meets any of the criteria in
10 CFR 54.4(a). Therefore, the staff finds that the applicant's response to RAI 2.4.1-2 is
acceptable, with the following exceptions with regard to the response to Items (i), (iii), (v), (vi)
and (ix) of RAI 2.4.1-2. In a follow-up RAI to RAI 2.4.1-2, dated May 12, 2008, the staff
requested the applicant to clarify/address these exceptions. The applicant provided clarification
responses to the follow-up RAI items by letter dated June 11, 2008. The follow-up RAI items
and their resolution are discussed below.

•   With regard to Item (i), the response stated that Primary Shield Wall is included as part of
    line item "Beams, columns, interior walls, slabs" in LRA Table 2.4-1. The staff noted that
    walls with lesser safety-significance, such as pressurizer shield, ring wall, and cylinder walls,
    have been listed as separate items in LRA Table 2.4-1. Considering that the primary shield
    wall is subjected to a more severe environment (high temperature and radiation exposure)
    and has a much higher safety-significance than the general interior wall, the staff requested,
    in a follow-up RAI dated May 12, 2008, that the applicant include the primary shield wall as
    a separate line item in LRA Table 2.4-1, to make its inclusion in the scope of license
    renewal and its consideration as being subject to AMR, explicitly clear.

    In its response, dated June 11, 2008, the applicant added the primary shield wall as a
    separate concrete component item in LRA Tables 2.4-1 and 3.5.2-1 with the appropriate
    intended functions. By doing so, the applicant has explicitly included the primary shield wall
    as a component subject to AMR. Therefore, the staff finds the response acceptable. The
    staff's evaluation of the AMR results for the primary shield wall is documented in SER
    Section 3.5.

•   With regard to Item (iii), the response stated that the retaining wall is included as part of line
    item "Beams, columns, interior walls, slabs" in LRA Table 2.4-1. The staff noted that the
    retaining wall at the equipment hatch entrance is an exterior wall and is subjected to a
    different environment than the interior wall. Therefore, in a follow-up RAI dated May 12,
    2008, the staff requested the applicant to explicitly include the retaining wall at the
    equipment hatch entrance in LRA Table 2.4-1 as a separate line item.

    In its response, dated June 11, 2008, the applicant added the equipment hatch entry
    retaining wall (exists for IP2 only) as a separate concrete component item in LRA
    Tables 2.4-1 and 3.5.2-1 with the appropriate intended functions. By doing so the applicant
    has explicitly included the IP2 retaining wall at the equipment hatch entrance as a
    component subject to AMR. Therefore, the staff finds the response acceptable.

- With regard to Item (v), the response stated that liner plate insulation is included with line item "Insulation Jacket" in LRA Table 2.4-1. The staff noted that materials for the insulation jacket and the insulation itself are not the same. The jacket is stainless steel but the insulation is PVC for IP2 and Urethane foam covered with gypsum board for IP3 (UFSAR Section 5.1). The insulation itself is not included in LRA Table 2.4-1 or LRA Table 2.4-4; nor are these materials identified in LRA Sections 3.5.2.1.1 or 3.5.2.1.4. They also were not addressed in the response to RAI 2.4.4-2. In a follow-up RAI dated May 12, 2008, the staff requested the applicant to appropriately address the scoping, screening, and AMR results for these in-scope insulation materials in the LRA.

  In its response, dated June 11, 2008, the applicant stated that the IP2 containment liner plate PVC insulation and IP3 containment liner urethane insulation are encapsulated within stainless steel jacketing (IP2 UFSAR Section 6C.8.4, and IP3 UFSAR Section 5.5) and are not exposed to containment atmosphere. The only visible and exposed parts of the insulation are the stainless steel jacketing. The aging management review results in LRA Table 3.5.2-1 for the liner plate insulation pertain to the stainless steel jacketing. The applicant added that the containment liner plate insulation within the jacketing is in scope and subject to aging management review for providing shelter and protection to the containment liner plate. The PVC and urethane insulation materials have no aging effects in the air-indoor environment and, therefore, no aging management program is necessary.

  In the above response, the applicant has clarified that, for both IP2 and IP3, the containment liner plate insulation within the jacketing is within the scope of license renewal and subject to AMR but does not need aging management since there are no aging effects in its protected environment. Based on the above response, it is the staff's understanding that the PVC and urethane insulation are encapsulated within the stainless steel insulation jacketing forming one composite unit, and the AMR results in LRA Table 3.5.2-1 for the line item "liner plate insulation jacket" includes the encapsulated insulation, which is exposed to an indoor air environment that does not promote aging effects. The staff finds that the applicant's response addressed the staff's concern with regard to scoping and screening of the liner insulation and, therefore, is acceptable.

- With regard to Item (vi), the response stated that protective coatings for the containment liner are not in scope because they do not perform an intended function. Staff noted that, although protective coating on the containment liner does not directly perform a license renewal function, it prevents degradation of the liner if properly maintained. Section XI.S8 of NUREG-1801, Volume 2, which is the AMP for protective coatings, recommends maintenance of the protective coatings to avoid clogging of the sumps. The GALL Report states that, if protective coatings are relied upon to manage the effects of aging, the structures monitoring program is to include provisions to address protective coating monitoring and maintenance (Item 25 in Table 5 of NUREG-1801, Volume 1). Therefore, in a follow-up RAI dated May 12, 2008, the staff requested the applicant to provide justification for excluding the protective coating on the containment liner from the scope of license-renewal and from being subject to an AMR.

  In its response, dated June 11, 2008, the applicant stated that the liner plates of IP2 and IP3 containment are provided with protective coatings. The applicant stated that, in response to Generic Safety Issue (GSI)-191, "Assessment of Debris Accumulation on PWR Sump Performance," the applicant's Civil/Structural Engineering group visually inspects coatings in the vapor containment building during refueling outages. Sump clogging for IP2 and IP3 was evaluated, and the evaluation results were provided by Entergy, Inc., in letter dated September 1, 2005, in response to NRC generic letter 2004-02, "Potential impact of debris blockage on emergency recirculation during design basis accidents at pressurized water reactors."

The applicant further added that the GALL Report states that, if protective coatings are relied upon to manage the effects of aging, the structures monitoring program should include provisions to address protective coating monitoring and maintenance. The applicant stated that, as indicated in LRA Table 3.5.1, Item 3.5.1-25, IP2 and IP3 containment liner protective coatings are not relied upon to manage the effects of aging. As shown in LRA Table 3.5.2-1, aging effects of liner plate and integral attachments are managed by the Containment Inservice Inspection-IWE and Containment Leak Rate Test programs for license renewal. Accordingly, the protective coating on the containment liner is not within the scope of license renewal and, therefore, is not subject to aging management review.

In the above response, the applicant clarified that inspection commitments of protective coatings and sump clogging evaluations were addressed as part of its response to the NRC's GSI-191 issue. The applicant reiterated that the aging effects of the liner plate are managed by the containment inservice inspection program per IWE and the Appendix J Containment Leakage Rate Testing Program, and that protective coatings are not relied upon to manage the effects of aging of the liner. Therefore, the staff accepts the applicant's determination that the protective coating on the containment liner may be considered outside the scope of license renewal and not subject to AMR. The staff finds the response acceptable.

• With regard to Item (ix), the response stated that the reactor cavity seal ring has no license renewal intended function. The staff notes that the reactor cavity seal ring is a flood barrier (FLB) to preclude borated water leaks through the seal and thereby prevent accumulation of borated water in the gap between the reactor vessel and the primary shield wall, which could induce corrosion of the reactor vessel and its supports as well as cause degradation of the primary shield wall concrete. Considering the above, in a follow-up RAI dated May 12, 2008, the staff requested the applicant to provide justification for excluding the reactor cavity seal from the scope of license-renewal and from being subject to an AMR.

In its response, dated June 11, 2008, the applicant stated that the reactor cavity seal ring is a nonsafety-related component and it has no license renewal intended function pursuant to 10 CFR 54.4(a). Therefore, the reactor cavity seal is not within the scope of license renewal nor subject to AMR. The applicant specifically explained that the reactor cavity seal ring is installed prior to filling the refueling cavity to allow for fuel handling operations and that plant procedures ensure proper installation to preclude leakage during refueling operations. The applicant added that, even if the seal were to leak during the time the refueling cavity is filled, sump pumps in the cavity beneath the reactor vessel would prevent water accumulation in the gap between the reactor vessel and the primary shield wall.

The applicant further stated that plant operating experience does not indicate that leakage from the reactor cavity seal ring has caused corrosion of the reactor vessel or its supports; nor has it caused degradation of primary shield wall concrete. Further, aging management programs shown in LRA Tables 3.1.2-1 and 3.5.2-1 will manage the effects of aging from corrosion, if any, of the reactor vessel and its supports and will manage degradation of the interior concrete walls from exposure to borated water leakage during refueling.

Based on the above response, the staff understands that the reactor cavity seal is a nonsafety-related component installed during each refueling outage prior to flooding of the reactor cavity for refueling operations using procedures to ensure a leaktight installation. Also, the applicant's operating experience has not indicated any degradation of the reactor vessel, its supports, and the primary shield wall attributable to leakage through the reactor cavity seal. Further, the applicant has procedures/programs in place to manage any effects even if the seal were to leak during refueling operations. Therefore, the staff accepts the applicant's determination that the seal does not perform a license renewal intended function pursuant to 10 CFR 54.4(a) and, therefore, the reactor cavity seal is not in scope of license renewal nor subject to AMR. The applicant's response resolved the staff's concern.

Based on the discussion above of the applicant's clarifying responses, the staff finds the applicant's response to RAI 2.4.1-2 acceptable.

In RAI 2.4.1-3, dated January 28, 2008, the staff requested that the applicant confirm whether the component identified as "Structural Steel: beams, columns, plates, trusses" in LRA Table 2.4-1 includes bracings, welds, and bolted connections. The applicant also was requested to confirm whether the pressurized channel shrouds used at liner welded joints (including those at penetrations) are included in a structure/commodity group, or to justify their exclusion from an AMR. In addition, the applicant was requested to confirm whether the components identified as "bellows penetrations" in LRA Table 2.4-1 include the refueling bellows, if refueling bellows are used at IP2 and IP3.

In its response, dated February 27, 2008, the applicant stated that the component identified as "Structural Steel: beams, columns, plates, trusses" in LRA Table 2.4-1 includes bracing and welds associated with the component. The applicant further clarified that bolted connections for structures/components are addressed in LRA section 2.4.4 and Table 2.4-4. The applicant stated that the pressurized channel shrouds associated with liner welded joints (including those at penetrations) are not addressed as a separate component group. They are considered integral to the components listed as "liner plate and integral attachments" and "Electrical penetration sleeves" and "Mechanical penetration sleeves" in LRA Table 2.4-1. The applicant stated that components identified as "bellows penetrations" in LRA Table 2.4-1 do not include "refueling bellows." The applicant further clarified that bellows penetrations in LRA Table 2.4-1 are associated with containment piping penetrations and that refueling bellows are not a feature of the IP2 or IP3 design.

The staff finds that the applicant's response adequately addressed the staff's questions with regard to the stated components and, the response to RAI 2.4.1-3 is acceptable, subject to resolution of the additional clarifications requested below with regard to bellows. With regard to bellows penetrations, the applicant's response stated that the bellows penetrations in LRA Table 2.4-1 are associated with containment piping penetrations and that refueling bellows are not a feature of the IP2 or IP3 design. In the follow-up RAI dated May 12, 2008, the staff requested the applicant to further describe the types of piping penetration bellows in each unit. Also, the staff requested the applicant to clarify if there are transfer canal bellows (with the number in each unit) at Indian Point and if they are in-scope of license renewal or not, with justification.

In its response, dated June 11, 2008, the applicant stated that IP2 and IP3 containment penetrations consist of a sleeve embedded in the concrete and welded to the containment liner. The applicant explained that differential expansion between a sleeve and one or more hot pipes passing through it is accommodated by using a nickel alloy or stainless steel bellows-type expansion joint between the outer end of the sleeve and the piping outside of the containment wall. The applicant added that details of the containment penetrations and bellows for each unit are shown in UFSAR Figures 5.1-30 (IP2) and 5.1-12 (IP3).

The applicant stated that, for each unit, a fuel transfer tube is provided for fuel movement between the refueling transfer canal in the reactor containment and the spent fuel pit. The fuel transfer tube consists of a 20-in. stainless steel pipe installed inside a 24-in. pipe. The applicant added that two bellows-type expansion joints (one inside containment and one in the spent fuel pit) are provided on the tubes to compensate for any differential movement between the two pipes and other structures. Figure 5.1-31 of IP2 UFSAR and Figure 5.1-14 of IP3 UFSAR show details of the fuel transfer tube and bellows for each unit. These penetration bellows are within the scope of license renewal and subject to an AMR. They are listed as "bellows penetration" in LRA Tables 2.4-1 and 3.5.2-1.

In its above response, the applicant confirmed that, in addition to the piping penetrations bellows, the two fuel transfer tube bellows for each unit were in scope of license renewal and subject to AMR and were included as part of line item "bellows penetration" in LRA Table 2.4-1. The staff finds that the response addressed the staff's question with regard to the types of bellows that were scoped and screened for license renewal. Therefore, the response is acceptable.

During its review of components listed as "Polar Crane, rails and girders" and "Manipulator Crane, crane rails and girders" in LRA Table 2.4-1, the staff determined that additional information was needed to complete its review. In RAI 2.4.1-4, dated January 28, 2008, the staff requested that the applicant confirm whether the column structure; bridge and trolley of the polar crane; and the bridge, trolley and mast of the manipulator crane were screened-in as subject to an AMR. The staff also requested that the applicant confirm whether fasteners and rail hardware associated with the polar crane and manipulator crane are within scope of license renewal and subject to an AMR; and if they were excluded, the staff requested that the applicant provide a justification. The staff also requested that the applicant indicate whether there were any other hoists and lifting devices (e.g. for the reactor vessel head and reactor internals) that should be included as components within the scope of license renewal and subject to an AMR; and if so, the staff requested that the applicant provide scoping, screening, and AMR results relevant to the LRA.

In its response, dated February 27, 2008, the applicant stated that the column structure; bridge and trolley of the polar crane; and the bridge, trolley and mast of the manipulator crane are screened-in as subject to an AMR. The applicant indicated that these components are subparts of "crane, rails and girders." The applicant stated that fasteners ("structural bolting") and rail hardware ("component support") associated with the polar crane and manipulator crane are within the scope of license renewal and subject to an AMR. The applicant indicated that these components are addressed in LRA Section 2.4.4, "Bulk Commodities." The applicant clarified that there were no hoists or lifting devices, other than those already identified in the LRA, that perform a license renewal intended function.

2-207

Because the applicant stated that the structures and components in question are subject to an AMR, the staff finds that the applicant adequately addressed the staff's questions; therefore, the response to RAI 2.4.1-4 is acceptable. The staff's concern described in RAI 2.4.1-4 is resolved.

Because of a lack of clarity in LRA Table 2.4-1 regarding components listed as Equipment Hatch and Personnel Lock, in RAI 2.4.1-5, dated January 28, 2008, the staff requested that the applicant clarify whether the flange double-gaskets, hatch locks, hinges, and closure mechanisms that help prevent loss of sealing/leak-tightness for these listed hatches were included within the scope of license renewal and subject to an AMR. The staff also requested that the applicant provide scoping, screening, and AMR results as appropriate or justify their exclusion.

In its response, dated February 27, 2008, the applicant stated that the flange double-gaskets, hatch locks, hinges, and closure mechanisms for the equipment hatch and personnel lock are within the scope of license renewal. The applicant clarified that the double gasket seals are included under the line item "equipment hatch and personnel lock seal" in LRA Table 2.4-1, and are subject to AMR. The AMR results are provided in Table 3.5.2-1. The applicant stated that hatch locks, hinges, and closure mechanisms are active components and are, therefore, not subject to aging management review as discussed in LRA Table 3.5.1, Line Item 3.5.1-17. The applicant added that satisfactory performance of these active components is demonstrated through routine testing under the Containment Leak Rate Program as required by Section 3.6.2 of the IP2 and IP3 Technical Specifications.

The staff finds that the applicant has adequately addressed the staff's concern with regard to the flange double-gaskets for the hatches in question. However, the response stated that the hatch locks, hinges, and closure mechanisms are active components and, therefore, not subject to AMR as discussed in LRA Table 3.5.1, Line Item 3.5.1-17. The staff noted that these components are passive during plant operation, during which time they are (and need to remain) in a closed position and are an integral part of the pressure boundary. Considering the above, in a follow-up RAI, dated May 12, 2008, the staff requested the applicant to provide the justification for excluding the hatch locks, hinges, and closure mechanisms from the scope of license-renewal and from being subject to an AMR.

In its response, dated June 11, 2008, the applicant stated that the IP2 and IP3 hatch locks, hinges, and closure mechanisms are in scope of license renewal. However, since they perform their functions with moving parts or change in configuration, they are not subject to AMR. The applicant added that consistent with NUREG-1801, Volume 1, Revision 1, Table 5, Item 17, their leaktightness in the closed position is demonstrated through routine testing under the containment leakage rate test program as required by IP2 and IP3 Technical Specifications (Reference LRA Table 3.5.1, Line Item 3.5.1-17). Since the applicant's response clarified that, in the closed position, the hatch locks, hinges, and closure mechanisms are considered integral to the hatch itself, whose leaktightness is demonstrated by routine local leak rate testing under the Containment Leakage Rate Test Program, the staff finds the response acceptable.

2.4.1.3 Conclusion

The staff reviewed the LRA, UFSAR, and RAI and follow-up RAI responses to determine whether the applicant failed to identify any SCs within the scope of license renewal. The staff found no omissions. In addition, the staff sought to determine if the applicant failed to identify any SCs subject to an AMR. The staff found no omissions. On the basis of its review, the staff concludes that the applicant has adequately identified the containment buildings SCs within the scope of license renewal, as required by 10 CFR 54.4(a), and those subject to an AMR, as required by 10 CFR 54.21(a)(1).

2.4.2 Water Control Structures

2.4.2.1 Summary of Technical Information in the Application

LRA Section 2.4.2 describes the water control structures, which include:

- discharge canal and outfall structure
- intake structure (IP1, IP2, IP3) and intake structure enclosure building (IP3)
- service water pipe chase (IP3)
- service water valve pit (IP2 and IP3)

The discharge canal and outfall structure, located west of the IP2 and IP3 turbine buildings, extends from the IP1 turbine building and carries SW system discharge to the river. Three IP3 backup SW pumps, which provide cooling water from the discharge canal in the unlikely event of damage to the SW intake structure, are supported on a slab spanning the walls of the canal. The SW pipe chase, a concrete structure enclosing the SW line, spans across the discharge canal. The discharge canal wall portion adjacent to the SW pipe chase is seismic Class I and part of the ultimate heat sink. The outfall structure enhances mixing of cooling water and river water to minimize thermal impact on the river. The discharge port gates can be adjusted mechanically to control fluid discharge velocity. The outfall structure does not support a license renewal function as defined by 10 CFR 54.4 and hence is not in the scope of license renewal.

The IP1 intake structure (also known as the screenwell house) is a seismic Class III structure located adjacent to the wharf and west of the station on the riverbank. It houses electrical components required for the alternate safe shutdown system, which is credited in the Appendix R safe shutdown analysis. The lower portion contains the IP1 intake, which houses the river water pumps that support IP2 SW. The structure is a reinforced concrete frame supported by a massive concrete substructure. Exterior walls of the intake structure are of concrete brick construction. The north and south ends of the structure are covered by a reinforced concrete roof slab.

The IP2 intake structure (also known as the screenwell structure) is west of the site, below grade at the Hudson River bank, and is open to the river on the west side. The IP3 intake structure (also known as the screenwell structure) is west of the containment structure. Each structure houses six CW pumps (each in a separate reinforced concrete compartment), six SW pumps (a SW bay enclosure protects the IP3 pumps), traveling and fixed screens, and screen wash equipment. On the east side of each structure, the SW strainer pit houses SW strainers, screen wash piping, and the strainer control panel. Both the SW strainer pit and the SW bay enclosure are seismic Class I.

The intake structure enclosure building located west of the containment structure provides an upper separate enclosure for the IP3 intake structure and protects CW and SW system components from the weather. Dampers located in the roof system release excess heat during normal operations. The intake structure enclosure consists of a single story steel-framed super-structure with exterior metal siding and ventilation panels.

The IP3 SW pipe chase, which protects SW lines that span the discharge canal and the SW valves and piping, is a reinforced concrete structure attached to the discharge canal wall. The discharge canal wall portion adjacent to the SW pipe chase is seismic Class I.

SW valve pits at the west side of the IP2 and IP3 heater bay buildings protect SW components in IP2 and IP3 intake structures. IP3 has an additional SW valve pit on the north end of the IP3 heater bay building to back up the SW pumps. The SW valve pits are underground reinforced concrete structures covered by structural steel plate welded to I-beams at ground level. The additional SW valve pit for IP3 has a precast concrete roof.

The water control structures contain safety-related components relied upon to remain functional during and following DBEs. The failure of nonsafety-related SSCs in the water control structures potentially could prevent the satisfactory accomplishment of a safety-related function. In addition, the water control structures perform functions that support fire protection.

LRA Table 2.4-2 identifies water control structures component types, grouped by material (steel/other metals, concrete), within the scope of license renewal and subject to an AMR as well as their intended functions.

2.4.2.2 Staff Evaluation

The staff reviewed LRA Section 2.4.2 and UFSAR Section 8.3 for IP2 using the evaluation methodology described in SER Section 2.4 and the guidance in SRP-LR Section 2.4.

During its review, the staff evaluated the structural component functions described in the LRA and UFSAR to verify that the applicant had not omitted from the scope of license renewal any SCs with intended functions, as required by 10 CFR 54.4(a). The staff then reviewed those SCs that the applicant identified as within the scope of license renewal to verify that the applicant had not omitted any passive and long-lived SCs subject to an AMR, in accordance with the requirements of 10 CFR 54.21(a)(1).

During its review of LRA Section 2.4.2, the staff identified areas in which additional information was necessary to complete the review of the applicant's scoping and screening results. The applicant responded to the staff's RAIs as discussed below.

In RAI 2.4.2-1, dated January 28, 2008, the staff noted that LRA Table 2.4-2 does not include the debris wall, fixed coarse screens, fine mesh traveling screens, and gates at the intake structure. Further, the table does not include metal decking, metal siding, grating, and ventilation panels for the intake structure enclosure; nor does it include manhole, ladder, and sump of the SW valve pit. The staff requested that the applicant confirm whether or not these components should be included within the scope of license renewal and subject to an AMR and, if so, to provide scoping, screening and AMR results. If not, the staff requested the applicant to justify their absence from LRA Table 2.4-2. The applicant also was requested to clarify whether the "structural steel" component in LRA Table 2.4- 2 includes, among other items, beams, plates, and welded/bolted connections.

In its response, dated February 27, 2008, the applicant stated that the debris wall, fixed coarse screens, fine mesh traveling screens, and gates at the intake structure are not safety-related and are not required to demonstrate compliance with 10 CFR 54.4(a)(3). The applicant stated that the system design is such that failure of these components will not prevent satisfactory accomplishment of a safety function. However, their support structures, being integral to the intake structure in some cases (e.g., embedded guides and steel supports), are included in the "structural steel" category listed in LRA Table 2.4.2. The applicant stated that metal siding for the intake structure enclosure is not safety-related and is not required to demonstrate compliance with 10 CFR 54.4(a)(3). The applicant added that failure of the metal siding component will not prevent satisfactory accomplishment of any safety function. The applicant stated that in-scope grating, decking, and ladders are bulk commodities addressed in LRA Table 2.4-4. The ventilation panels for the intake structure enclosure are addressed as "vents and louvers" and listed in LRA Table 2.4-4. Furthermore, the applicant stated that manholes are included in LRA Table 2.4-3. The sump of the SW valve pit is integral to the in-scope SW valve pit; thus, it is not listed as a separate item. The applicant clarified that the "structural steel" component type in LRA Table 2.4-2 includes columns, beams, plates, and their welded connections. Structural bolting is included as a bulk commodity and listed in LRA Table 2.4-4.

In reviewing the response to RAI 2.4.2-1, the staff also reviewed the discussion on the "Service Water System" and "Tornado Design Criteria" in Sections 9.6.1 and 16.2, respectively, of the IP3 UFSAR. Based on the description in these UFSAR sections, the SW supply is assured by redundancy of two supply lines, four intakes and screens, and six pumps, of which only two pumps, one intake and screen, and one supply line are required for prolonged shut-down. Further, the backup SW system provides an additional source of service water independent of the intake structure. The existence of these redundancies in the SW system confirms the applicant's statement, in the RAI response, that failure of the intake structure components noted in the RAI, which are part of the SW system, will not prevent satisfactory accomplishment of the safety function of the SW system. However, in the response, the applicant stated that in-scope grating, decking, and ladders are bulk commodities addressed in LRA Table 2.4-4. Since this is a generic statement, in a follow-up RAI, dated May 12, 2008, the staff requested the applicant to clarify if the specific components in question that were identified in the RAI (i.e. metal decking and grating of the intake structure enclosure and ladder of the service water valve pit) are included in the scope of license renewal, and subject to AMR as bulk-commodities addressed in LRA Table 2.4-4.

In its response, dated June 11, 2008, the applicant stated that metal decking and grating of the intake structure enclosure and ladder of the service water valve pit have license renewal intended functions as defined by 10 CFR 54.4(a)(2) and, therefore, they are in scope of license renewal and subject to an AMR. The applicant added that these structural components are included in LRA Table 2.4-4, line item "Stairway, handrail, platform, grating, decking, and ladders." Since the applicant explicitly clarified that the specific structural components identified in the RAI were subject to an AMR, the staff finds the response acceptable.

Based on the above response to RAI 2.4.2-1 and to the follow-up RAI, and the descriptions in Section 9.6.1 and Section 16.2 of the IP3 UFSAR, the staff finds that the applicant has adequately addressed and/or clarified the scoping and screening of the specific structural components identified in the RAI. Therefore, the applicant's response to RAI 2.4.2-1 is acceptable.

The staff also requested additional information in RAI 2.4.2-2, dated May 12, 2008, regarding other structural components. In Part (a) of RAI 2.4.2-2, the staff noted that LRA Table 2.2-3 and LRA Section 2.4.2 include "discharge canal and outfall structure" as being within the scope of license renewal. The description in LRA Section 2.4.2, in the second paragraph under the subtitle "Discharge Canal and Outfall Structure," states that the outfall structure does not support a license renewal function and, therefore, is not in scope. The staff requested the applicant to explain why the "outfall structure" was included in LRA Table 2.2-3 and LRA Section 2.4.2. The staff requested the applicant to discuss this inconsistency and take appropriate action in scoping the outfall structure.

In Part (b) of RAI 2.4.2-2, because of a lack of clarity in the description in LRA Section 2.4.2 with regard to the discharge canal, the staff requested the applicant to confirm/clarify if (i) the entire discharge canal is considered within the scope of license renewal and subject to AMR or (ii) only the portion adjacent to/supporting the service water pipe chase, and the portion supporting and including the slab on which the Unit 3 service water backup pumps are mounted, are within the scope of license renewal and subject to an AMR.

In its response to Part (a) of RAI 2.4.2-2, dated June 11, 2008, the applicant stated that the "outfall structure" is included in LRA Table 2.2-3 and LRA Section 2.4.2 as part of line item "discharge canal and outfall structure" because this line item is the name of one continuous structure that includes the discharge canal and the outfall structure. The only portion that is within the scope of license renewal is the discharge canal. The applicant reiterated that the description in LRA Section 2.4.2, in the second paragraph under the subtitle "Discharge Canal and Outfall Structure," states that "[t]he outfall structure does not support a license renewal function as defined by 10 CFR 54.4 and, therefore, is not in scope." The applicant added that this statement specifically addresses exclusion of the outfall structure portion of the structure from the scope of license renewal and AMR. The staff finds the response acceptable because the applicant clarified that only the discharge canal is within the scope of license renewal; the outfall structure portion of the "discharge canal and outfall structure" is not within the scope of license renewal.

In its response to Part (b) of RAI 2.4.2-2, dated June 11, 2008, the applicant stated that the entire discharge canal is within the scope of license renewal and subject to AMR. Since the response clarified that the entire discharge canal is conservatively included as being in scope of license renewal and subject to AMR, the staff finds the clarification provided by the applicant acceptable.

2.4.2.3 Conclusion

The staff reviewed the LRA, UFSAR, RAI and follow-up RAI responses, and description of related structural components to determine whether the applicant failed to identify any SCs within the scope of license renewal. The staff found no omissions. In addition, the staff sought to determine if the applicant failed to identify any SCs subject to an AMR. Again, the staff found no omissions. On the basis of its review, the staff concludes that the applicant has adequately identified the water control structures SCs within the scope of license renewal, as required by 10 CFR 54.4(a), and those subject to an AMR, as required by 10 CFR 54.21(a)(1).

2.4.3 Turbine Buildings, Auxiliary Buildings, and Other Structures

2.4.3.1 Summary of Technical Information in the Application

LRA Section 2.4.3 describes the turbine buildings, auxiliary buildings, and other structures:

- Appendix R diesel generator foundation, fuel oil tank vault, switchgear and enclosure (IP3)
- auxiliary feedwater pump building (IP2, IP3)
- boric acid evaporator building (IP2)
- city water storage tank foundation and meter house
- condensate storage tanks foundation (IP2, IP3)
- containment access facility and annex (IP3)
- control buildings (IP2, IP3)
- diesel generator buildings (IP2, IP3)
- electrical tunnels (IP2, IP3)
- emergency lighting poles and foundations
- fan houses (IP2, IP3)
- fire pump house (IP2)/fire protection pump house (IP3)
- fire water storage tank foundation (IP2, IP3)
- fuel storage buildings (IP2, IP3)
- gas turbine generator Nos. 1, 2, and 3 enclosure and fuel tank foundation (includes gas turbine substation switchgear structures and foundation)
- maintenance and outage building elevated passageway (IP2)
- manholes and duct banks
- new station security building
- nuclear service building (IP1)
- power conversion equipment building (IP3)
- PABs (IP2, IP3)
- primary water storage tanks foundation (IP2, IP3)
- radiation monitoring enclosure (IP2)
- refueling water storage tanks foundation (IP2, IP3)
- security access and office building (IP3)

- superheater building (IP1)

- superheater stack (IP1)

- transformer/switchyard support structures

- transmission towers (SBO recovery path) and foundations

- turbine building (IP1, IP2, IP3) and heater bay (IP2, IP3)

- utility tunnel

- waste holdup tank pit (IP2, IP3)

The Appendix R diesel generator, fuel oil tank vault, and switchgear are located in separate, adjacent enclosures in the yard area north of the AFW pump room. The Appendix R diesel generator, fuel oil tank vault, and switchgear support a power supply sufficient to allow the plant to be brought to cold shutdown in a loss of offsite power coincident with a fire causing the loss of all three EDGs or their distribution systems.

The IP2 AFW pump building in the shield wall area between the shield wall and the IP2 containment building is a seismic Class I structure that protect the Class I AFW pumps. The MS lines also located in this building are supported by the structural steel framing.

The IP3 AFW pump building in the shield wall area between the shield wall and the IP3 containment building also includes the shield wall area enclosure. It is a seismic Class I structure that protects the Class I AFW pumps and MS lines located in this area.

The boric acid evaporator building is a seismic Class I reinforced concrete structure supported by the roof slab of the IP2 waste hold-up tank pit. The exterior walls are of concrete and concrete block construction. Portions of the concrete walls are removable. Over the concrete block portion is light-weight roofing over metal decking and over the concrete walls is a concrete slab.

The city water storage tank and meter house is a source of water for the AFW system for both IP2 and IP3 and of emergency water for SI, RHR, and charging pumps. The city water storage tank foundation supports the storage tank safety function. The meter house shelters and protects the storage tank components. A free-standing, 1,500,000-gallon vertically cylindrical carbon steel city water storage tank is supported by a reinforced concrete spread footing foundation on rock. The meter house is a single-story concrete brick and steel structure with a concrete roof slab.

Two separate reinforced concrete slab foundations support the condensate storage tanks for IP2 and IP3.

The containment access facility and annex adjacent to the PAB is a handling area for contaminated material and a personnel access to containment. The containment access facility and annex is Class III except for the seismic Class I structural steel portion interfacing with the PAB. The containment access facility and annex has structural steel framing supported on the PAB roof floor slab and insulated metal siding.

The control buildings house the central control room, cable spreading room, and other safety-related equipment and components. The IP2 control building adjacent to the IP2 turbine building on the west and the superheater building on the south contains both the IP1 and IP2 control rooms. It is a multi-story Class I steel framed structure with north and east exterior walls of insulated metal-sandwich panels. Floor slabs are composite-type construction, concrete over steel beam. The IP3 control building is a multi-story Class I concrete structure with concrete and concrete brick exterior is adjacent to the IP3 turbine building on one end and the diesel generator building on the south. Both structures are founded on bedrock.

The seismic Class I IP2 diesel generator building consists of a reinforced concrete foundation on bedrock, a prefabricated rigid steel superstructure with exterior insulated metal siding, and a solid, corrugated metal roof. The diesel generators rest on reinforced concrete foundations supported by the structure's main slab. A concrete shield wall on the west side serves as missile protection between the control panel and diesels. The IP3 diesel generator building is a single-story reinforced concrete structure on a concrete slab supported on bedrock. Each diesel generator building houses three safety-related diesel generators. Each diesel has separate underground storage vaults, integral to its building, for fuel oil tanks. Foundations for the fuel oil tanks are the same as for the structure.

The electrical tunnels are partially below-grade, seismic Class I reinforced concrete structures that contain electrical cable, conduit, and cable trays that support plant operations. The IP2 electrical tunnel running eastward from the east side of the control building is attached to the south side of an east-west retaining wall. The elevation of the lower slab of the tunnel slopes from the control building up to the PAB. The tunnel then turns northward past the west side of the PAB to the electrical penetration area adjacent to the IP2 containment building. The IP3 electrical tunnels run from the control building past the PAB to the containment penetration vault. The electrical tunnels consist of two seismic Class I reinforced concrete conduits, one above the other. Both the upper and lower tunnels are eight feet wide by eight feet high.

Pole-mounted security lighting around the perimeter of the plant site provides emergency lighting in an Appendix R fire and a loss of offsite power by illuminating ingress and egress. Each emergency light pole is a single-pole steel structure supported by a reinforced concrete foundation.

Each fan house is a seismic Class I structure containing the piping penetration area. Safety-related valves in the piping penetration area may be used to achieve safe shutdown. Each fan house building is a multi-story reinforced concrete and masonry block wall structure founded on bedrock. A steel superstructure on top of each building supports the roof framing system. The IP2 fan house southeast of the IP2 containment structure and between the IP2 containment, the IP2 PAB, and the IP2 fuel storage building is isolated from the containment structure and the PAB. Its east wall is common with the west wall of the fuel storage building. The IP3 fan house southeast of the IP3 containment structure and between the IP3 containment, the IP3 PAB, containment access facility, and the IP3 fuel storage building is isolated from the containment structure and the PAB. Its east wall is common with the west wall of the fuel storage building and its south wall is common to the containment access facility annex.

The IP2 fire protection pump house (also known as diesel fire pump house) houses the main diesel firewater pump and protects fire protection system components. The structure is of structural steel framing with exterior insulated metal siding and a composite metal roof. The foundation is a reinforced concrete slab on grade. The IP3 fire protection pump house contains the electric motor-driven fire pump, the diesel-driven fire pump, and equipment for an adequate source of fire water. The structure is a reinforced concrete and concrete block wall construction with a concrete roof slab. The foundation is a reinforced concrete slab on bedrock.

The IP2 fire water storage tank (also known as suction tank) foundation is the main support for the 300,000-gallon fire water storage tank. Water for the dedicated diesel-driven fire pump for normal operations comes from the tank. The IP3 fire water storage tank foundations are the main supports for two 350,000-gallon fire water storage tanks. The tanks and their piping, electrical, and instrumentation systems are the source of fire protection system water and IP3 makeup water treatment.

For IP2 and IP3, the fuel storage building is designed to handle and store both spent and new fuel and supports the spent fuel crane and other fuel-handling equipment. In addition, the floor of IP2 provides support for a single-failure-proof gantry crane. Each structure is located adjacent to but separate from its containment building.

The gas turbine generator No. 1 enclosure and tank foundation are seismic Class III structures providing shelter and protection from the elements for gas turbine No. 1 and its associated equipment. Gas turbine No. 1 is located adjacent to the Unit 1 turbine building and supports no license renewal function; however, the associated switchgear components and fuel supply tank provide support for the SBO/Appendix R diesel generator set. The gas turbine No. 1 enclosure consists of structural steel framing with exterior metal siding on a reinforced concrete slab. The fuel tank foundation is a reinforced concrete spread footing which supports the fuel tank supplying the SBO/Appendix R diesel.

The gas turbine generators Nos. 2 and 3 enclosure is a seismic Class III structure that shelters and protects the equipment from the elements. The gas turbine Nos. 2 and 3 enclosure located at the Buchanan substation houses gas turbine generators Nos. 2 and 3 and their switchgear equipment. The switchgear and associated components within the structure support offsite power recovery following station blackout. The gas turbine Nos. 2 and 3 fuel tank foundation supports the fuel tank, an alternate source of EDG fuel. These fuel tanks shared by IP2 and IP3 are credited for minimum EDG fuel oil inventory. If the EDGs require the reserves in these tanks, the contents can be transported by tanker truck.

The gas turbine substation switchgear structures and foundation support equipments required to support offsite power recovery following station blackout. It consists of a reinforced concrete slab that supports the substation and switchgear support structures. Component equipment is anchored by welding or bolting to embedments in the concrete slab.

The maintenance and outage building and elevated passageway are seismic Class II structures used by maintenance and outage personnel. The structures are southeast of the IP2 containment structure, across from the PAB, and adjacent to the fuel storage building. The building has two major floors and an elevated passageway for access to the PAB. A safety-related conduit routed through one end of the building near the bridge connects the maintenance and outage building to the PAB.

Manholes and duct banks throughout the applicant's yard allow underground routing of cables and piping. These structural components are of reinforced and non-reinforced concrete.

The new station security building east of the IP1 containment structure provides offices for personnel and contains the security generator credited as a source of backup power to the station security lighting system. For IP2, this lighting illuminates exterior ingress and egress in an Appendix R fire and a loss of offsite power.

The IP1 nuclear service building adjacent to but separated from the IP1 containment structure protects alternate safe shutdown system components in support of IP2. These components consist of cables in conduit for various systems: chemical and volume control, CCW, RHR, and SI systems. The structure contains treatment and decontamination facilities and examination rooms for site personnel.

The IP3 power conversion equipment building houses power conversion system components.

The IP2 PAB is a seismic Class I structure housing safety injection pumps, component cooling pumps, heat exchangers, and RHR pumps. The IP3 PAB houses components required for recirculation (e.g., component cooling pumps, heat exchangers, and SI and RHR pumps).

The IP2 and IP3 primary water storage tank foundations are the main supports for the 165,000-gallon primary water storage tank for each unit. The tanks supply demineralized water for the primary water makeup systems.

The IP2 radiation monitoring enclosure houses radiation monitors R46, R49 and R53. Monitors R46 and R53 monitor the SW return from all containment fan cooler units.

For both IP2 and IP3, the RWST foundation is the main support for the 350,000-gallon RWST. The tank supplies borated water to the refueling canal, SI pumps, RHR pumps, and the containment spray pumps for a LOCA.

The IP3 security access and office building located west of the service admin complex provides offices for personnel and contains the security generator credited as a source of backup power to the station security lighting system. For IP3, this lighting illuminates exterior ingress and egress in an Appendix R fire or a loss of offsite power.

The IP1 superheater building is adjacent to but physically separated from the control building. The superheater stack is located on top of the superheater building. The structure contains the technical support center, provides office area for personnel, supports alternate safe shutdown system components, and houses a safety-related battery room.

The IP1 superheater stack on top of the superheater building carries exhaust from the superheaters and also supports a ventilation duct carrying exhaust from the containment structure. Failure of the stack could result in damage to the IP2 control building, the EDG building, and in-scope IP3 structures. To minimize this risk, the applicant shortened the stack and reinforced its support structure to satisfy IP3 tornado protection criteria.

The offsite power source required to support SBO recovery actions is fed through one of the station auxiliary transformers. Specifically, the path includes the 138kV and 345kV switchyard circuit breakers feeding either station auxiliary transformers.

The transformer/switchyard support structures physically support the station auxiliary transformers and the other switchyard components in the SBO recovery path. These support structures include the transformer foundations and support steel, transformer pothead foundations and support steel, and switchyard breaker foundations.

Transmission towers (SBO recovery path) and foundations are parts of the path to restore offsite power.

The IP1 turbine building is an extension of the IP2 turbine building and is integrally attached to the superheater building and the IP2 turbine building. The structure is classified as seismic Class III but was analyzed to ensure that there is no potential for gross structural collapse as a result of a design basis event. Equipment and components on the IP1 operating floor have been removed and the supporting systems for these components are not in service. The facility houses the station blackout/Appendix R diesel and two fire water pumps, along with their associated components relied upon in the site's safe shutdown analysis. The building is constructed of heavy structural steel framing with steel supported reinforced concrete slabs forming the floor area. Crane rails located within IP1 extending the entire length of the structure also provide support for IP2. The building's exterior face is constructed of metal-sandwich panels and concrete brick.

The IP2 turbine building and heater bay extension of the IP1 turbine building is similar to IP1 and is seismic Class III. Although the turbine building and heater bay are seismic Class III structures, they were analyzed for potential gross structural collapse as a result of a design-basis event. Attached to the superheater building and the IP1 turbine building, the building houses the IP2 turbine generator, FW heaters, and their supporting systems as well as cabling, switchgear, and other SBO/Appendix R diesel equipment.

The IP3 turbine building and heater bay is a seismic Class III structure that houses the turbine generator and its auxiliaries. The structure is designed not to affect Class I structures.

The utility tunnel is a seismic Class III structure. The tunnel shelters and protects the city water supply piping for AFW backup water and other miscellaneous functions. The utility tunnel is a rectangular reinforced concrete structure founded on rock.

The IP2 waste holdup tank pit is adjacent to the refueling water tank and its top slab supports the boric acid evaporator building. The IP3 waste holdup tank pit, two structures joined to form a single structure, is adjacent to the primary water storage tank and the radioactive machine shop. The waste holdup tank pits house liquid waste holdup tanks which are the collection points for liquid radwaste. A sump services the water tanks.

The turbine buildings, auxiliary buildings, and other structures contain safety-related components relied upon to remain functional during and following DBEs. The failure of nonsafety-related SSCs in the turbine buildings, auxiliary buildings, and other structures potentially could prevent the satisfactory accomplishment of a safety-related function. In addition, the turbine buildings, auxiliary buildings, and other structures perform functions that support fire protection and SBO.

LRA Table 2.4-3 identifies turbine buildings, auxiliary buildings, and other structure component types, grouped by material (steel/other metals, concrete), within the scope of license renewal and subject to an AMR as well as their intended functions.

2.4.3.2 Staff Evaluation

The staff reviewed LRA Section 2.4.3 and IP2 UFSAR Sections 1.3.8, 1.11.4.12, 1.11.6, 7.2.4.1.4, and 9.5.2, and IP3 UFSAR Sections 8.4, 9.6.2, 9.6.2.9, and 11.1.2.1, using the evaluation methodology described in SER Section 2.4 and the guidance in SRP-LR Section 2.4.

During its review, the staff evaluated the structural component functions described in the LRA and UFSAR to verify that the applicant had not omitted from the scope of license renewal any SCs with intended functions, as required by 10 CFR 54.4(a). The staff then reviewed those SCs that the applicant identified as within the scope of license renewal to verify that the applicant had not omitted any passive and long-lived SCs subject to an AMR in accordance with the requirements of 10 CFR 54.21(a)(1).

During its review of LRA Section 2.4.3, the staff identified areas in which additional information was necessary to complete the review of the applicant's scoping and screening results. The applicant responded to the staff's RAIs as discussed below.

In the second part of RAI 2.4-1, dated January 28, 2008 (the first part of RAI 2.4-1 is addressed in SER Section 2.4), the staff noted that the structure identified as "Gas Turbine Substation Switchgear Structures and Foundation" in LRA Table 2.2-3 was not included in the structures listed at the beginning of the subsection "Description" of LRA Section 2.4.3. The staff requested that the applicant address the scoping and screening of these structures or clarify where they were addressed in the LRA.

In the last paragraph of its response to RAI 2.4-1, dated February 27, 2008, the applicant stated that the "Gas Turbine Substation Switchgear Structures and Foundation" area is addressed in LRA Section 2.4.3, subsection titled "Description" under "Gas Turbine Generator No. 1, 2 and 3 Enclosure and Fuel Tank Foundation." The staff verified that a description of switchgear structures and foundation was included in the subsection in Section 2.4.3 describing the gas turbine generators No. 1, 2 and 3 enclosure and fuel tank foundations, as stated by the applicant. The staff finds the applicant's response acceptable subject to further clarification as requested in the follow-up RAI, dated May 12, 2008.

Because of a lack of clarity in LRA Table 2.4-3, and the applicant's response to RAI 2.4-1 with regard to switchgear structures and foundation, the staff sought clarification regarding which specific structural components in Table 2.4-3 cover the switchgear structures and foundation. The staff noted that the component line item "foundations" in LRA Table 2.4-3 does not list "switchgear structures" in the structure list provided within parenthesis.

In its response, dated June 11, 2008, the applicant stated that the switchgear foundation is listed in LRA Table 2.4-4, as equipment pads/foundations. Since the applicant clarified that the switchgear foundations are included as a concrete bulk commodity item as part of line item "equipment pads/foundation" in LRA Table 2.4-4, and the embedments to which the switchgear equipment is anchored are included as part of bulk commodity line item "anchorages/embedments" in LRA Table 2.4-4, the staff finds the response acceptable.

In RAI 2.4.3-1, dated January 28, 2008, the staff noticed the following in the LRA with regard to the fuel storage buildings:

(i)   LRA Section 2.4.3 states that the fuel storage buildings have the following intended functions pursuant to 10 CFR 54.4(a)(1) and (a)(2): "Maintain integrity of nonsafety-related components such that safety functions are not affected by maintaining pool water inventory (Units 2 and 3)."

(ii)   LRA Section 2.1.2.2, "Screening of Structures," states that the screening of structural components and commodities was based primarily on whether they perform an intended function.

(iii)   LRA Table 3.5.2-3, "Turbine Building, Auxiliary Building, and Other Structures, Structural Components and Commodities (IP2 and IP3)," identifies structural components subject to aging management based on materials of construction and intended functions for components of structures, including the fuel storage buildings.

(iv)   The intended functions listed in LRA Table 3.5.2-3 (e.g., pressure boundary, missile barrier, and shelter or protection) agree with the intended functions listed in LRA Table 2.0-1, "Intended Functions: Abbreviations and Definitions." However, the intended functions for the fuel storage building listed in LRA Section 2.4.3 do not agree with the listed intended functions in LRA Tables 2.0-1 and 3.5.2-3.

With reference to the above, the staff noted in the RAI that, pursuant to 10 CFR Part 54.21, the LRA must identify and list those SCs subject to an AMR. The staff requested that the applicant clarify the LRA Section 2.4.3 description of the intended function(s) of the fuel storage building components, using the list of intended functions from LRA Table 2.0-1. The staff added that, to satisfy the requirements of 10 CFR Part 54.21, the clarification must be adequate to reasonably identify the fuel storage building structural components subject to an AMR by the component or commodity, material of construction, and intended functions listed in LRA Table 3.5.2-3.

In its response, dated February 27, 2008, as supplemented in LRA Amendment No. 3, dated March 24, 2008, the applicant stated that the intended functions listed in LRA Tables 2.0-1 and 3.5.2-3 are component intended functions, which are determined during the screening process. The intended functions in LRA Section 2.4.3, in contrast, are the intended functions of the structure in its entirety and are determined during the scoping process. The applicant explained that the scoping process determines whether or not the structure has an intended function (i.e., providing containment or isolation to mitigate post-accident offsite doses, or providing support or protection to safety-related equipment), whereas the screening process identifies those components that support the structure intended function(s) via specific component intended functions (i.e., providing shelter and protection or providing support for safety-related equipment). The structure and system level functions that are assessed against the scoping requirements of 10 CFR Part 54.4 are not intended to match the component level functions defined in LRA Table 2.0-1. While similarities exist between the terminology used for component intended functions versus structure intended functions, a direct correlation between the structure intended functions in LRA Section 2.4 and the component intended functions in the tables in LRA Section 3.5 does not exist. The applicant clarified that the structure level intended functions of the fuel storage buildings are to: (a) maintain integrity of nonsafety-related components such that safety functions are not affected by maintaining pool water inventory, and (b) provide support and protection for safety-related equipment within the scope of license renewal. The applicant also provided a tabulation of component level intended functions (as defined in LRA Table 2.0-1) supporting each of the two structure level intended functions for the fuel storage buildings.

In its response, the applicant used a broader structure level intended function concept in the scoping process and supplemented that by more detailed component level intended functions for the structural components during the screening process. Because the applicant 1) has clarified the structure level intended functions of the fuel storage buildings, and 2) provided a tabulation of the structural component intended functions for each of the two structure level intended functions (as defined in LRA Table 2.0-1), the staff finds the applicant's response acceptable. Therefore, the staff's concern described in RAI 2.4.3-1 is resolved.

In RAI 2.4.3-2, dated January 28, 2008, the staff noted that, in LRA Section 2.4.3, the top of the spent fuel pit wall forms the north wall of each unit's fuel building. The staff further noted that UFSAR Figure 1.2-4 (IP2), "Cross Section of Plant," indicates that at least part of the fuel building exterior wall is below grade. LRA Table 2.4-3 lists pressure boundary as an intended function for the concrete component "exterior walls" but does not list pressure boundary as an intended function of the concrete component "exterior walls-below grade," representing the fuel building wall. The staff requested that the applicant update LRA Table 2.4-3 to include the pressure boundary intended function for the spent fuel pit wall that is below grade or provide justification for excluding this intended function.

In its response, dated February 27, 2008, the applicant stated that it agrees that the spent fuel pit wall below grade also performs a pressure boundary intended function. The applicant revised LRA Tables 2.4-3 and 3.5.2-3 to include the pressure boundary intended function for exterior walls below-grade which includes the spent fuel pit wall. The staff finds the applicant's response adequately addresses the staff's concerns raised in the RAI and, therefore, is acceptable. The staff's concern described in RAI 2.4.3-2 is resolved.

In RAI 2.4.3-3, dated January 28, 2008, the staff noted that LRA Table 2.4-3 does not include the leak chase channel of the IP3 spent fuel pit as a component subject to an AMR. The staff requested the applicant to include this as a component subject to an AMR or provide a justification for its exclusion.

In its response, dated February 27, 2008, the applicant stated that the leak chase channel is an integral attachment to the liner plate, which is subject to AMR and included in line item "Spent fuel pool liner plate and gate" in LRA Table 2.4-3. The staff agrees with the applicant's position that the leak chase channel, which is welded to the liner plate, can be considered an integral attachment to the liner plate and included as part of the liner plate component. The staff finds the applicant's response adequately addresses the staff's concerns raised in the RAI and, therefore, is acceptable. The staff's concern described in RAI 2.4.3-3 is resolved.

In RAI 2.4.3-4, dated January 28, 2008, the staff noted that, although LRA Table 2.4-3 lists "Crane rails and girders" as a component type subject to an AMR, it is not clear whether this component refers to just crane rails and girders or also refers to the cranes themselves. If it includes the cranes, the applicant was requested to clarify whether all relevant subcomponents ("...including bridge and trolley, rails, and girders") of these in-scope crane systems have been screened in as items requiring an AMR. The staff also requested that the applicant identify the specific cranes in each of these structures that are included within the above component type as within the scope of license renewal and subject to an AMR, and those that are excluded, with technical bases. The applicant also was requested to confirm whether fasteners and rail hardware associated with this component type are within the scope of license renewal and subject to an AMR or provide the technical bases for their exclusion. The staff also requested that the applicant confirm whether there are other hoists and lifting devices that should be included within the scope of license renewal (and subject to an AMR) and, if so, provide their scoping, screening, and AMR results, relevant to the LRA.

In its response, dated February 27, 2008, the applicant stated that the component type "crane rails and girders" in LRA Table 2.4-3 includes bridge and trolley and also refers to the cranes themselves. The applicant further stated that there are no hoists or lifting devices that perform an intended function that would place them in scope and subject to an AMR. The applicant clarified that the specific cranes in scope and subject to an AMR are discussed in LRA Section 2.4-1 for containment buildings and in Section 2.4-3 for turbine building(s) and fuel storage building(s). The applicant confirmed that fasteners and rail hardware are in scope and subject to an AMR. They are, however, considered bulk commodities and are included in LRA Table 2.4-4, line item "structural bolting." Since the language of the line item as currently written could be misleading, in a follow-up RAI, dated May 12, 2008, the staff requested the applicant to correct the line item "crane rails and girders" in LRA Table 2.4-3 to read "cranes, rails and girders."

In its response to the follow-up RAI, dated June 11, 2008, the applicant stated that the line item "crane rails and girders" LRA Table 2.4-3 and LRA Table 3.5.2-3 is corrected to read "cranes, rails and girders". Since the applicant corrected the line item, the staff finds the response acceptable.

In RAI 2.4.3-5, dated January 28, 2008, the staff requested that the applicant confirm whether the component identified as "Structural Steel: beams, columns, plates" in LRA Table 2.4-3 includes bracings, welds, and bolted connections or indicate where they were included. The staff also requested that the applicant include "Battery Racks" (e.g., for emergency diesels), turbine generator pedestals and their structural bearing pads, and diesel generator pedestals and the concrete curb around diesel generator foundations as components subject to an AMR.

In its response, dated February 27, 2008, the applicant clarified that the component identified as "Structural Steel: beams, columns, plates, trusses" in LRA Table 2.4-3 includes bracings and welds associated with the component. The applicant added that bolted connections are addressed in LRA Section 2.4.4 and LRA Table 2.4-4. The applicant further clarified that battery racks (e.g., for emergency diesel) are within the scope of license renewal and subject to an AMR and are included as bulk commodities within line item "component and piping support" in LRA Table 2.4-4. The applicant further clarified that the turbine generator pedestals, diesel generator pedestals, and the concrete curb around diesel generator foundations are included within the LRA Table 2.4-3 as part of line item "Floor slabs, interior walls and ceiling" and line item "Foundations." The applicant stated that structural bearing pads associated with the turbine generator pedestal are not within the scope of license renewal because they are not safety-related and not required to demonstrate compliance with 10 CFR 54.4(a)(3). Failure of the bearing pads will not prevent satisfactory accomplishment of a safety function. Based on this response, the staff finds that the applicant has adequately clarified the inclusion or justified the exclusion, as applicable, of each of the structural components noted in the RAI. The staff finds the applicant's response adequately addresses the staff's concerns raised in the RAI and, therefore, is acceptable. The staff's concern described in RAI 2.4.3-5 is resolved.

2.4.3.3 Conclusion

The staff reviewed the LRA, UFSAR, and RAI and follow-up RAI responses to determine whether the applicant failed to identify any SCs within the scope of license renewal. The staff found no omissions. In addition, the staff sought to determine if the applicant failed to identify any SCs subject to an AMR. Again, the staff found no omissions. On the basis of its review, the staff concludes that the applicant has adequately identified the turbine buildings, auxiliary buildings, and other structures SCs within the scope of license renewal, as required by 10 CFR 54.4(a), and those subject to an AMR, as required by 10 CFR 54.21(a)(1).

2.4.4 Bulk Commodities

2.4.4.1 Summary of Technical Information in the Application

LRA Section 2.4.4 describes bulk commodities, the structural components or commodities that perform or support intended functions of in-scope SSCs. Bulk commodities unique to a specific structure are included in the review for that structure (LRA Sections 2.4.1 through 2.4.3). Bulk commodities common to Indian Point in-scope SSCs (*e.g.*, anchors (including rock bolts), embedments, pipe and equipment supports, instrument panels and racks, cable trays, and conduits) are addressed in this section.

Insulation may have the specific intended functions of (1) controlling the heat load during DBAs in areas with safety-related equipment, (insulation and Insulation jacket) or (2) maintaining integrity such that falling insulation does not damage safety-related equipment (reflective metallic-type reactor vessel insulation).

Bulk commodities have the following intended functions for 10 CFR 54.4(a)(1), (a)(2), and (a)(3): Provide support, shelter, and protection for safety-related equipment and nonsafety-related equipment within the scope of license renewal.

LRA Table 2.4-4 identifies bulk commodities' component types, grouped by material (steel/other metals, concrete, other materials), within the scope of license renewal and subject to an AMR as well as their intended functions.

2.4.4.2 Staff Evaluation

The staff reviewed LRA Section 2.4.4 using the evaluation methodology described in SER Section 2.4 and the guidance in SRP-LR Section 2.4.

During its review, the staff evaluated the structural component functions described in the LRA and UFSAR to verify that the applicant had not omitted from the scope of license renewal any SCs with intended functions, as required by 10 CFR 54.4(a). The staff then reviewed those SCs that the applicant identified as within the scope of license renewal to verify that the applicant had not omitted any passive and long-lived SCs subject to an AMR, in accordance with the requirements of 10 CFR 54.21(a)(1).

During its review of LRA Section 2.4.4, the staff identified areas in which additional information was necessary to complete the review of the applicant's scoping and screening results. The applicant responded to the staff's RAIs as discussed below.

In LRA Section 2.4.4 and LRA Table 2.4-4, the applicant discussed and listed the structural bulk commodities components common to in-scope structures that are subject to an AMR. Because of a lack of clarity in LRA Table 2.4-4, in RAI 2.4.4-1, dated January 28, 2008, the staff requested that the applicant confirm or clarify and appropriately address whether the following bulk commodities have been screened in as components subject to an AMR, in LRA Table 2.4-4:

(i)   expansion anchors
(ii)   vibration isolation elements
(iii) flood curbs
(iv)   waterproofing membrane
(v)   sliding support bearings and sliding support surfaces

The applicant also was requested to explicitly state the specific materials that are classified as "Other Materials" in LRA Table 2.4-4.

In its response, dated February 27, 2008, the applicant clarified the screening of each component identified in the RAI as follows:

(i)   Expansion Anchors are addressed in LRA Table 2.4-4 under line item "anchorages/embedments."

(ii)   There are no vibration isolation elements identified as within the scope of license renewal and subject to AMR.

(iii) Flood curbs are included in the review of structures. Considered integral to floor slabs, they are included in the review for those line items identified in LRA Tables 2.4-1 as "beams, columns, interior walls, slabs," Table 2.4-2 as "beams, columns, floor slabs and walls" and Table 2.4-3 as "floor slabs, interior walls, ceilings."

(iv)   Waterproofing membranes are not in-scope. Waterproofing membranes are not safety-related and are not required to demonstrate compliance with 10 CFR 54.4(a)(3). Failure of these membranes will not prevent satisfactory accomplishment of a safety function.

(v)   The sliding support bearings and sliding support surfaces identified as within the scope of license renewal are documented in LRA Table 2.4-1, line item "Lubrite sliding surfaces."

The applicant also stated that materials classified as "Other Materials" in LRA Table 2.4-4 are those materials that were not captured by what is considered basic structural materials (i.e., steel or concrete) and that the material make-up of these commodities is specifically identified in LRA Section 3.5.2.1.4.

The staff finds that the applicant adequately clarified the issues related to the screening of the five specific structural components identified in the RAI. The staff also verified that, in LRA

Section 3.5.2.1.4, the applicant identified the bulk commodity component materials that make up the line item "Other Materials." These other materials, identified in LRA Section 3.5.2.1.4 are aluminum, cera blanket, cerafiber, elastomer, fiberglass and/or calcium silicate, mineral wool, and pyrocrete. The staff finds that the applicant's response adequately addresses the staff's concerns raised in the RAI and, therefore, is acceptable. The staff's concern described in RAI 2.4.4-1 is resolved.

In RAI 2.4.4-2, dated January 28, 2008, with regard to the components "insulation" and "insulation jacket" identified in LRA Table 2.4-4, the staff pointed out that it was unclear as to which insulation (and material) and insulation jacket within the scope of license renewal were included in these items. The applicant was requested to clarify whether the insulation and jacketing on the containment liner, reactor vessel, RCS, MS and FW systems are included.

The applicant also was requested to provide the following information with regard to insulation that is used to control the maximum temperature of safety-related structural elements:

(a)   Identify the structures and structural components designated as within the scope of license renewal that have insulation and/or insulation jacketing, and identify their location in the plant. Identify locations of the thermal insulation that serve an intended function in accordance with 10 CFR 54.4(a)(2) and describe the scoping and screening results of thermal insulation, and provide the technical basis for its exclusion from the scope of license renewal.

(b)   For insulation and insulation jacketing materials associated with item (a) above that do not require aging management, submit the technical basis for this conclusion, including plant-specific operating experience.

(c)   For insulation and insulation jacketing materials associated with item (a) above that require aging management, indicate the applicable LRA sections that identify the AMP(s) credited to manage their aging.

In its response, dated February 27, 2008, the applicant addressed each of the items in the RAI as follows:

(a)   The applicant stated that structures and structural components within the scope of license renewal that have insulation and/or insulation jacketing that serves an intended function pursuant to 10 CFR 54.4(a)(2) are the containment liner and high-temperature piping at containment piping penetrations. The applicant stated that the containment liner insulation is listed in LRA Table 2.4-1, and the insulation associated with hot containment penetrations is addressed in LRA Section 2.4.4 and in LRA Table 2.4-4.

(b)   The applicant clarified that insulation and insulation jacketing materials associated with item (a) do not require an AMP because these insulation materials are exposed to indoor air environment and the containment liner insulation is encapsulated in a stainless steel jacket and is not subject to external environments. The applicant further stated that, in these environments, these materials have no aging effects requiring management. The operating experience review specifically considered plant-specific information related to the effects of aging on insulation materials, and that review confirmed that no aging effects requiring management are applicable to the insulation materials that are subject to an AMR at IP2 and IP3.

(c)   The applicant stated that aging management review results for insulation and insulation jacketing materials are shown in LRA Tables 3.5.2-1 and 3.5.2-4.

The applicant reiterated that, since there are no aging effects requiring management for insulation, no AMP is credited, noting that insulation materials in an indoor air environment are not susceptible to degradation from the effects of aging.

In its response, and in the context of insulation that serves to limit the temperature of safety-related structural components, the applicant confirmed that the structures and structural components, within the scope of license renewal and subject to an AMR, that have insulation and/or insulation jacketing are the containment liner and high-temperature piping at the containment penetrations. The applicant concluded that none of the in-scope insulating material used at IP2 and IP3 requires any management for aging effects because of its favorable operating experience and the fact that it is only exposed to an indoor air environment and encapsulated in metallic jacketing. The staff finds that this conclusion is consistent with the GALL Report, Volume II. The staff further finds that the applicant's response to RAI 2.4.4-2 adequately addressed the staff's question with regard to insulation and, therefore, is acceptable. The staff's concern described in RAI 2.4.4-2 is resolved.

2.4.4.3 Conclusion

The staff reviewed the LRA, UFSAR, and RAI responses to determine whether the applicant failed to identify any SCs within the scope of license renewal. The staff found no omissions. In addition, the staff sought to determine whether the applicant failed to identify any SCs subject to an AMR. The staff found no omissions. On the basis of its review, the staff concludes that the applicant has adequately identified the bulk commodities SCs within the scope of license renewal, as required by 10 CFR 54.4(a), and those subject to an AMR, as required by 10 CFR 54.21(a)(1).

2.5 Scoping and Screening Results: Electrical and Instrumentation and Control Systems

This section documents the staff's review of the applicant's scoping and screening results for electrical and I&C systems.

In accordance with the requirements of 10 CFR 54.21(a)(1), the applicant must list passive, long-lived SCs within the scope of license renewal and subject to an AMR. To verify that the applicant properly implemented its methodology, the staff's review focused on the implementation results. This focus allowed the staff to confirm that there were no omissions of electrical and I&C system components that meet the scoping criteria and are subject to an AMR.

The staff's evaluation of the information in the LRA sought to determine whether the applicant had identified, in accordance with 10 CFR 54.4, components and supporting structures for

2-226

electrical and I&C systems that appear to meet the license renewal scoping criteria. Similarly, the staff evaluated the applicant's screening results to verify that all passive, long-lived components were subject to an AMR in accordance with 10 CFR 54.21(a)(1).

In its scoping evaluation, the staff reviewed the applicable LRA sections, focusing on components that had not been identified as within the scope of license renewal. The staff reviewed relevant licensing basis documents, including the UFSAR, for each electrical and I&C system to determine whether the applicant had omitted from the scope of license renewal components with license renewal intended functions in accordance with 10 CFR 54.4(a). The staff also reviewed the licensing basis documents to determine whether the LRA specified all license renewal intended functions in accordance with 10 CFR 54.4(a). The staff requested additional information to resolve any omissions or discrepancies identified.

After its review of the scoping results, the staff evaluated the applicant's screening results. For those SCs with intended functions, the staff sought to determine whether (1) the functions are performed with moving parts or a change in configuration or properties, or (2) the SCs are subject to replacement after a qualified life or specified time period, as described in 10 CFR 54.21(a)(1). For those meeting neither of these criteria, the staff sought to confirm that these SCs were subject to an AMR, as required by 10 CFR 54.21(a)(1). The staff requested additional information to resolve any omissions or discrepancies identified.

2.5.1 Electrical and Instrumentation and Control Systems

2.5.1.1 *Summary of Technical Information in the Application*

LRA Section 2.5 describes the electrical and instrumentation and control systems. As stated in LRA Section 2.1.1, plant electrical and instrument and control (I&C) systems are included in the scope of license renewal as are electrical and I&C components in mechanical systems. The default inclusion of plant electrical and I&C systems in the scope of license renewal reflects the method for the integrated plant assessment (IPA) of electrical systems. This method is different from the methods used for mechanical systems and structures.

The applicant stated that the basic philosophy of the electrical and I&C components IPA is that components are included in the review unless specifically screened out. In the plant spaces approach, this method eliminates the need for unique identification of every component and its specific location so components are not excluded improperly from an AMR. The electrical and I&C IPA began by grouping all components into commodity groups of similar electrical and I&C components with common characteristics and by determining component level intended functions of the commodity groups.

The IPA eliminated commodity groups and specific plant systems from further review as the intended functions of commodity groups were examined. In addition to the plant electrical systems, certain switchyard components required to restore offsite power following SBO were included conservatively within the scope of license renewal even though those components are not relied on in safety analyses or plant evaluations to perform a function that demonstrates compliance with the Commission's regulations for SBO (10 CFR 50.63).

The applicant further stated that the offsite power system provides the electrical interconnection between IPEC and the offsite transmission network. The offsite power sources required to support SBO recovery actions supply the station auxiliary transformers. Specifically, the offsite power recovery path includes the station auxiliary transformers, the 138 kV and 13.8 kV switchyard circuit breakers supplying the station auxiliary transformers, the circuit breaker-to-transformer and transformer-to-onsite electrical distribution interconnections, control circuits, and structures.

The electrical and instrumentation and control systems perform functions that support SBO and EQ.

LRA Table 2.5-1 identifies electrical and instrumentation and control systems component types within the scope of license renewal and subject to an AMR:

- cable connections (metallic parts)

- electrical cables and connections not subject to 10 CFR 50.49 EQ requirements

- electrical cables not subject to 10 CFR 50.49 EQ requirements used in instrumentation circuits

- electrical connections not subject to 10 CFR 50.49 EQ requirements exposed to borated water leakage

- fuse holders (insulation material)

- high-voltage insulators for SBO recovery

- inaccessible medium-voltage (2kV to 35kV) cables not subject to 10 CFR 50.49 EQ requirements

- metal-enclosed bus (non-segregated) and connections for SBO recovery

- metal-enclosed bus (non-segregated), insulation/insulators for SBO recovery

- metal-enclosed bus (non-segregated) enclosure assemblies for SBO recovery

- switchyard bus and connections for SBO recovery

- transmission conductors and connections for SBO recovery

- 138 kV direct burial insulated transmission cables

The intended functions of the electrical and instrumentation and control systems component types within the scope of license renewal include the following functions:

- connect specified electrical circuit portions to deliver voltage, current, or signals

- insulate and support electrical conductors

- structurally or functionally support equipment required for the 10 CFR 54.4(a)(3) regulated events

2.5.1.2 Staff Evaluation

The staff reviewed LRA Section 2.5 and the UFSAR using the evaluation methodology described in SER Section 2.5 and the guidance in SRP-LR Section 2.5, "Scoping and Screening Results: Electrical and Instrumentation and Controls Systems."

During its review, the staff evaluated the system functions described in the LRA and UFSAR to verify that the applicant had not omitted from the scope of license renewal any components with intended functions delineated under 10 CFR 54.4(a). The staff then reviewed those components that the applicant had identified as within the scope of license renewal to verify that the applicant had not omitted any passive and long-lived components subject to an AMR in accordance with the requirements of 10 CFR 54.21(a)(1).

During its review of LRA Section 2.5, the staff identified several areas in which additional information was necessary to complete the review of the applicant's scoping and screening results. The applicant responded to the staff's RAIs as discussed below.

The staff noted that, according to LRA Section 2.5, two independent paths from the safety-related buses to the first circuit breaker from the offsite transmission line were not included within the scope of license renewal. General Design Criterion 17 of 10 CFR Part 50, Appendix A, requires that electric power from the transmission network to the onsite electric distribution system be supplied by two physically independent circuits to minimize the likelihood of their simultaneous failure. In addition, the staff noted that the guidance provided by letter dated April 1, 2002, "Staff Guidance on Scoping of Equipment Relied on to Meet the Requirements of the Station Blackout Rule (10 CFR 50.63) for License Renewal (10 CFR 54.4(a)(3))," and later incorporated in SRP-LR Section 2.5.2.1.1, states:

> For purposes of the license renewal rule, the staff has determined that the plant system portion of the offsite power system that is used to connect the plant to the offsite power source should be included within the scope of the rule. This path typically includes switchyard circuit breakers that connect to the offsite system power transformers (startup transformers), the transformers themselves, the intervening overhead or underground circuits between circuit breaker and transformer and transformer and onsite electrical system, and the associated control circuits and structures. Ensuring that the appropriate offsite power system long-lived passive SCs that are part of this circuit path are subject to an AMR will assure that the bases underlying the SBO requirements are maintained over the period of extended license.

According to this guidance, the NRC staff position is that, for the purposes of license renewal, the specified offsite power recovery path elements should be included in the scope of license renewal. In RAI 2.5-1, dated October 24, 2007, the staff conveyed its position that both paths from the safety-related 480 V buses to the first circuit breaker from the offsite line used to control the offsite circuits to the plant should be included within the scope of license renewal. Therefore, the staff requested that the applicant provide a detailed explanation of which high voltage breakers and other components in the switchyard will be connected from the startup transformers up to the offsite power system for the purpose of SBO recovery.

In its response, dated November 16, 2007, the applicant stated that the Buchanan substation, which includes the 345 kV, 138 kV, and 13.8 kV sections, provides for the interconnection of multiple sources of power and constitutes the offsite power source for IP2 and IP3.

In the LRA, Figure 2.5-2, "IP2 Offsite Power Scoping Diagram," shows the IP2 primary offsite power source, the 6.9 kV source from the station auxiliary transformer which is connected to the 138 kV Buchanan substation through circuit breaker F2. The applicant's November 16, 2007 response revised the scoping boundary for both offsite power sources for IP2. First, the station auxiliary transformer is connected to the 138 kV Buchanan substation via switchyard bus, overhead transmission conductors, and underground transmission conductors through motor-operated disconnect F3A (primary path). The staff determined that this change to a motor-operated disconnect is not consistent with the staff guidance and, therefore, is unacceptable. Secondly, the November 16, 2007 response delineated the secondary offsite power source (alternate path). The gas turbine (GT) autotransformer is connected to the 13.8 kV Buchanan substation via underground medium voltage cable through 13.8 kV circuit breaker F2-3.

LRA Figure 2.5-3, "IP3 Offsite Power Scoping Diagram," was modified in the applicant's November 16, 2007, response to add the secondary offsite power feeder, indicating that the 6.9 kV buses receive power from two independent sources: the 138 kV/6.9 kV station auxiliary transformer and the 13.8 kV/6.9 kV GT autotransformer. The station auxiliary transformer is connected to the 138 kV Buchanan substation via switchyard bus and overhead transmission conductors through circuit breaker BT2-6, and the GT autotransformer is connected to the 13.8 kV Buchanan substation via underground medium voltage cable through 13.8 kV circuit breaker F3-1.

During a telephone conference, documented in a conference call summary dated December 4, 2007, the staff requested that Entergy explain its response to RAI 2.5-1 with regard to why the connection point for offsite power (for the purpose of station blackout recovery) changed from circuit breaker F2 to a motor-operated disconnect for IP2. The staff informed the applicant that this change is not consistent with the staff's guidance and, therefore, is unacceptable.

In a letter dated March 24, 2008, the applicant modified its scoping boundary for the primary offsite power path for IP2, as shown in modified Figure 2.5-2, "IP2 Offsite Power Scoping Diagram." The station auxiliary transformer is connected to the 138 kV Buchanan substation via switchyard bus, overhead transmission conductors, and underground transmission conductors through switchyard breakers F2 and BT 3-4. The change from motor-operated disconnects to 138 kV circuit breakers addresses the staff's concern for the scoping boundary for the primary offsite power path and provides closure for Open Item 2.5-1.

By letter dated May 20, 2009, the staff requested that the applicant explain why the secondary offsite circuit (the delayed access circuit) path, from the first inter-tie with the offsite distribution systems at the Buchanan substations to the safety buses, was not included in the scope of license renewal.

By letter dated June 12, 2009, the applicant stated that the components up to and including either the 138 kV circuit breaker F1 or 345 kV circuit breaker F7 for IP2, and either the 138 kV circuit breaker F3 or 345 kV circuit breaker F7 for IP3 were not included in the scope of license renewal because they do not meet the scoping criteria specified in 10 CFR 54.4. The staff finds the response acceptable as it is in accordance with the IP2 and IP3 current licensing basis and applicable regulatory requirements. This closes Open Item 2.5-1.

The applicant did not specifically exclude the associated control circuits and structures for the circuit breakers and thus, it was unclear if these components are included in the scope of license renewal. In RAI 2.5-5, the staff requested that the applicant confirm whether the associated control cables and structures for the circuit breakers have been included in the scope of license renewal. In letter dated August 14, 2008, the applicant clarified its response to RAI 2.5-1 and confirmed that the associated control cables and structures for the circuit breakers have been included in the scope of license renewal. Therefore, the staff finds the response acceptable.

In RAI 2.5-2, dated October 24, 2007, the staff requested the applicant to clarify why elements such as resistance temperature detectors (RTDs), sensors, thermocouples, and transducers are not included in the list of components and/or commodity groups subject to an AMR if a pressure boundary is applicable. In its response, dated November 16, 2007, the applicant stated that RTDs, sensors, thermocouples, and transducers associated with the pressure boundary are evaluated in mechanical systems. Examples are thermowells and flow elements. LRA Section 2.1.2.3.1 states that the pressure boundary function that may be associated with some electrical and I&C components was considered in the mechanical aging management reviews. The staff verified through a sampling of mechanical systems that the applicant had scoped and screened the passive mechanical components (e.g., thermowells and flow elements) associated with the electrical elements in question. Therefore, the staff finds the response acceptable.

In RAI 2.5-3, dated October 24, 2007, the staff requested clarification as to why Section 2.5 of the LRA did not include splices, terminal blocks, control cables, and isolated-phase bus in the commodity group of "cables & connections, bus, electrical portions of electrical and I&C penetration assemblies." In its response, dated November 16, 2007, the applicant stated that electrical splices, terminal blocks, and control cables were included in the commodity group "electrical cables and connections not subject to 10 CFR 50.49 EQ requirements." Thus, these components are subject to an aging management review. The isolated-phase bus is not subject to an AMR because it does not perform an intended function. Since the applicant clarified that the electrical splices, terminal blocks, and control cables are subject to an AMR, the staff finds the response acceptable.

2.5.1.3 Conclusion

The staff reviewed the LRA, UFSAR, and RAI responses to determine whether the applicant failed to identify any SSCs within the scope of license renewal. In addition, the staff sought to determine whether the applicant failed to identify any components subject to an AMR. The staff found no such omissions. On the basis of its review, the staff concludes that the applicant has adequately identified the electrical and I&C component commodity groups components within the scope of license renewal, as required by 10 CFR 54.4(a), and those subject to an AMR, as required by 10 CFR 54.21(a)(1).

2.6 Conclusion for Scoping and Screening

The staff reviewed the information in LRA Section 2, "Scoping and Screening Methodology for Identifying Structures and Components Subject to Aging Management Review and Implementation Results" and determines that the applicant's scoping and screening methodology is consistent with the requirements of 10 CFR 54.4(a) and 10 CFR 54.21(a)(1), except as noted above. Accordingly, the staff concludes that the applicant has adequately identified those systems and components within the scope of license renewal, as required by 10 CFR 54.4(a), and those subject to an AMR, as required by 10 CFR 54.21(a)(1).

With regard to these matters, the staff concludes that reasonable assurance exists that the activities authorized by the renewed licenses will continue to be conducted in accordance with the CLB and that any changes made to the CLB, in order to comply with 10 CFR 54.29(a), are in accordance with the Atomic Energy Act of 1954, as amended, and NRC regulations.

NRC FORM 335 (9-2004)
NRCMD 3.7

U.S. NUCLEAR REGULATORY COMMISSION

BIBLIOGRAPHIC DATA SHEET

(See instructions on the reverse)

1. REPORT NUMBER
(Assigned by NRC, Add Vol., Supp., Rev., and Addendum Numbers, if any.)

NUREG-1930, Volume 1

2. TITLE AND SUBTITLE

Safety Evaluation Report
Related to the License Renewal of Indian Point Nuclear Generating Unit Nos. 2 and 3

3. DATE REPORT PUBLISHED

MONTH	YEAR
November	2009

4. FIN OR GRANT NUMBER

5. AUTHOR(S)

Kimberly Green

6. TYPE OF REPORT

Technical

7. PERIOD COVERED *(Inclusive Dates)*

04/23/07-08/06/09

8. PERFORMING ORGANIZATION - NAME AND ADDRESS *(If NRC, provide Division, Office or Region, U.S. Nuclear Regulatory Commission, and mailing address; if contractor, provide name and mailing address.)*

Division of License Renewal
Office of Nuclear Reactor Regulation
U.S. Nuclear Regulatory Commission
Washington, DC 20555-0001

9. SPONSORING ORGANIZATION - NAME AND ADDRESS *(If NRC, type "Same as above"; if contractor, provide NRC Division, Office or Region, U.S. Nuclear Regulatory Commission, and mailing address.)*

Same as above

10. SUPPLEMENTARY NOTES

11. ABSTRACT *(200 words or less)*

This safety evaluation report (SER) documents the technical review of the Indian Point Nuclear Generating Unit Nos. 2 and 3 (IP2 and IP3) license renewal application (LRA) by the U.S. Nuclear Regulatory Commission (NRC) staff. By letter dated April 23, 2007, and as supplemented by letters dated May 3 and June 21, 2007, Entergy Nuclear Operations, Inc. (Entergy or the applicant), submitted the LRA in accordance with Title 10, Part 54, of the Code of Federal Regulations. Entergy requests renewal of the IP2 and IP3 operating licenses for a period of 20 years beyond the current expirations at midnight on September 28, 2013, for IP2, and at midnight on December 12, 2015, for IP3.

Indian Point is located approximately 24 miles north of the New York City boundary line. The NRC issued operating licenses on September 28, 1973, for IP2, and on December 12, 1975, for IP3. IP2 and IP3 employ a pressurized water reactor design with a dry ambient containment. Westinghouse Electric Corporation supplied the nuclear steam supply system. The licensed output of each unit is 3216 megawatts thermal with a gross electrical output of approximately 1080 megawatts electric.

On January 15, 2009, the staff issued an SER with open items, in which it identified 20 open items necessitating further review. This SER presents the staff's review of information submitted through August 6, 2009. The staff resolved the 20 open items before it made its final determination on the LRA.

12. KEY WORDS/DESCRIPTORS *(List words or phrases that will assist researchers in locating the report.)*

10 CFR Part 54, license renewal, Indian Point, scoping and screening, aging management, aging effects, time-limited aging analysis, safety evaluation report

13. AVAILABILITY STATEMENT

unlimited

14. SECURITY CLASSIFICATION

(This Page)
unclassified

(This Report)
unclassified

15. NUMBER OF PAGES

16. PRICE

Printed
on recycled
paper

Federal Recycling Program